Composites forming technologies

Related titles:

Geosynthetics in civil engineering
(ISBN-13: 978-1-85573-607-8; ISBN-10: 1-85573-607-1)
Geosynthetics are essential to civil engineering and have a multitude of applications. The first part of the book looks at design principles for geosynthetics, their material properties and durability, and the range of national and international standards governing their use. Part II reviews the range of applications for synthetics as well as quality assurance issues. There are chapters on geosynthetic applications as filters, separators and barrier materials, in improving building foundations and landfill sites, and as limited design life materials.

Multi-scale modelling of composite material systems
(ISBN-13: 978-1-85573-936-9; ISBN-10: 1-85573-936-4)
This book focuses on the fundamental understanding of composite materials at the microscopic scale, from designing microstructural features to the predictive equations of the functional behaviour of the structure for a specific end-application. The papers presented discuss stress- and temperature-related behavioural phenomena based on knowledge of physics of microstructure and microstructural change over time.

Durability of composites for civil structural applications
(ISBN-13: 978-1-84569-035-9; ISBN-10: 1-84569-035-4)
This comprehensive book on the durability of FRP composites will make it easier for the practising civil engineer and designer to use these materials on a routine basis. It addresses the current lack, or inaccessibility, of data related to the durability of these materials, which is proving to be one of the major challenges to the widespread acceptance and implementation of FRP composites in civil infrastructure. The book should help further the acceptance of composites for civil structural applications by providing a source for practising engineers, decision makers, and students involved in architectural engineering, construction and materials, disaster reduction, environmental engineering, maritime structural technology, transportation engineering and urban planning.

Details of this book and a complete list of Woodhead's titles can be obtained by:

- visiting our website at www.woodheadpublishing.com
- contacting Customer Services (e-mail: sales@woodhead-publishing.com; fax: +44 (0) 1223 893694; tel.: +44 (0) 1223 891358 ext. 130; address: Woodhead Publishing Limited, Abington Hall, Abington, Cambridge CB21 6AH, England)

If you would like to receive information on forthcoming titles in this area, please send your address details to: Francis Dodds (address, tel. and fax as above; e-mail: francisd@woodhead-publishing.com). Please confirm which subject areas you are interested in.

Composites forming technologies

Edited by
A. C. Long

The Textile Institute

CRC Press
Boca Raton Boston New York Washington, DC

WOODHEAD PUBLISHING LIMITED
Cambridge England

Published by Woodhead Publishing Limited in association with The Textile Institute
Woodhead Publishing Limited
Abington Hall, Abington
Cambridge CB21 6AH, England
www.woodheadpublishing.com

Published in North America by CRC Press LLC
6000 Broken Sound Parkway, NW
Suite 300, Boca Raton FL 33487, USA

First published 2007, Woodhead Publishing Limited and CRC Press LLC
© 2007, Woodhead Publishing Limited
The authors have asserted their moral rights.

This book contains information obtained from authentic and highly regarded sources. Reprinted material is quoted with permission, and sources are indicated. Reasonable efforts have been made to publish reliable data and information, but the authors and the publishers cannot assume responsibility for the validity of all materials. Neither the authors nor the publishers, nor anyone else associated with this publication, shall be liable for any loss, damage or liability directly or indirectly caused or alleged to be caused by this book.

Neither this book nor any part may be reproduced or transmitted in any form or by any means, electronic or mechanical, including photocopying, microfilming and recording, or by any information storage or retrieval system, without permission in writing from Woodhead Publishing Limited.

The consent of Woodhead Publishing Limited does not extend to copying for general distribution, for promotion, for creating new works, or for resale. Specific permission must be obtained in writing from Woodhead Publishing Limited for such copying.

Trademark notice: product or corporate names may be trademarks or registered trademarks, and are used only for identification and explanation, without intent to infringe.

British Library Cataloguing in Publication Data
A catalogue record for this book is available from the British Library

Library of Congress Cataloging in Publication Data
A catalog record for this book is available from the Library of Congress

Woodhead Publishing Limited ISBN-13: 978-1-84569-033-5 (book)
Woodhead Publishing Limited ISBN-10: 1-84569-033-8 (book)
Woodhead Publishing Limited ISBN-13: 978-1-84569-253-7 (e-book)
Woodhead Publishing Limited ISBN-10: 1-84569-253-5 (e-book)
CRC Press ISBN-13: 978-0-8493-9102-6
CRC Press ISBN-10: 0-8493-9102-4
CRC Press order number: WP9102

The publishers' policy is to use permanent paper from mills that operate a sustainable forestry policy, and which has been manufactured from pulp which is processed using acid-free and elementary chlorine-free practices. Furthermore, the publishers ensure that the text paper and cover board used have met acceptable environmental accreditation standards.

Project managed by Macfarlane Production Services, Dunstable, Bedfordshire, England (macfarl@aol.com)
Typeset by Godiva Publishing Services Ltd, Coventry, West Midlands, England
Printed by TJ International Limited, Padstow, Cornwall, England

Contents

Contributor contact details — xi

Introduction — xv

1 Composite forming mechanisms and materials characterisation — 1
A C LONG and M J CLIFFORD, University of Nottingham, UK

1.1 Introduction — 1
1.2 Intra-ply shear — 3
1.3 Axial loading — 9
1.4 Ply/tool and ply/ply friction — 10
1.5 Ply bending — 12
1.6 Compaction/consolidation — 14
1.7 Discussion — 19
1.8 References — 19

2 Constitute modelling for composite forming — 22
R AKKERMAN and E A D LAMERS, University of Twente, The Netherlands

2.1 Introduction — 22
2.2 Review on constitutive modelling for composite forming — 22
2.3 Continuum based laminate modelling — 29
2.4 Multilayer effects — 34
2.5 Parameter characterisation — 35
2.6 Future trends — 43
2.7 References — 44

3		**Finite element analysis of composite forming**	**46**
		P BOISSE, INSA de Lyon, France	
3.1		Introduction: finite element analyses of composite forming, why and where?	46
3.2		The multiscale nature of composite materials and different approaches for composite forming simulations	48
3.3		The continuous approach for composite forming process analysis	50
3.4		Discrete or mesoscopic approach	57
3.5		Semi-discrete approach	59
3.6		Multi-ply forming and re-consolidation simulations	70
3.7		Conclusions	75
3.8		References	75
4		**Virtual testing for material formability**	**80**
		S V LOMOV, Katholieke Universiteit Leuven, Belgium	
4.1		Introduction	80
4.2		Mechanical model of the internal geometry of the relaxed state of a woven fabric	82
4.3		Model of compression of woven fabric	84
4.4		Model of uniaxial and biaxial tension of woven fabric	89
4.5		Model of shear of woven fabric	93
4.6		Parametric description of fabric behaviour under simultaneous shear and tension	96
4.7		Conclusions: creating input data for forming simulations	111
4.8		References	112
5		**Optimization of composites forming**	**117**
		W-R YU, Seoul National University, Korea	
5.1		Introduction	117
5.2		General aspects of optimization	118
5.3		Optimization of composite forming	126
5.4		Conclusions	142
5.5		References	142
6		**Simulation of compression moulding to form composites**	**144**
		E SCHMACHTENBERG, Universität Erlangen-Nürnberg, Germany and K SKRODOLIES, Institut für Kunststoffverarbeitung, Germany	
6.1		Introduction	144
6.2		Theoretical description of the simulation	145

6.3	Examples of use of the simulation	161
6.4	Measurement of the material data	172
6.5	References	174
6.6	Symbols	175

7 Understanding composite distortion during processing 177
M R WISNOM and K D POTTER, University of Bristol, UK

7.1	Introduction	177
7.2	Fundamental mechanisms causing residual stresses and distortion	177
7.3	Distortion in flat parts	181
7.4	Spring-in of curved parts	186
7.5	Distortion in more complex parts	192
7.6	Conclusions	194
7.7	References	195

8 Forming technology for composite/metal hybrids 197
J SINKE, Technical University Delft, The Netherlands

8.1	Introduction	197
8.2	Development of composite/metal hybrids	198
8.3	Properties of fibre metal laminates	201
8.4	Production processes for fibre metal laminates	205
8.5	Modelling of FML	213
8.6	Conclusions	218
8.7	References	219

9 Forming self-reinforced polymer materials 220
I M WARD and P J HINE, University of Leeds, UK and D E RILEY, Propex Fabrics, Germany

9.1	Introduction	220
9.2	The hot compaction process	220
9.3	Commercial exploitation	224
9.4	Postforming studies	225
9.5	Key examples of commercial products	232
9.6	Future developments	235
9.7	Acknowledgements	236
9.8	References	236

10 Forming technology for thermoset composites 239
R PATON, Cooperative Research Centre for Advanced Composite Structures Ltd, Australia

10.1	Introduction	239
10.2	Practicalities of forming thermoset prepeg stacks	240
10.3	Deformation mechanisms in woven fabric prepeg	241
10.4	Tape prepreg	247
10.5	Forming processes	248
10.6	Tooling equipment	250
10.7	Diaphagm forming tooling	251
10.8	Potential problems	252
10.9	Process capabilities	253
10.10	Future trends	253
10.11	References	254

11 Forming technology for thermoplastic composites 256
R BROOKS, University of Nottingham, UK

11.1	Introduction	256
11.2	Thermoplastic composite materials (TPCs) for forming	256
11.3	Basic principles of TPC forming technologies	262
11.4	Forming methods	264
11.5	Some recent developments	273
11.6	Conclusions	275
11.7	References	275

12 The use of draping simulation in composite design 277
J W KLINTWORTH, MSC Software Ltd, UK and
A C LONG, University of Nottingham, UK

12.1	Introduction	277
12.2	Zone and ply descriptions	277
12.3	Composites development process	278
12.4	Composites data exchange	281
12.5	Draping and forming simulation	282
12.6	Linking forming simulation to component design analysis	284
12.7	Conclusions	291
12.8	References	292

13	Benchmarking of composite forming modelling techniques	293

J L GORCZYCA-COLE and J CHEN, University of Massachusetts Lowell, USA and J CAO, Northwestern University, USA

13.1	Introduction	293
13.2	Forming process and fabric properties	295
13.3	Experimental	297
13.4	Numerical analyses	313
13.5	Conclusions and future trends	315
13.6	Acknowledgements	316
13.7	References and further reading	317
	Index	318

Contributor contact details

(* = main contact)

Editor
A.C. Long
School of Mechanical Materials and Manufacturing Engineering
University of Nottingham
University Park
Nottingham NG7 2RD
UK

E-mail: Andrew.Long@nottingham.ac.uk

Chapter 1
A.C. Long* and M.J. Clifford
School of Mechanical Materials and Manufacturing Engineering
University of Nottingham
University Park
Nottingham NG7 2RD
UK

E-mail: mike.clifford@nottingham.ac.uk

Chapter 2
R. Akkerman* and E.A.D. Lamers
Construerende Technische Wetenschappen
Universiteit Twente – CTW
Postbus 217
7500AE Enschede
The Netherlands

E-mail: r.akkerman@ctw.utwente.nl

Chapter 3
P. Boisse
Laboratoire de Mécanique des Contacts et des Solides
UMR CNRS 5514
INSA de Lyon
France

E-mail: Philippe.Boisse@insa-lyon.fr

Chapter 4
S. Lomov
Department of Metallurgy and Materials Engineering
Kasteelpark Arenberg 44
BE-3001 Heverlee
Belgium

E-mail: stepan.lomov@mtm.kuleuven.be

Chapter 5
W.-R. Yu
Dept. of Materials Science and Engineering
College of Engineering

Seoul National University
San 56-1 Silim dong
Gwanak-gu
Seoul 151-744
Korea

E-mail: woongryu@snu.ac.kr

Chapter 6

E. Schmachtenberg
Universität Erlangen-Nürnberg
Lehrstuhl für Kunststofftechnik
Am Weichselgarten 9
91058 Erlangen-Tennenlohe
Germany

E-mail: Schmachtenberg@lkt.uni-erlangen.de

K. Skrodolies
Institute of Plastics Processing at RWTH Aachen University
Pontstraße 49
52062 Aachen
Germany

E-mail: zentrale@ikv.rwth-aachen.de

Chapter 7

M. R. Wisnom* and K. D. Potter
Professor of Aerospace Structures
University of Bristol
Advanced Composites Centre for Innovation and Science
Queens Building 0.64
University Walk
Bristol BS8 1TR
UK

E-mail: M.Wisnom@bristol.ac.uk

Chapter 8

J. Sinke
Faculty of Aerospace Engineering
Technical University Delft

Aerospace Materials and Manufacturing
Kluyverweg 1
2629HS, Delft
The Netherlands

E-mail: j.sinke@tudelft.nl

Chapter 9

I.M. Ward*, P.J. Hine and D.E. Riley
IRC in Polymer Science & Technology
School of Physics and Astronomy
University of Leeds
Leeds LS2 9JT
UK

E-mail: I.M.Ward@leeds.ac.uk
p.j.hine@leeds.ac.uk
Derek.Riley@propexfabrics.com

Chapter 10

R. Paton
Cooperative Research Centre for Advanced Composite Structures Ltd
506 Lorimer St, Fishermens Bend
Port Melbourne, 3207
Australia

E-mail: r.paton@crc-acs.com.au

Chapter 11

R. Brooks
School of Mechanical Materials and Manufacturing Engineering
University of Nottingham
University Park
Nottingham NG7 2RD
UK

E-mail: richard.brooks@nottingham.ac.uk

Chapter 12

J.W. Klintworth* and A.C. Long
MSC Software Ltd
MSC House
Lyon Way
Frimley
Camberley
Surrey GU16 7ER
UK

E-mail:
    john.klintworth@mscsoftware.com

Chapter 13

J.L. Gorczyca-Cole and J. Chen*
University of Massachusetts Lowell
One University Avenue
Lowell, MA 01854
USA

E-mail: Julie_chen@uml.edu

J. Cao
Department of Mechanical
    Engineering
Northwestern University
2145 Sheridan Road
Evanston, IL 60208-3111
USA

E-mail: jcao@northwestern.edu

Introduction

Composite materials are available in many forms and are produced using a variety of manufacturing methods. A range of fibre types is used – primarily carbon and glass – and these can be combined with a variety of polymer matrices. This book concentrates on 'long' fibre composites, including fibres from a few centimetres in length (i.e. excluding injection moulding compounds). So the processing methods of interest include compression moulding of thermoplastic or thermoset moulding compounds; resin transfer moulding based on dry fibre preforms; forming and consolidation of thermoset prepreg and thermoplastic sheets; and forming of new material forms including composite/metal laminates and polymer/polymer (self-reinforced) composites.

Whatever the material form or manufacturing process, there is one common step: forming of initially planar material into a three dimensional shape. This is the focus of 'Composite Forming Technologies'. The book includes descriptions of industrial forming processes, case studies and applications, and methods used to simulate composite forming. This description is intended for manufacturers of polymer composite components, end-users and designers, researchers in the fields of structural materials and manufacturing, and materials suppliers. Whilst the bulk of the text is devoted to modelling tools, the intention is to provide useful guidance and to inform the reader of the current status and limitations of both research and commercial tools. It is hoped that this will form essential reading for the users of such modelling tools, whilst encouraging others to 'take the plunge' and adopt a simulation approach to manufacturing process design.

This text may be considered broadly in two halves, with Chapters 1–7 covering the fundamental aspects of modelling and simulation, and Chapters 8–13 describing practical aspects including manufacturing technologies and modern practices in composites design. The first chapter provides a comprehensive introduction to the range of deformation mechanisms that can occur during forming for a range of materials, along with appropriate test methods and representative data. Chapter 2 describes fundamental constitutive models as required for composite forming, including the bases for commercial kinematic (draping) and mechanical (forming) simulations. The latter topic is continued in Chapter 3, including a detailed description of finite element simulation techniques for forming of dry fabric preforms. The methodology here can be considered similar to that used for sheet metal forming, albeit with a more complex material model. Chapter 4 continues the modelling theme, with a description of 'virtual testing', whereby materials input data for forming simulation are predicted from the material structure. This topic is of particular interest, as it may offer the opportunity to select materials that are fit-for-forming, or even to design new materials with a specific component in mind. Chapter 5 details the use of modern simulation techniques for composite forming within an optimisation scheme, with the aim of selecting materials and process parameters to eliminate such defects as wrinkling or undesirable fibre orientations. Chapter 6 describes the methodology and current status of simulation tools for compression moulding, including applications to sheet moulding compound (SMC) and glass mat thermoplastic (GMT). The following chapter completes the initial treatment of simulation and modelling, with a description of composite distortion – notably the common phenomenon of 'spring-in' – caused by manufacturing induced stresses.

The second half of the book begins with four chapters describing forming technologies for a range of materials. This begins with a relatively new family of materials – composite/metal hybrids – which have recently found applications in the aerospace sector (notably as fuselage panels for the Airbus A380). Another new family is covered next, referred to as 'self-reinforced polymers'. These materials include fibre and matrix from the same polymer material, addressing one of the current concerns for polymer composites – recycling. The next two chapters cover more conventional materials – thermoset prepreg and thermoplastic composite sheet. Prepreg forming technologies are described in detail, from the traditional hand lay-up and autoclave cure approach to current developments in automated tape placement and diaphragm forming. The thermoplastics chapter includes a detailed description of the range of material forms, along with their appropriate forming and consolidation techniques. Chapter 12 describes the current state-of-the-art in simulation software for composite forming within an industrial context, detailing the use of modern software tools to design the material lay-up, and describing how these tools can be integrated within the manufacturing environment. Finally Chapter 13 covers the issue of benchmarking of composite forming. This topic is particularly timely, drawing on current worldwide efforts to compare both formability characterisation tests and forming simulation tools for benchmark materials. It is hoped that this will lead to standardisation of formability testing – a key requirement for more widespread use of analysis tools – and guidelines on the accuracy of the range of simulation approaches that are currently available.

1
Composite forming mechanisms and materials characterisation

A C LONG and M J CLIFFORD, University of Nottingham, UK

1.1 Introduction

This chapter describes the primary deformation mechanisms that occur during composites forming. Experimental procedures to measure material behaviour are described, and typical material behaviour is discussed. The scope of this description is reasonably broad, and is relevant to a variety of manufacturing processes. While other materials will be mentioned, the focus here is on forming materials based on continuous, aligned reinforcing fibres. Specifically, materials of interest here include:

- Dry fabrics, formed to produce preforms for liquid composite moulding.
- Prepregs, comprising aligned fibres (unidirectional or interlaced as a textile) within a polymeric (thermoset or thermoplastic) matrix.

While other materials are also formed during composites processing, the above have received by far the most attention amongst the research community. The techniques described here can also be applied to polymer/polymer composites, although these materials present a number of challenges (see Chapter 9). Moulding compounds such as glass-mat thermoplastics (GMTs) and thermoset sheet moulding compounds (SMCs) are formed by a compression (flow) moulding process; here formability is usually characterised by rheometry (see Chapter 6).

Focusing on continuous, aligned fibre materials, a number of deformation mechanisms during forming can be identified (Table 1.1). The remainder of this chapter will focus on methods for characterising materials behaviour. Materials testing typically has a number of objectives. Often the primary motivation is simply to understand materials behaviour during forming, and in particular to rank materials in terms of formability. If this can be related to the material structure, then this understanding may facilitate design of new materials or optimisation of manufacturing process conditions. Another aim may be to obtain materials data for forming simulation. For the most advanced codes, this may

2　Composites forming technologies

Table 1.1 Deformation mechanisms for continuous, aligned fibre based materials during forming

Mechanism	Schematic	Characteristics
Intra-ply shear		• Rotation of between parallel tows and at tow crossovers, followed by inter-tow compaction • Rate and temperature dependent for prepreg • Key deformation mode (along with bending) for biaxial reinforcements to form 3D shapes
Intra-ply tensile loading		• Extension parallel to tow direction(s) • For woven materials initial stiffness low until tows straighten; biaxial response governed by level of crimp and tow compressibility • Accounts for relatively small strains but represents primary source for energy dissipation during forming
Ply/tool or ply/ply shear		• Relative movement between individual layers and tools • Not generally possible to define single friction coefficient; behaviour is pressure and (for prepreg) rate and temperature dependent
Ply bending		• Bending of individual layers • Stiffness significantly lower than in-plane stiffness as fibres within tows can slide relative to each other; rate and temperature dependent for prepreg • Only mode required for forming of single curvature and critical requirement for double curvature
Compaction/ consolidation		• Thickness reduction resulting in increase in fibre volume fraction and (for prepreg) void reduction • For prepreg behaviour is rate and temperature dependent.

require a full mechanical characterisation of the material under axial, shear and bending loads. The use of such data is described in detail in Chapter 3.

In almost all cases, test methods are non-standardised and have been developed by designers or researchers with a particular material and process in mind. This means that test methods, specimen dimensions, data treatment and presentation differ between practitioners. Here we will give a description of what we believe to be 'best practice', although this is clearly a subjective assessment. Benchmarking and comparison of results between laboratories is being addressed within an international exercise; this is discussed in detail in Chapter 13.

1.2   Intra-ply shear

This mechanism occurs when the material is subjected to in-plane shear. This essentially corresponds to relative sliding of parallel tows within a fabric layer or composite ply, and (for textile-based materials) rotation of tows at their crossovers. Intra-ply shear is usually considered to be the primary deformation mechanism for aligned fibre-based materials. Coupled with low bending resistance, the ability of materials to shear in this way allows them to be formed to three dimensional shapes without forming folds or wrinkles. A good analogy here is to compare a woven fabric to a sheet of paper. Both may have a similar bending stiffness, but unlike paper the ability of the fabric to shear allows it to be formed over shapes with double curvature.

Various experimental methods exist to characterise the shear resistance of dry textiles and aligned or woven composite materials. Early developments here were for apparel fabrics; of particular relevance is the 'Kawabata Evaluation System for Fabrics (KES-F)', a series of test methods and associated testing equipment for textile mechanical behaviour including tensile, shear, bending, compression and friction.[1] However whilst this system has been used widely for clothing textiles, its application to reinforcement fabrics has been limited.[2] This is probably due to the fact that KES-F provides single point data at relatively low levels of deformation, coupled with the limited availability of the (expensive) testing equipment.

Amongst the composites forming community, two widely used test methods are the *picture frame test*[3–9] and the *bias extension test*.[8–12] In this section we present a guide to the use of these test methods and how to make good use of the output data.

1.2.1   Picture frame test

The picture frame (or rhombus) test can be used to measure the force generated by shearing technical textiles and textile composites, including thermoplastic and thermoset based materials. Cross-shaped test samples can be cut or stamped

4   Composites forming technologies

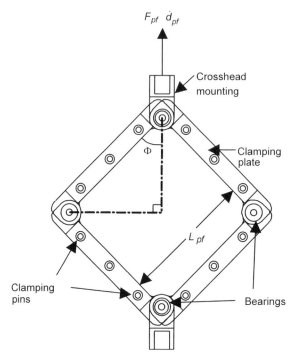

1.1 Schematic of picture frame shear rig. L_{pf} is the side length measured between the centres of the bearings, F_{pf} is the axial force measured by the load cell, $\dot{d}_{pf}$ is the rate of crosshead displacement and Φ is the frame angle.

from rolls or sheets of material using a template. Great care should be taken to ensure that the fibres are perfectly aligned with the edges of the template. Test samples are held in a purpose-built square frame, hinged at each corner (see Fig. 1.1). The frame is loaded into a tensile test machine, and two diagonally opposite corners are extended, imparting pure and uniform shear in the test specimen on a macroscopic scale. There is no uniformly applied standard test procedure. Depending on the material to be tested, various methods can be used to restrain test specimens. For dry fabric, impaling samples on a number of pins may improve repeatability. This approach avoids imparting tensile strain in the fibres and reduces bending of tows.[9] Other materials, such as thermoplastic composites, may need to be tightly clamped to prevent fibres from slipping.[7]

During the test, the axial force required to deform the sample is recorded. Since many impregnated composite materials are based on polymers with viscosities that depend on shear rate, it may be useful to perform tests at different cross-head displacement rates. It is important to observe the surface of the test samples during the test, as misaligned or poorly clamped samples can wrinkle almost from the start. These results should be discarded. It is usual for samples to wrinkle towards the end of the test (between 50° and 70° of shear deformation) and by careful observation the test can be used to estimate the 'locking angle' of a material. For successful tests, shear force can be calculated from the cross-head force using:

$$F_s = \frac{F_{pf}}{2\cos\Phi} \quad\quad 1.1$$

where Φ is the frame angle and F_{pf} is the measured axial picture frame force. Test data can be normalised by dividing the shear force by the length of the picture frame, L_{pf}. Graphs of shear force against shear angle can be produced, where the shear angle is defined as:

$$\theta = \pi/2 - 2\Phi \quad\quad 1.2$$

The shear angle can be calculated from the cross-head displacement, d_{pf}, by

$$\theta = \frac{\pi}{2} - 2\cos^{-1}\left[\frac{1}{\sqrt{2}} + \frac{d_{pf}}{2L_{pf}}\right] \quad\quad 1.3$$

The picture frame test procedure is relatively simple to perform and results should be reasonably repeatable if sufficient care is taken in cutting test samples and the correct clamping technique is used. The major benefit of the test is that shear angle and angular shear rate can be easily calculated from the cross-head displacement and displacement rate.

Typical results from picture frame experiments conducted on a range of textile reinforcements and prepregs are shown in Figs 1.2 and 1.3. For clarity these graphs represent single experiments, although it should be noted that even well controlled experiments exhibit scatter of up to $\pm 20\%$ in shear force for a particular angle. Most materials exhibit some similarities in terms of their shear force curve. Initially the shear resistance is low – for dry fabrics this represents dry friction at tow crossovers, whilst for prepreg this corresponds to lubricated friction or viscous shear of polymer between fibres or yarns. Towards the end of the test the resistance increases significantly – this happens once adjacent yarns come into contact, representing yarn compaction (fabrics) or squeeze flow (prepreg). If the test were continued, the curve would tend towards an asymptote corresponding to maximum possible deformation. Non-crimp fabrics (Fig. 1.2b) exhibit more complex behaviour, as the shear resistance depends on the direction of shear. These materials consist of perpendicular layers of tows held together by a stitching thread. This thread restricts the movement of the tows, so that materials exhibit higher resistance to shear when they are sheared parallel to the stitch.

Based on the data given in Figs 1.2 and 1.3, at the simplest level the curve may be approximated using a bi-linear model:

$$F_s/L_{pf} = E_0 \tan\theta \quad\quad (\theta < \theta_0) \quad\quad 1.4$$
$$F_s/L_{pf} = E_0 \tan\theta_0 + E_\infty \tan(\theta - \theta_0) \quad\quad (\theta \geq \theta_0)$$

Clearly the material response (and hence the constants θ_0, E_0 and E_∞ in equation 1.4) depends on material type and (for prepreg) experimental conditions such as

6   Composites forming technologies

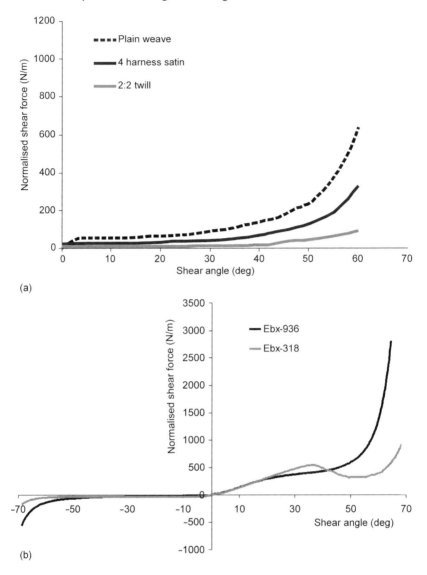

1.2 Picture frame shear data for dry glass fabric reinforcements. (a) Three woven fabrics with superficial density 800 g/m². (b) Non-crimp fabrics retained with tricot (Ebx-936) and chain (Ebx-318) stitch oriented at 45° to the tows in each case. Negative shear angle represents deformation perpendicular to the stitch.

rate and temperature. However, as a rough guide, θ_0 is likely to be around 40° for dry fabrics and <30° for prepreg, and is indicative of the point at which the material starts to lock as adjacent tows come into contact. The ratio E_0/E_∞ is of the order 3–4 for woven fabrics and closer to unity for prepreg. The most

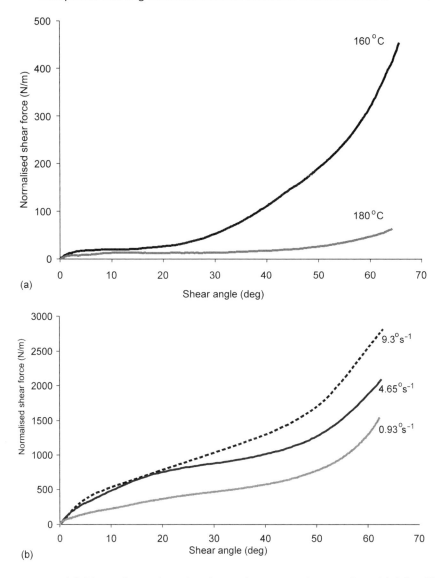

1.3 Picture frame shear data for pre-impregnated composites. (a) 2:2 twill weave glass/polypropylene thermoplastic composite at two temperatures. (b) 5 harness satin weave carbon/epoxy prepreg at various angular shear rates (room temperature).

rigorous approach to intra-ply shear characterisation would require every material to be characterised under all possible forming conditions. As this is clearly not a practical proposition, researchers have attempted to develop models to predict materials formability from textile structure[13] and matrix rheology.[14] This approach is discussed further in Chapter 4.

8　Composites forming technologies

1.2.2　Bias extension test

The bias extension test involves clamping a rectangular piece of bidirectional material such that the tows are orientated initially at $\pm 45°$ to the direction of the applied tensile force. The material sample can be characterised by the aspect ratio, $\lambda = L_o/w_o$, where the sample width w_o is usually greater than 100mm. Figure 1.4 shows an idealised bias extension test sample with $\lambda = 2$. The sample is divided into a number of regions which deform at different rates as the test proceeds. Generally it can be shown that the shear angle in region A is always twice that in regions denoted B, while region C remains un-deformed. The deformation in region A is the same as the deformation produced by the picture frame test, as long as intra-ply shear is the only mechanism; in practice the angle will be somewhat lower than in an equivalent picture frame sample as intra-ply slip will occur (i.e. tow spacing will increase) particularly as the material approaches locking. The sample aspect ratio must be at least two for the three different deformation regions to exist. Increasing the length/width ratio, λ, to higher values serves to increase the area of region A.

Bias extension tests are simple to perform and can provide reasonably repeatable results. Axial force and cross-head displacement are recorded during a test. The test provides a useful method to estimate the locking angle of a material; once the material in region A reaches the locking angle, it usually ceases to shear. As with the picture frame test, different clamping conditions have been suggested by various researchers, but the boundary conditions tend to affect the data much less than for picture frame tests. The method may be preferred for gaining shear data at elevated temperatures for thermoplastic composites, since the influence of relatively cool material adjacent to the metal clamps during high temperature testing is of less importance than in picture frame tests.

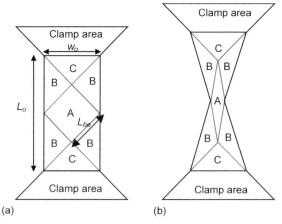

1.4 Idealised bias extension test sample with $\lambda = L_0/w_0 = 2$, where L_0 and w_0 are respectively the initial length and width of the specimen.

Producing graphs of shear force against shear angle can be achieved by following the same data analysis procedure as for picture frame test data, considering region A to be equivalent to a picture frame specimen with side length L_{be} (see Fig. 1.4). However, for increased accuracy, it may be necessary to measure the shear deformation rather than rely on the material following idealised deformation kinematics. This can be achieved by visual analysis, which can prove time consuming without an automated image acquisition and analysis approach.

Forces from bias extension tests can be normalised by dividing by a characteristic dimension, such as sample width. However, this does not allow data from samples with different aspect ratios to be compared directly. Recent work by Harrison[9] considered the energy dissipated within regions A, B and C to develop a more sophisticated normalisation technique for bias extension test data. In addition to allowing results from different aspect ratio samples to be compared, this also allows picture frame and bias extension test data to be correlated directly. Whilst this might allow mechanical data to be extracted in terms of shear force versus shear strain in a form suitable for simulation software, the data analysis procedure is extremely complex and hence at present the picture frame test is preferred for this purpose.

1.3 Axial loading

Loading of aligned fibre based materials along the fibre axis or axes typically results in very large forces and very low maximum strains in comparison to intra-ply shear. This might suggest that deformation under axial loading is of secondary importance, and indeed this is reflected in the relatively limited attention received by this topic. Boisse[15] has long argued that this behaviour cannot be neglected, since the high magnitude of the axial stiffness indicates that tensile loading of the fibres accounts for the majority of energy dissipated during forming.

Axial loading of textiles and composites can be conducted using standard tensile testing equipment, although as for the bias extension test (Section 1.2.2) wide samples are usually used. Unidirectional fibre materials will typically exhibit a linear force-displacement response when loaded parallel to the fibre axis. This is not the case for textile based materials, which exhibit an initial, non-linear stiffening due to crimp in the tows. As the fibres become aligned with the direction of loading, the response becomes linear and is determined by the fibre modulus and volume fraction. The importance of this 'de-crimping' depends on the properties of the transverse tows, and in particular their resistance to bending and compaction. If the transverse tows are also loaded, then the de-crimping zone will decrease in magnitude. Boisse[16] has analysed a wide range of fabrics using a specially designed biaxial loading frame. Some typical results are given in Fig. 1.5 for a plain weave fabric. When loaded

10   Composites forming technologies

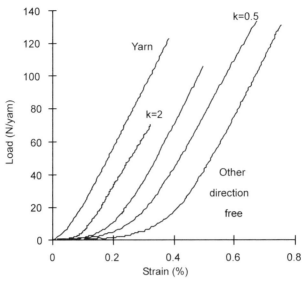

1.5 Results of biaxial tensile tests for a balanced glass plain weave fabric.[17] The load is measured along one tow direction, with the constant *k* determining the ratio between strains in the loading and transverse directions.

uniaxially (denoted 'other direction free'), the non-linear force region extends to a strain of approximately 0.5%, but the force to completely straighten the tows is low. As the ratio between strains in the tested (warp) and transverse (weft) directions increases, the force curve tends towards the behaviour of an individual tow (yarn). The testing procedure and relevance of this behaviour to composite forming simulation are described in detail in Chapter 3.

1.4   Ply/tool and ply/ply friction

During an automated forming operation, friction between the material and forming tools governs the transfer of loads to the material. In multi-layer forming processes, friction between individual layers of material is also of importance. For example, when forming layers of prepreg at different orientations to each other, compressive forces generated by intra-ply shear in one layer may be transferred into adjacent layers, causing compression along the fibre direction and hence wrinkling of the form. Hence to model forming processes accurately, measurement of friction at ply/tool and ply/ply interfaces is important.

A number of test methods are possible for measurement of friction. The simplest approach is the so-called 'inclined plane method'. Here a block of tooling material is placed on a piece of fabric/prepreg mounted on a rigid plate. The plate is then inclined until the block starts to move, with the tangent of the angle of inclination defining the friction coefficient. More sophisticated variations on this approach exist, for example pulling the block along the surface of fabric/prepreg and measuring the force required to maintain a constant velocity. However, these techniques are best suited to materials that exhibit a constant coefficient of friction. Unfortunately this does not tend to be the case for fabrics or prepregs.

Several alternatives to the above have been developed to allow variables such as normal pressure, testing rate and temperature to be varied. Ply pull out tests have been used to measure ply/ply friction.[18] Data from such tests are discussed in Chapter 10. Murtagh[19] developed a device whereby a layer of thermoplastic prepreg or tooling material was sandwiched between two prepreg layers, held together with a controlled normal pressure. The whole apparatus was heated to evaluate the effect of temperature on behaviour, and the tooling material was withdrawn at various rates with the required force measured using a load cell. Wilks[20] designed apparatus based on a similar principle to Murtagh, although here a layer of fabric/prepreg was sandwiched between two layers of tooling material (Fig. 1.6). For friction between dry fabric and tooling materials, some dependence on normal pressure is observed, with friction reducing marginally with increasing pressure as the fabric surface was flattened against the tool. Friction coefficients of between 0.2 and 0.4 have been measured between glass fabrics and tooling materials.

Typical results are shown in Fig. 1.7 for friction between a glass/polypropylene thermoplastic composite and a steel tool. Shear stress increases

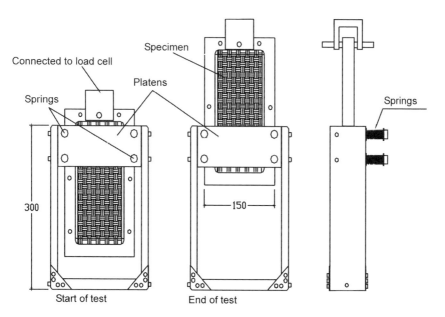

1.6 Schematic of ply/tool friction measurement apparatus.

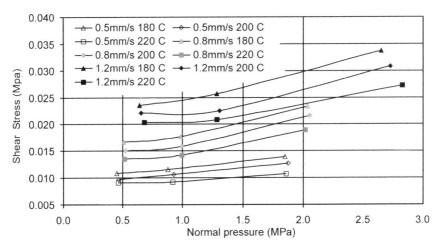

1.7 Shear stress at the interface between steel tools and glass/polypropylene under applied normal pressure.[20] Results are for velocities of 0.5, 0.8 and 1.2 mm/s at 180°C, 200°C and 220°C.

with normal pressure, although the relationship is not linear. At a given normal pressure, the shear stress increases with increasing rate and with decreasing temperature.

Along with other published studies, the results in Fig. 1.7 suggest that for prepreg, the friction coefficient (ply/tool and ply/ply) should be defined as a function of rate, temperature and pressure. Wilks[20] suggested a phenomenological model for shear stress at the interface of the form:

$$\tau = \eta\dot{\gamma} + \mu P \qquad 1.5$$

where the first term represents shearing of a polymer film at the prepreg surface (with viscosity η and shear strain rate $\dot{\gamma}$) and the second term representing Coulomb friction caused by fibre reinforcement penetrating the polymer film (with friction coefficient μ and normal pressure P). In practice the values of μ and the polymer film thickness used to define $\dot{\gamma}$ must be determined empirically.

1.5 Ply bending

The ability of fabric or prepreg to bend out of plane is of course critical for forming of curved components. It is perhaps surprising then that this topic has received relatively little attention, at least for fabric reinforcements and composites. One reason may be that the bending resistance is usually orders of magnitude lower than intra-ply shear resistance, which in turn is significantly lower than tensile stiffness in the fibre direction(s). However, in terms of modelling, this in fact presents a problem as traditional shell element formulations will have a bending stiffness related to the in-plane stiffness of the material. Hence it is necessary to understand the relative magnitude of bending stiffness so that the forces associated with out of plane bending can be scaled appropriately.

Bending stiffness has long been measured for apparel fabrics.[1,21] A standardised test can be performed for bending resistance of fabric under its own weight.[22] This involves sliding a strip of fabric off the edge of a platform until the self-weight causes it to bend to a specified angle (41.5° is specified in the standard). The length of strip necessary to reach this threshold value is recorded, from which the bending rigidity is approximated. Results for woven apparel fabrics indicate that the bending rigidity in the fibre directions is significantly higher than in the bias (±45°) direction. Young et al.[23] have applied this technique to calibrate a finite element model for composite bending. The procedure involved simulating the experiment and adjusting a bending scale factor so that the predicted bending behaviour matched experimental observations.

More informative techniques measure the mechanical resistance to bending using, for example, cantilever, three-point bending or axial buckling experiments, and such tests have been applied recently to prepreg. For example Martin et al.[24] measured the three-point bending behaviour of unidirectional glass/polypropylene composites using a V-shaped punch. At elevated temperature this resulted in an increase in bending force with increasing displacement rate. All tests exhibited an initial increase up to a peak value, after which the force plateaued or reduced gradually. One issue with this approach is that the material has to be supported during bending to stop it from deforming under its own weight. To avoid this problem a simple buckling test can be used.[25] Wang et

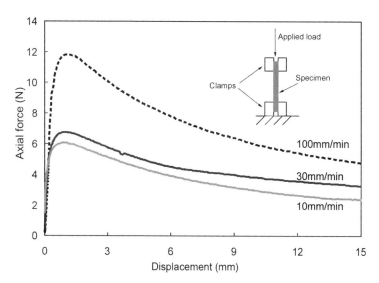

1.8 Bending/buckling behaviour of unidirectional carbon/epoxy thermoset prepreg under axial compressive loading at different rates.

al.[26] have applied this technique to thermoset prepreg, including unidirectional and woven carbon/epoxy composites. Typical data are included in Fig. 1.8 for unidirectional prepreg, illustrating the effect of rate on bending/buckling behaviour. The force response increases linearly with increasing rate, suggesting that the phenomenon is dominated by flow of the polymer between fibre layers. The curve shape illustrates a peak load corresponding to buckling, followed by a reduction in force as displacement is increased. Similar tests for woven fabrics show an initial peak followed by a small reduction and then a plateau in the force, illustrating that the fibre architecture has a clear effect on the bending behaviour. Both material response curves fall within the range observed for dry textiles,[25] although for such materials no clear rate effect is observed.

1.6 Compaction/consolidation

At the end of forming, the material must be compacted or consolidated to increase the fibre volume fraction and (for prepreg) eliminate voids. An understanding of compressibility, typically in terms of compaction pressure versus thickness or fibre volume fraction, allows the required pressure to be determined for the target fibre content. For multi-layer, multi-material reinforcement preforms, the compressibility of each material type is likely to be different, so that each layer attains a different fibre volume fraction under the imposed compaction pressure (or at the desired laminate thickness). Compaction has received a great deal of attention, particularly for dry fibre mats and fabrics. Robitaille[27] has published an extensive review of both experimental methods and modelling approaches for reinforcement compaction, whilst Garcia[28] has reviewed similar techniques for thermoplastic and thermoset prepreg. Testing procedures appear relatively simple, with material compacted between two parallel platens within a universal testing machine. The platens are moved together usually at constant rate to either a pre-determined load or thickness. At this point either the thickness or force can be held constant to measure relaxation or compaction creep. Care must be taken to ensure that the platens are as flat and as parallel as possible, and their relative displacement must be measured carefully – best practice here is to attach an LVDT (linear variable displacement transducer) between the platens.

1.6.1 Compaction behaviour of reinforcements

A great deal of experimental data exists for fabric reinforcements, and it is beyond the scope of this chapter to provide a comprehensive review. Large data sets have published by Robitaille[27] and Correia,[29] and these form the basis for the present discussion. Typical data are included in Fig. 1.9. The graph shows the evolution of average fibre volume fraction (V_f) as a function of compaction pressure (P) for multiple layer stacks of a range of materials. All results show

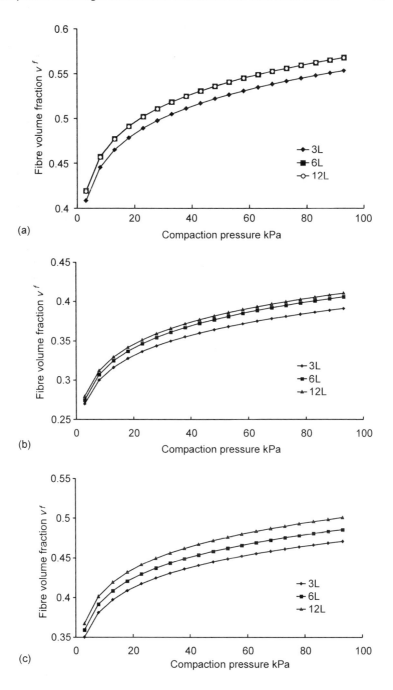

1.9 Typical dry glass fabric compaction behaviour at low pressures (up to 1 bar) for 3, 6 and 12 material layers: (a) triaxial non-crimp, (b) unidirectional non-crimp, (c) plain weave.

that fibre volume fraction initially builds up rapidly with pressure, tending towards a plateau defining the maximum practical fibre content. Each graph shows that for low pressure compaction (as illustrated here), increasing the number of layers within the stack can ease compaction (i.e. lower pressure to attain the required fibre volume fraction). This applies particularly to woven fabrics (e.g. Fig. 1.9c) and may be explained by nesting between the layers.

Other effects observed experimentally include:

- Number of compaction cycles – if the platens are moved apart, the unloading curve does not superimpose on the loading curve and the unloaded material will generally have a higher fibre volume fraction than before compaction due to unrecovered compaction within the tows. Subsequent loading cycles will each attain a higher fibre volume fraction for the same applied pressure, converging on a maximum value after a number of cycles. This is relevant for example in liquid composite moulding, where two compaction cycles may be applied – firstly during preform manufacture and again on mould tool closure.
- Saturation – at low rates (where fluid flow effects are negligible), the compaction curve is usually shifted to the right if the material is lubricated, i.e. the material is more compressible, so that lower pressures are required to achieve a given fibre volume fraction. This is relevant in vacuum infusion (a.k.a. VARTM), where the material is compacted under atmospheric pressure as the resin front advances, with the reinforcement behind the flow front lubricated.

Several authors have presented models for compaction behaviour of fabric reinforcements. These fall into two groups: phenomenological models based on solid mechanics principles, and empirical models to provide a simple representation of the data. Phenomenological models are based typically on representation of reinforcement fibres as a series of beams contacting at a finite number of points along their length. The number of contacts typically increases during compaction, so that the bridging fibre sections gradually stiffen. For example, Cai and Gutowski[30] have proposed a series of models, providing valuable insight into compaction behaviour. However, the models are not strictly predictive, as whilst they consist of physically meaningful parameters, the values of these must be adjusted to fit to experimental data.

For convenience simple empirical models may be preferred. Power law relationships have been proposed by several authors, for example:

$$V_f = V_{f0} \cdot P^B \qquad 1.6$$

Here B is an empirical factor often referred to as the stiffening index, and V_{f0} is equivalent to the fibre volume fraction at a compaction pressure of 1 Pa (although V_{f0} is also usually determined empirically). This type of equation has been found to fit well to experimental data for a wide range of materials.[27,29]

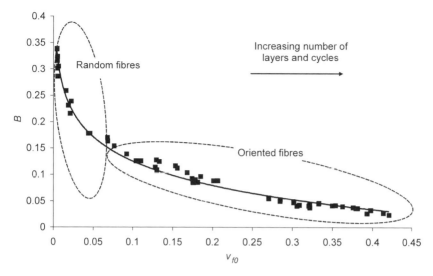

1.10 Compaction master curve, showing the relationship between stiffening index (B) and initial fibre volume fraction (V_{f0}) from equation (1.6) for a range of dry reinforcements.

Correia[29] recently showed that the two parameters in equation (1.6) appear to be related, as indicated in Fig. 1.10. This also provides a good way of characterising material behaviour, with highly compressible materials (such as random mats) towards the left of the curve and less compressible materials (based on highly aligned fibres, e.g. non-crimp fabrics) towards the right.

1.6.2 Consolidation behaviour of prepreg

The majority of thermoset and thermoplastic prepreg materials undergo limited reductions in thickness during consolidation (typically <20%). This is because materials in semi-finished (i.e. as supplied) form are usually relatively well consolidated, and all that is required is the reduction of void content to an acceptable level (typically <1% for aerospace applications). Unfortunately as the materials have no clear paths for entrapped voids to escape, very high pressures may be required to achieve the desired degree of consolidation. Hence reasonably powerful hydraulic presses or high-pressure autoclaves are usually employed to consolidate these materials. A number of authors have conducted consolidation experiments for both thermoset and thermoplastic materials,[31–33] using a similar approach as described above for dry fabric. As would be expected behaviour is highly dependent on temperature and rate, with increasing time at pressure resulting in a reduction in void content to a limiting value.

In an attempt to reduce the pressure levels required for consolidation, various partially impregnated or 'semi-preg' materials have been developed in recent years. The general idea is to minimise the required polymer flow distances and provide continuous channels for air removal within the mould. For example, materials can be assembled as layers of resin film and dry fibre, with the resulting process often referred to as resin film infusion. One popular family of thermoplastic materials is based on co-mingled yarns, where reinforcement fibres are intimately mixed with polymeric fibres. Such materials can be processed by heating, pressure application and cooling. Pressure application can be undertaken within a hydraulic press at modest pressures or using a vacuum bag.

A number of studies have been published on consolidation behaviour of co-mingled fabrics, following initial work by Van West.[32] Wilks[20] analysed consolidation behaviour of co-mingled glass/polypropylene between parallel platens at various rates and temperatures. Typical results are included in Fig. 1.11. Consolidation pressure increases with reducing thickness (increasing normalised thickness) and this increase becomes steeper as the material approaches full consolidation. The pressure to achieve a given thickness increases approximately linearly with compaction rate. In the tests shown here, voids could not be eliminated completely as the compaction pressure was limited to 0.7 MPa. Garcia Gil[28] performed a similar analysis for vacuum consolidation, demonstrating that void contents of <2% could be achieved under atmospheric compaction pressure within 300 seconds at 180°C. Here the vacuum-based process is advantageous as it removes most of the air from the material prior to significant polymer flow. A matched mould process in contrast may allow voids to become entrapped within the material before the target thickness is reached.

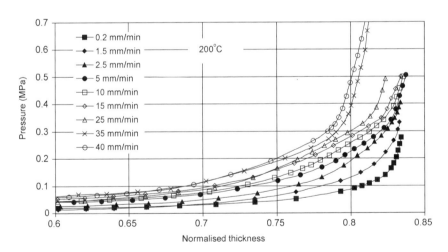

1.11 Typical consolidation behaviour for a commingled glass/polypropylene material, showing required pressure as a function of normalised thickness (or degree of consolidation) at 200°C for various consolidation rates. Normalised thickness is the fully consolidated thickness (zero voidage) divided by the current thickness.

Various models have been proposed for prepreg consolidation, based typically on a combination of fabric compaction and fluid flow.[28,31–33] Where voids are entrapped within the material prior to consolidation (e.g. for traditional prepregs or co-mingled fabrics consolidated without a vacuum) it is important to consider the pressure generated within entrapped voids, which can be done conveniently using the ideal gas law.

1.7 Discussion

This chapter has described the deformation behaviour of reinforcements and composites, focusing on the key deformation mechanisms. Clearly the majority of attention amongst the research community has been on intra-ply shear and compaction behaviour, and here a wealth of data are available in the literature. Less attention has been applied to tensile and bending loads for materials used within composites, although much may be learned here from the large body of work on conventional textiles. The majority of the tests used are non-standard, and as materials data are required increasingly for manufacturing process simulations, this issue must be addressed as a matter of urgency. Some initial efforts here are discussed in Chapter 13.

Almost all of the effort in the field of materials characterisation for composites forming has involved analysis of a single deformation mechanism in isolation. In practice of course several modes will occur simultaneously, for example intra-ply shear and in-plane tension. Coupling between these mechanisms is not clear at present, and may provide insights allowing increased accuracy from forming simulations. Given the wide variety of materials available, predictive modelling to determine material behaviour from constituent properties becomes highly desirable. Such 'virtual testing' tools would allow a wide range of materials to be analysed prior to component manufacture, facilitating selection and design of new materials with formability in mind. This is the subject of Chapter 4.

1.8 References

1. Saville B P, *Physical testing of textiles*, Woodhead Publishing Ltd, Cambridge, 1999.
2. Lomov S V, Verpoest I, Barburski M and Laperre J, 'Carbon composites based on multiaxial multiply stitched preforms. Part 2. KES-F characterisation of the deformability of the preforms at low loads', *Composites Part A*, 2003 **34**(4) 359–370.
3. McGuiness G B and O'Bradaigh C M, 'Development of rheological models for forming flows and picture frame testing of fabric reinforced thermoplastic sheets', *Journal of Non-Newtonian Fluid Mechanics*, 1997 **73** 1–28.
4. Long A C, Rudd C D, Blagdon M and Johnson M S, 'Experimental analysis of fabric deformation mechanisms during preform manufacture', *Proc. 11th International*

Conference on Composite Materials (ICCM-11), Gold Coast, Australia, July 1997, 238–248.
5. Prodromou A G and Chen J, 'On the relationship between shear angle and wrinkling of textile composite preforms', *Composites Part A*, 1997 **28A** 491–503.
6. Mohammad U, Lekakou C, Dong L and Bader M G, 'Shear deformation and micromechanics of woven fabrics', *Composites Part A*, 2000 **31** 299–308.
7. Milani A S, Nemes J A, Pham X T and Lebrun G, 'The effect of fibre misalignment on parameter determination using picture frame test', *Proc. 14th International Conference on Composite Materials (ICCM-14), San Diego, USA*, July, 2003.
8. Lebrun G, Bureau M N and Denault J, 'Evaluation of bias-extension and picture-frame test methods for the measurement of intraply shear properties of PP/glass commingled fabrics', *Composite Structures*, 2003 **61** 341–352.
9. Harrison P, Clifford M J and Long A C, 'Shear characterisation of woven textile composites: a comparison between picture frame and bias extension experiments' *Composites Science and Technology*, 2004 **64**(10–11) 1453–1465.
10. Potter K D, 'The influence of accurate stretch data for reinforcements on the production of complex structural mouldings', *Composites*, July 1979 161–167.
11. Murtagh A M and Mallon P J, 'Shear characterisation of unidirectional and fabric reinforced thermoplastic composites for pressforming applications', *Proc. 10th International Conference on Composite Materials (ICCM-10), Whistler, Canada*, August 1995, 373–380.
12. Wang J, Page R and Paton R, 'Experimental investigation of the draping properties of reinforcement fabrics', *Composites Science and Technology*, 1998 **58** 229–237.
13. Souter B J, *Effects of fibre architecture on formability of textile preforms*, PhD Thesis, University of Nottingham, 2001.
14. Harrison P, Clifford M J, Long A C and Rudd C D, 'A constituent based predictive approach to modelling the rheology of viscous textile composites', *Composites Part A* 2005 **37**(7–8) 915–931.
15. Boisse P, Borr M, Buet K and Cherouat A, 'Finite element simulations of textile composite forming including the biaxial fabric behaviour', *Composites Part B*, 1997 **28**(4), 453–464.
16. Boisse P, Gasser A and Hivet G, 'Analyses of fabric tensile behaviour: Determination of the biaxial tension-strain surfaces and their use in forming simulations', *Composites Part A*, 2001 **32**(10) 1395–1414.
17. Long A C (editor), *Design and manufacture of textile composites*, Woodhead Publishing Ltd, Cambridge, 2005. Chapter 2.
18. Scherer R and Friedrich K, 'Inter- and intraply-slip flow processes during thermoforming of CF/PP-laminates', *Composites Manufacturing*, 1999 **2**(2) 92–96.
19. Murtagh A M, *Characterisation of shearing and frictional behaviour in sheetforming of thermoplastic composites*, PhD Thesis, University of Limerick, 1995.
20. Wilks C E, *Processing technologies for woven glass/polypropylene composites*, PhD Thesis, University of Nottingham, 2000.
21. Hearle J W S, Backer S, Grosberg P, *Structural mechanics of fibers, yarns and fabrics*, Wiley, New York, 1969.
22. BS 3356:1990, 'Determination of bending length and flexural rigidity of fabrics', British Standards Institution, June 1991.
23. Young M, Cartwright B, Paton R, Yu X, Zhang L and Mai Y-W, 'Material

characterisation tests for finite element simulation of the diaphragm forming process', *Proc. 4th Int. ESAFORM Conference on Material Forming, University of Liege*, April 2001.
24. Martin T A, Bhattacharyya D and Collins I F, 'Bending of fibre-reinforced thermoplastic sheets', *Composites Manufacturing*, 1995 **6**(3–4) 177–187.
25. Lindberg J, Behre B and Dahlberg B, 'Shearing and buckling of various commercial fabrics', *Textile Research Journal*, Feb 1961 99–122.
26. Wang J, Lin H, Long A C, Clifford M J and Harrison P, 'Predictive modelling and experimental measurement of the bending behaviour of viscous textile composites', *Proc. 9th Int. ESAFORM Conference, Glasgow*, April 2006.
27. Robitaille F and Gauvin R, 'Compaction of textile reinforcements for composites manufacturing. I: Review of experimental results', *Polymer Composites*, 1198 **19**(2) 198–216.
28. Garcia Gil R, *Forming and consolidation of textile composites*, PhD Thesis, University of Nottingham, 2003.
29. Correia N, *Analysis of the vacuum infusion moulding process*, PhD Thesis, University of Nottingham, 2004.
30. Cai Z and Gutowski T G, 'The 3D deformation behaviour of a lubricated fibre bundle', *Journal of Composite Materials*, 1992 **26**(8) 1207–1237.
31. Hubert P and Poursartip A, 'A method for the direct measurement of the fibre bed compaction curve of composite prepregs', *Composites Part A*, 2001 **32** 179–187.
32. Van West B P, Pipes R B and Advani S G, 'The consolidation of commingled thermoplastic fabrics', *Polymer Composites*, 1991 **12**(6) 417–427.
33. Bernet N, Michaud V, Bourban P-E and Manson J-A E, 'Commingled yarn composites for rapid processing of complex shapes', *Composites Part A*, 2001 **32**(11) 1613–1626.

2
Constitutive modelling for composite forming

R AKKERMAN and E A D LAMERS, University of Twente, The Netherlands

2.1 Introduction

Fibre reorientation occurs when forming a reinforced structure, such as a fabric onto a doubly curved surface. This leads to a change in the angle between warp and fill yarns. The composite properties change inhomogeneously, corresponding to the varying angle between warp and fill yarns. Many composite properties are determined by the angle between the warp and fill yarns, such as the mechanical properties, the coefficients of thermal expansion, the local fibre volume fractions, the local thickness and the permeability. The extent of the fibre reorientation is affected by the product shape and the forming process. The forming process may cause tensile stresses in the fabric yarns, causing subsequent product distortions. Also, wrinkling risks are present due to the incapability of the fabric to deform beyond a maximum shear deformation. The local change in composite properties must be taken into account in order to predict the properties of a product. Drape modelling can predict the process induced fibre orientations and stresses, which can speed up the product development process compared with trial-and-error development. The constitutive model for these biaxially reinforced composite materials is the primary element of these process simulations.

2.2 Review on constitutive modelling for composite forming

Both dry and pre-impregnated fabrics can be draped over a mould during the composite forming process. When forming a dry fabric over the mould, the result is a preform. This preform can be impregnated subsequently with a polymer, for instance in one of the Liquid Composite Moulding (LCM) processes.

When draping pre-impregnated composites, the fabric is embedded in the matrix material. In the case of thermoplastics, several plies can be stacked into a pre-consolidated laminate preform. This preform is heated above the glass transition or melting temperature of the polymer matrix, formed on the tool and subsequently cooled or cured until the product is form stable.

Various drape models for dry and pre-impregnated fabrics have been proposed in the past. Lim and Ramakrishna published a review in 2002 on the forming of composite sheet forming. Two approaches are distinguished in their review: the mapping approach and the mechanics approach. The use of mapping approaches is discussed in Chapter 12, whilst the mechanical modelling of forming is described in Chapter 3. Here, we will concentrate on the underlying constitutive models, using a different classification: the discrete approach and the continuum approach. This classification is based on the representation of the material by the models. Draping multi-layered composites gives rise to additional complexity which will be discussed subsequently.

2.2.1 Discrete models

Three schemes are distinguished in the discrete drape approach: the mapping schemes, the particle based schemes and the truss based schemes.

Mapping based schemes

Mapping schemes are most commonly employed in commercial packages for drape predictions. A layer of fabric is represented by a square mesh which is fitted onto the drape surface. The mapping scheme is based on the assumption that the fabric only deforms due to shear deformation, and fibre extension can be neglected. The resin, if present, is also neglected during the simulation. The fabric always remains in a fixed position on the draping surface after having been mapped. The shape of the product must be represented in algebraic expressions when modelling draping with a mapping scheme.

Several methods are used to predict the fibre reorientation of the fabric. The geometrical model, also referred to as the kinematics or fishnet model, is a widely used model to predict the resulting fibre reorientation for doubly curved fabric reinforced products. This model was initially described by Mack and Taylor in 1956, based on a pinned-joint description of the weave. The model assumes inextensible fibres pinned together at their crossings, allowing free rotation at these joints. An analytic solution of the fibre redistribution was presented for a fabric oriented in the bias direction on the circumference of simple surfaces of revolution, such as cones, spheres and spheroids. The resulting fibre orientations were solved as a function of the constant height coordinate of the circumference.

From the early 1980s up to the late 1990s many authors presented numerically based drape solutions, based on the same assumptions as Mack and Taylor (see, for example, Robertson *et al.*, 1984; Smiley and Pipes, 1988; Heisey and Haller, 1988; Long and Rudd, 1994; Bergsma, 1995; and Trochu *et al.*, 1996). Typically, these drape models start from an initial point and two initial fibre directions. Further points are then generated at a fixed equal distance from the previous points creating a mesh of quadrilateral cells. There is no unique solution for this geometrical drape method. This problem is generally solved by defining two fibre paths on the drape surface. Bergsma (1995) introduced 'strategies' in order to find solutions for the drape algorithm, without pre-defining fibre paths. Bergsma also included a mechanism to incorporate the locking phenomenon in his drape simulations.

Alternatively to the fishnet model, Van der Weën (1991) presented a computationally efficient energy based mapping method in 1991. Rather than creating a new cell on a geometric basis, the cell in the mesh is mapped onto the drape surface by minimising the elastic energy in the drape cell, while only accounting for the deformation energy used to extend the fibres. Long *et al.* (2002) presented a similar approach based on minimisation of shear strain energy, demonstrating the capability to predict different fibre patterns depending on material type.

The mapping scheme is quite simple in its application and implementation, and requires very limited computational efforts. The results of the mapping scheme agree well with reality if the product shape is convex.

However, the mapping schemes do not predict unique solutions. User interference or 'strategies' are required to solve the drape problem. Inaccurate drape predictions are obtained for products where bridging occurs or when the preform slides over the mould during forming. The scheme is not suited to incorporate the processing conditions accurately during draping or to give an accurate representation of the composite properties. Especially in tight weaves, the error of assuming a zero in-plane fabric shear stiffness during draping leads to errors. From the late 1970s it was shown experimentally that the resin material also affects the deformation properties (Potter, 1979). In addition, the geometrical approach might find infeasible solutions when draping products with holes.

Forming of multi-layered composites is simulated by repeatedly draping single layers of fabric, since the model only represents one layer of fabric. The through-thickness shear interaction between the individual layers is not accounted for.

Particle based schemes

From the first half of the 1990s particle based schemes were used to predict the fabric drape behaviour. The fabric, or cloth, is represented as a discontinuous sheet using micro-mechanical structural elements. These elements, also called particles, interact and must be chosen to be small enough to still represent the weave's behaviour.

An interacting particle model was developed by Breen *et al.* in 1994. Energy functions define the interaction between the particles, placing the particles at the crossings of the yarns in the fabric. The energy contribution in the particles consists of thread repelling, thread stretching, thread bending, thread trellising and gravity. The total energy in the cloth is simply the sum of the energy of all particles. The modelling strategy for particle based solutions is generally time dependent. In the first time step, the model accounts for the gravity and the collision between the cloth and the drape surface. In the next step a stochastic energy-minimising technique is used to find the local energy minima for the cloth. Finally, permutations are introduced to produce a more asymmetric final configuration.

Similarly to the energy based functions, force based functions were also developed for the interactions of the particles (Colombo *et al.*, 2001). This representation method is applied in commercially available software, since it is computationally more attractive than the energy based particle interaction functions.

Cordier and Magnenat-Thalmann (2002) simulated the cloth behaviour on dressed virtual humans in real time. They proposed a hybrid drape algorithm combining the advantages of physically (particle) based and geometric deformations, avoiding the computationally expensive collision calculations as much as possible. The cloth is segmented into three sections in their simulations. Cloths that remain at constant distance to the drape body are modelled in the first section. Typically, these are stretch cloths. In the second layer, the loose cloth follows predefined discs, representing the limbs. Finally, floating cloth such as skirts is represented in the third section. A force based particle method is used for modelling the floating cloth, incorporating the collision with the underlying body and the cloth itself. Real-time modelling of cloth behaviour is feasible with this approach using middle range computers (up to 1 GHz PCs).

The method requires the mechanical properties of the cloth and the product shape as input. Typically, the method is used for modelling the shape of hanging cloth on objects or humans in the fashion industry. The emphasis is therefore not the deformation and stresses within the fabric but the resulting shape of the cloth as a whole. Possibly, this is why no implementation in the technical industry has been found for this method in the literature.

Truss based schemes

Fabrics are woven using a periodic arrangement of fibre bundles. These periodic arrangements are called Representative Volume Elements (RVE), or unit cells. The fibres in these unit cells can be represented using trusses. The fibre interaction, such as shear-locking of the fabric, is modelled with diagonal stringers. Kato et al. (1999) proposed a unit cell representation based on such a fabric lattice model in 1999.

In 2003, Tanov and Brueggert modelled the inflation of a car side airbag in an FE (Finite Element) simulation, using a loosely woven fabric model. The yarns in the fabric were represented by pinned-joined bars with two locking springs on

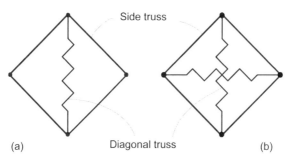

2.1 Schematic representation of the unit cell by truss elements: (a) one diagonal spring element, (b) two diagonal spring elements.

the diagonal of their unit cell. A schematic representation of this unit cell is depicted in Fig. 2.1(b).

Sharma and Sutcliffe (2003) also represented the unit cell in a network of pinned-joined trusses. The edge trusses represent the fibres in the fabric. To introduce shear stiffness in the unit cell, one diagonal spring is introduced, as presented in Fig. 2.1(a). An FE analysis was performed with this pinned-joined net of trusses to predict the draping process.

The advantages of these models are the simple mechanics and ease of use. Little input is required and the simulation time is relatively short compared to continuum based approaches. The resin behaviour is not incorporated in these models. Forming multi-layered composites can be simulated just as with the mapping based schemes since these models are also based on the representation of a single layer of fabric. Through-thickness shear interaction between the fabric layers during forming is not accounted for.

2.2.2 Continuum models

Several constitutive models have been proposed for fabric drape modelling. A distinction can be made between elastic material models, viscous material models and multi-component models. Most of these constitutive models are formulated in a plate or shell theory and implemented in Finite Element formulations.

Elastic models

Finite Element drape simulations by means of elastic models have been applied since the mid 1990s. One of the earliest models was presented by Chen and Govindaraj in 1995. They developed an elastic orthotropic continuum based model to represent the fabric drape behaviour. The material model was based on a flexible shell theory. A non-linear FE formulation was used to predict the forming of a fabric onto a table.

At the same time, Kang and Yu (1995) developed a similar shell based drape model. They used a convective coordinate system in a total Lagrangian formulation with an orthotropic continuum based elastic material model. The nonlinear incremental formulation of the total Lagrangian scheme was solved using Newton's method. Again, the draping of a cloth onto a table was simulated.

A few years later Boisse *et al.* (1997) modelled the bi-axial fabric behaviour in forming processes. The undulation of the yarns in the fabric was accounted for in the bi-axial weave model, assuming fibres with stiffness in the fibre direction only. A refined version of this material model was presented by Boisse *et al.* (2001).

Ivanov and Tabiei (2002) developed an elastic material model based on the RVE of a plain weave fabric. A homogenisation technique accounts for the weave's microstructure, where the yarns are assumed to be transversely isotropic. The shear properties of the fabric are neglected up to the locking angle. On further shear the yarn shear properties are used for the fabric response.

Elastic material laws are fairly simple. The implementation of these models in FE packages is therefore reasonably simple as well, compared to more advanced material models. Generally, the matrix material behaviour is viscoelastic. The properties of the matrix cannot be taken into account accurately with a purely elastic material law. Only single layers of fabric were draped using these models in the literature.

Viscous models

Spencer (2000) modelled the behaviour of impregnated woven fabrics as a viscous fluid. The fibres were assumed inextensible in his model, effectively restraining the deformation of the fluid in the fibre directions. The fluid was also assumed incompressible. In the plane stress situation, the model simplifies to a single parameter model and is able to simulate the draping behaviour of the fabric. Similarly, Spencer (2001) proposed a viscoplasticity model for draping fabric reinforced composites.

The elastic behaviour of the fabric itself is not incorporated in these material models. Therefore, processing-induced fibre stresses are unaccounted for in these models. The models account for the drape behaviour of one layer of fabric only.

Multi-component models

Multi-component models are a combination of several material models. The fibres are often represented as elastic materials in these models. Some models incorporate the fabric shear behaviour, others account for the resin behaviour using viscous material laws.

Sidhu *et al.* (2001) proposed a bi-component FE analysis to model dry fabric forming. The yarns were represented as trusses, using a linear elastic material law. A layer of shell elements accounted for the friction between the yarns and for the locking of the fabric. The material behaviour in the shells was non-linearly elastic and orthotropic. The nodes of the truss and the shell elements were connected in the FE formulation. Friction between the fabric and the tooling was not incorporated in the simulation of the stamping of a spherical shape.

Cherouat and Billoët (2001) presented a model for draping thermoplastic composite materials. The yarns in the fabric were represented using truss elements. The material law representing the yarns was non-linear elastic and based on three deformation mechanics: straightening, relative rotation and tensile stretching. The resin was modelled as an isotropic viscoelastic medium in a layer of shell elements. Again the nodes of the two meshes are connected in the FE simulations. Coulomb friction between the tooling and the fabric was assumed in the forming simulation of a hemispherical product.

McEntee and Ó Brádaigh (1998) modelled the drape behaviour of a multi-layered thermoplastic composite on tools with a single curvature. Two-dimensional elements were stacked through the thickness of the sheet, each ply represented by a row of elements. The constitutive relation in each ply was based on the 'ideal fibre reinforced fluid' model by Rogers (1989). A layer of contact elements was placed between the ply elements. Experiments demonstrated the presence of a resin rich layer between the individual plies during forming, justifying a viscous contact behaviour.

De Luca *et al.* (1998) modelled the drape behaviour of composite laminates in a dynamic explicit FE scheme. Each single fabric layer in the laminate was represented by a layer of elements. Per layer, a bi-phase material model was applied, decoupling the behaviour of the elastic fibres and the viscous matrix. The shell element layers were stacked in the thickness direction of the sheet. Between the shell elements, a 'specialised viscous-friction law' was applied. The model predicts a clear interaction between the laminate lay-up and the drapeability. Experimental results confirm the importance of this interlaminar shear effect. The method provides good results but becomes quite slow by expanding the problem computationally.

Recapitulating, draping can be modelled using the combination of continuum based material models and the FE method. A drape simulation is non-linear due to the large deformations of the fabric during draping and the evolving contact conditions.

The required input consists of the tool and laminate geometry definition, the material property data and the appropriate boundary conditions. The results of the simulation combine the information on the material deformation with the loading required for shaping. The interlaminar shear effects during forming can play a significant role in the drapeability of multi-layered composite components. The drape behaviour of multi-layered composites can be modelled by stacking multiple element layers through the thickness of the sheet and connecting them by friction laws.

2.2.3 Multi-layered models

Current drape predictions are based on single fabric layer models or an assembly of single fabric layer models. In some production processes the fabric layers are formed sequentially onto the mould. Interlaminar shear between the individual fabric layers is small in such a production process. Modelling each single fabric layer sequentially suffices for such a process.

However, the interaction between the layers in the sheet is important when draping multi-layered composites. FE simulations with multiple elements through-the-thickness of the sheet are used to account for this interaction between the layers. A friction law accounts for the interlaminar shear between the individual fabric layers. The drawback of using multiple sheet elements on top of each other is the increase of the complexity of the FE model. The total number of degrees of freedom (DOFs) grows linearly with the number of layers in the model and so does the number of contact conditions to be evaluated. A non-linear system of equations has to be solved in the FE representation. The computation times to solve such a non-linear system of equations will easily increase at least quadratically with the increasing number of DOFs. As a result the computation time with increasing layers in the drape simulation behaves correspondingly.

A computationally more efficient method is preferred to predict the drape behaviour of multi-layered fabric composites. This can be achieved in a multi-layer drape material model as presented by Lamers in 2004. This drape model incorporates the inter-ply and intra-ply shear behaviour of multi-layered fabric reinforced composites. The use of multiple elements through-the-thickness of the laminate (and the corresponding contact logic) is avoided by accounting for the interlaminar shear within the multi-layer model. The same type of FE element can be used for the single layer and the multi-layer material model. The number of DOFs in an FE simulation with this multi-layer material model will therefore be equal to the number DOFs of the single layer model. Hence, the computation time for solving the non-linear system of equations will be comparable.

2.3 Continuum based laminate modelling

Continuum mechanics defines the kinematics, stresses, strains and the conservation laws for arbitrary continuous media. Forming simulations require a constitutive relation in addition to these conservation laws.

Here, constitutive relations are presented for single layer and multi-layered composites, based on linear elastic fibres and a Newtonian viscous resin. An isothermal approach is used in both models, if necessary incorporating an interaction between the warp and fill fibre families. First, a single layer intra-ply material model is presented. This will be extended for multi-layered woven fabric composite material, incorporating inter-ply shear behaviour.

2.3.1 Kinematics

The single layer drape material constitutive equation derived here is an extension to the 'Fabric Reinforced Fluid' (FRF) model by Spencer (2000). Extending the FRF model with elastic components enables the model to incorporate fibre stresses and an elastic fabric shear response in drape predictions. The composite components, the woven fabric and the resin, are subjected to an affine deformation (see Fig. 2.2). The warp and fill fibre families of the fabric are represented by vectors a and b; $\varphi(X,t)$ maps the original configuration of a material particle at position X to the current configuration. A continuous distribution of the fibre families is assumed which can vary spatially and in time,

$$a(X,t) = F(X,t) \cdot a_o(X), \qquad b(X,t) = F(X,t) \cdot b_o(X) \qquad 2.1$$

where $a_o(X)$, $b_o(X)$ are the original fibre orientation unit vectors at $t = 0$ and $F(X,t)$ is the deformation gradient. The length of a fibre may change, leading to a fibre stretch defined as:

$$\lambda_a = \frac{a}{a_o} = \sqrt{a_o \cdot F^T \cdot F \cdot a_o}. \qquad 2.2$$

Allowing fibre extension leads to a change in length of the characteristic vectors a and b of the fibre families. The corresponding unit vectors are introduced as:

$$a^* = \frac{1}{\|a\|} a, \qquad b^* = \frac{1}{\|b\|} b. \qquad 2.3$$

The two bias directions of the weave, s_1 and s_2, are depicted in Fig. 2.3. The angle ϕ is the angle between the s_1 direction and the fibre families a and b,

$$\phi = \frac{1}{2} \arccos(a^* \cdot b^*). \qquad 2.4$$

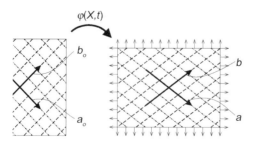

2.2 Affine deformation of fibres and matrix.

Constitutive modelling for composite forming   31

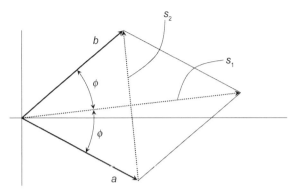

2.3 Directions of the two fibre families a and b and the bias directions s1,2 of the weave.

Often, 2ϕ is referred to as the enclosed fibre angle. The material shear angle θ depends on ϕ as:

$$\theta = 2(\phi_o - \phi), \qquad 2.5$$

where again subscript o refers to the original configuration. Generally, ϕ_o is 45° for woven fabrics.

Constitutive equation for a single layer

The composite material consists of fibres in a textile structure embedded in a matrix, with volume fractions V_f and V_m respectively. The fibre volume is distributed over the fibre families a and b. The sum of the volume fractions evidently equals 1,

$$V_{f_a} + V_{f_b} + V_m = V_f + V_m = 1. \qquad 2.6$$

The matrix will typically have a viscous or viscoelastic response to deformations whereas the fibres have a high tensile stiffness. The fabric structure causes interactions between the fibre families. The amount of fabric shear is restricted and stresses in one fibre direction affect the tensile response in the other direction as well as the shear response. Here, the total Cauchy stress is assumed to be caused by elastic and viscous effects working in parallel,

$$\sigma = \sigma_e + \tau, \qquad 2.7$$

where σ_e is the elastic stress contribution and τ is an extra viscous stress contribution. The elastic and viscous contributions to the stress are solved separately.

The fibres dominate the elastic response. They are modelled as linear elements as a first approximation, having no stiffness properties in any direction but the fibre longitudinal direction. The stress contribution of the fibre is given as:

$$\sigma_a = V_{f_a} E_a (\lambda_a - 1) \boldsymbol{A}^*, \qquad 2.8$$

where $\boldsymbol{A}^*$ is the dyadic product of the unit vector $\boldsymbol{a}^*$,

$$\boldsymbol{A}^* = \boldsymbol{a}^* \boldsymbol{a}^* \qquad 2.9$$

and E_a is the fibre longitudinal modulus.

An analogous relation holds for fibre family b, resulting in:

$$\sigma_b = V_{f_b} E_b (\lambda_b - 1) \boldsymbol{B}^*, \qquad 2.10$$

where

$$\boldsymbol{B}^* = \boldsymbol{b}^* \boldsymbol{b}^* \qquad 2.11$$

and E_b is the longitudinal modulus of fibre family b.

Further elastic effects are found from matrix compression and fabric shear, leading to

$$\sigma_e = \sigma_a + \sigma_b + \sigma_h + \sigma_f, \qquad 2.12$$

where σ_h is the hydrostatic stress and σ_f is the fabric elastic stress. The hydrostatic stress originates from to compression of the matrix:

$$\sigma_h = K \, tr(\epsilon) \boldsymbol{I}, \qquad 2.13$$

where K is the bulk modulus of the matrix, ϵ is the strain tensor and tr is the trace operation.

The fabric elastic stress σ_f is caused by the interaction between the fibres, ultimately leading to shear-locking and wrinkling. This non-linear shear response is modelled as an exponential function of the material shear angle θ as:

$$\sigma_f = V_f m (e^{n\theta} - e^{-n\theta}) \frac{1}{2} (\boldsymbol{C}^* + \boldsymbol{C}^{*T}), \qquad 2.14$$

with

$$\boldsymbol{C}^* = \boldsymbol{a}^* \boldsymbol{b}^*, \qquad 2.15$$

and constants m, n depending on the fabric architecture, which can be evaluated from shear experiments.

The viscous stress contribution is attributed to the matrix, which is described as a linear viscous medium. The stress response of incompressible viscous fluids is represented by an isotropic tensor-valued function of the rate of deformation tensor (Schowalter, 1978). The rate of deformation tensor $\boldsymbol{D}$ is the symmetric part of the velocity gradient,

$$\boldsymbol{D} = \frac{1}{2} \left(\vec{\nabla} \boldsymbol{v} + \boldsymbol{v} \overleftarrow{\nabla} \right) \qquad 2.16$$

When $\boldsymbol{D}$ meets the incompressibility condition, the most general representation for the viscous stress contribution τ of incompressible viscous fluids can be written as a power series expansion, or:

$$\tau = \tau(\boldsymbol{D})$$
$$= -p\boldsymbol{I} + \psi_1 \boldsymbol{D} + \psi_2 \boldsymbol{D}^2 \qquad 2.17$$

where p is the hydrostatic pressure, $\boldsymbol{I}$ is the unit tensor and $\psi_{1,2}$ are functions of the invariants $tr\ \boldsymbol{D}^2$ and $tr\ \boldsymbol{D}^3$. In Spencer's FRF model, the viscous stress contribution depends on the rate of deformation and also on the fibre directions:

$$\tau = \tau(\boldsymbol{D}, \boldsymbol{a}^*, \boldsymbol{b}^*). \qquad 2.18$$

Spencer derived an expression for the linear anisotropic viscous response of an incompressible matrix material with inextensible fibres. In this case, the viscous stress contribution is given in its most general form by:

$$\tau(\boldsymbol{D}, \boldsymbol{a}^*, \boldsymbol{b}^*) = 2\eta \boldsymbol{D} + 2\eta_1(\boldsymbol{A}^* \cdot \boldsymbol{D} + \boldsymbol{D} \cdot \boldsymbol{A}^*) + 2\eta_2(\boldsymbol{B}^* \cdot \boldsymbol{D} + \boldsymbol{D} \cdot \boldsymbol{B}^*) +$$
$$= 2\eta_3(\boldsymbol{C}^* \cdot \boldsymbol{D} + \boldsymbol{D} \cdot \boldsymbol{C}^{*T}) + 2\eta_4(\boldsymbol{C}^{*T} \cdot \boldsymbol{D} + \boldsymbol{D} \cdot \boldsymbol{C}^*), \qquad 2.19$$

where $\eta, \eta_1, \eta_2, \eta_3$ and η_4 are characteristic anisotropic viscosities generally being functions of the angle between the fibre families a and b. In general, these viscosities will be temperature dependent.

Here, the inextensibility condition is relieved somewhat by assigning a finite stiffness to the fibres. In this way it is not necessary to use Lagrange multipliers in subsequent FE calculations. The fibre stresses are now constitutively determined by the strain in the fibre direction. As a result, $\boldsymbol{D}$ does not necessarily meet the inextensibility conditions in the fibre directions. A modified rate of deformation tensor $\boldsymbol{D}^*$ is introduced to overcome this discrepancy:

$$\boldsymbol{D}^* = \boldsymbol{D} + c_1\boldsymbol{A}^* + c_2\boldsymbol{B}^* + c_3\boldsymbol{I} \qquad 2.20$$

where $\boldsymbol{D}^*$ is the deviatoric part, satisfying the inextensibility and incompressibility conditions:

$$\boldsymbol{A}^* : \boldsymbol{D}^* = 0,$$
$$\boldsymbol{B}^* : \boldsymbol{D}^* = 0, \qquad 2.21$$
$$\boldsymbol{I} : \boldsymbol{D}^* = 0.$$

With these conditions it can be found that the unknowns $c_{1,2,3}$ are:

$$c_1 = \frac{1}{d}\left[(3g_B - I_B^2)\boldsymbol{A}^* : \boldsymbol{D} - (3g_{AB} - I_A I_B)\boldsymbol{B}^* : \boldsymbol{D} + (g_{AB}I_B - I_A g_B)I_D\right]$$

$$c_2 = \frac{1}{d}\left[(3g_A - I_A^2)\boldsymbol{B}^* : \boldsymbol{D} - (3g_{AB} - I_A I_B)\boldsymbol{A}^* : \boldsymbol{D} + (g_{AB}I_A - g_A I_B)I_D\right] \qquad 2.22$$

$$c_3 = \frac{1}{d}\left[(g_{AB}I_B - I_A g_B)\boldsymbol{A}^* : \boldsymbol{D} + (g_{AB}I_A - g_A I_B)\boldsymbol{B}^* : \boldsymbol{D} - (g_{AB}^2 - g_A g_B)I_D\right]$$

where

$$d = g_A I_b^2 + 3g_{AB}^2 - 2g_{AB}I_A I_B + g_B I_A^2 - 3g_A g_B,$$
$$I_A = tr\ \boldsymbol{A}^*, \quad I_B = tr\ \boldsymbol{B}^*, \quad I_D = tr \boldsymbol{D}$$

and

$$g_A = A^* : A^*, \quad g_B = B^* : B^*, \quad g_{AB} = A^* : B^*.$$

The total stress in (2.7) is found by using (2.20) in (2.19) and combining the terms in equations (2.8, 2.10, 2.12, 2.14, 2.19) to:

$$\begin{aligned}\sigma = &-p\boldsymbol{I} + V_{f_a}E_a(\lambda_a - 1)\boldsymbol{A}^* + V_{f_b}E_b(\lambda_b - 1)\boldsymbol{B}^* + \\ &+ \frac{1}{2}V_f m(e^{n\theta} - e^{-n\theta})(\boldsymbol{C}^* + \boldsymbol{C}^{*T}) + \\ &+ V_m[2\eta\boldsymbol{D}^* + 2\eta_1(\boldsymbol{A}^* \cdot \boldsymbol{D}^* + \boldsymbol{D}^* \cdot \boldsymbol{A}^*) + 2\eta_2(\boldsymbol{B}^* \cdot \boldsymbol{D}^* + \boldsymbol{D}^* \cdot \boldsymbol{B}^*) + \\ &+ 2\eta_3(\boldsymbol{C}^* \cdot \boldsymbol{D}^* + \boldsymbol{D}^* \cdot \boldsymbol{C}^{*T}) + 2\eta_4(\boldsymbol{C}^{*T} \cdot \boldsymbol{D}^* + \boldsymbol{D}^* \cdot \boldsymbol{C}^*)],\end{aligned} \quad 2.23$$

The constitutive relation is symmetric with respect to the fibre families a and b when the fibre families are mechanically equivalent. This symmetry in the constitutive relation reduces the number of independent parameters in the extra viscous stress contribution as $\eta_1 = \eta_2$ and $\eta_3 = \eta_4$ (Spencer, 2000). Three parameters η, η_1 and η_3, generally being functions of $\cos 2\phi$, are required to determine the extra viscous stress response of the fabric. They can either be even or odd functions of $\cos 2\phi$.

2.4 Multilayer effects

A composite laminate consists of multiple layers, generally with different fibre orientations. A multilayer model must allow these individual plies to slide with respect to each other and deform individually. Figure 2.4 illustrates the deformation of a three ply laminate due to a load on the centre ply. The laminate deforms from its original configuration to a current configuration. The layers are stacked on each other in the original configuration, having independent fibre orientations a_i and b_i, where i is the ply index. The plies are allowed to deform individually, conforming to their fibre orientations. The strains ϵ_i, rotation ω_i, the velocities v_i and the rates of deformation D_i will generally also be non-uniform over the laminate thickness as a result.

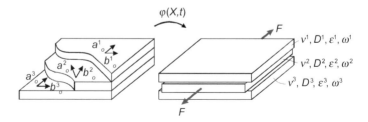

2.4 Deformation of the individual layers during composite forming.

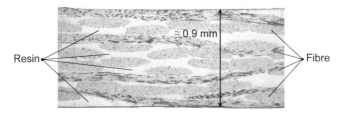

2.5 Cross-sectional view of an 8H satin glass fibre weave reinforced PPS, illustrating the resin rich layers between the fabric plies.

Since the individual plies in the laminate can have different velocities, the interface between these plies must deform correspondingly. Therefore, a slip law needs to be defined which is able to describe the sliding of the individual plies.

Figure 2.5 shows a microscopic view of the cross-section of a compression moulded glass poly(phenylene sulphide) (PPS) laminate, reinforced with four layers of 8H satin fabric. The horizontal glass fibre yarns in dark grey indicate the separate fabric layers. Some yarn contact is noticed in Fig. 2.5. However, resin is observed between most of the plies. These plies are separated by the lighter coloured thin PPS layers. Based on the observations of these resin rich layers between the plies and following the work of McEntee and Ó Brádaigh (1998), the interlaminar behaviour is assumed to be viscous. This leads to a viscous slip law expressed in the velocity differences between adjacent plies,

$$v^j_{rel} = v^{i+1} - v^i, \qquad 2.24$$

where the suffix i indicates the ply and j indicates the interface layer between the plies i and $i+1$. When assuming a friction law which depends linearly on the velocity difference between the plies, the interface traction γ^j is defined as:

$$\gamma^j = \frac{1}{\beta_j} v^j_{rel}, \qquad 2.25$$

with a slip factor β_j. This constant friction factor is derived as

$$\beta_j = \frac{h_j}{\eta_j}, \qquad 2.26$$

where h_j is the averaged thickness of the interface layer and η_j the viscosity of the interface layer.

2.5 Parameter characterisation

The material property data need to be determined before the constitutive relations (2.23) and (2.25) can be employed in composite forming simulations. Both the intra-ply and inter-ply composite properties need to be characterised. Materials characterisation is discussed in Chapter 1.

There is little standardisation in the experimental characterisation of pre-impregnated technical fabrics at production temperature. Some standardisation for measuring the intra-ply response of dry textiles is available in the garment industry. The Kawabata Evaluation System for Fabrics (KES-F) (Kawabata, 1975) is used to measure the formability of fashion cloths at room temperature. Unfortunately, KES-F is not applicable for testing pre-impregnated composite materials since it is not designed to test at high temperatures, to apply large shear deformations or to measure at relatively high loads. The topic of standardisation is discussed further in Chapter 13.

The number of intra-ply material parameters required for the single layer material model is quite extensive. The fibre properties are usually available in the literature (e.g., Peters, 1998). The other material properties, m, n, η, η_1, η_2, η_3 and η_4, need to be quantified experimentally. Picture frame experiments can be used to determine these intra-ply properties. The inter-ply composite properties can be determined using pull-out tests (Murtagh and Mallon, 1997).

2.5.1 Picture frame

McGuinness and Ó Brádaigh (1997) presented the picture or trellis frame experiment in 1997 in order to measure the in-plane response of woven fabric reinforced composites. Since then, several authors (Prodromou and Chen, 1997; Mohammed et al., 2000; Peng et al., 2002) applied the picture frame experiment to determine the properties of dry and pre-impregnated fabrics. An initially square frame deforms into a rhombus, imposing a shear deformation onto the fabric. The fabric is placed in the frame with the fibre directions parallel to the arms of the frame, as shown in Fig. 2.6. In this manner the fabric deforms without extension in the fibre directions. Left or right shear deformations, denoted with the subscripts l and r, are imposed on the fabric by extending the frame in the vertical or horizontal direction respectively. The shear response of the composite can be determined by measuring the load response $F_{l,r}$ of the frame. The load,

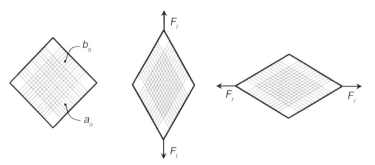

2.6 Schematic deformation of the composite in the picture frame experiment; from left to right: initial configuration, left shear and right shear.

Constitutive modelling for composite forming 37

required to deform the fabric, is measured as a function of the crosshead displacement in a tensile testing machine. The shear response of the weave can be unsymmetric, when it different for left and right shear (McGuinness and Ó Brádaigh, 1997). Impregnated composites are tested above the glass transition or melt temperature of the matrix material. The entire picture frame, including the specimen, is placed in an oven in order to perform these high-temperature tests. Ideally, the temperature dependent viscosities are determined for a range of temperatures, appropriate for the composite forming conditions.

Picture frame kinematics

The crosshead displacement of the tensile testing machine causes the frame to shear. The velocity at which the crosshead travels is usually constant during the test, resulting in a variation of the shear rate of the fabric. The fabric deformation can be expressed in terms of the crosshead displacement and velocity.

A schematic view of the frame in the undeformed and deformed state is shown in Fig. 2.7, using the bias directions as the *xy*-coordinate system. The frame with side length l_f and initial frame angle ϕ_o is shown on the left-hand side, using one line of symmetry. The deformed frame (after right shear) is presented on the right-hand side, where also the displacement Δu_r and its velocity v_r for right shear are indicated. The length l_r of the frame in the deformed state is given as a function of the displacement by:

$$l_r = l_f \cos \phi_o + \Delta u_r. \qquad 2.27$$

Hence the frame angle ϕ can be expressed in terms of the displacement Δu_r and the length of the sides of the picture frame l_f,

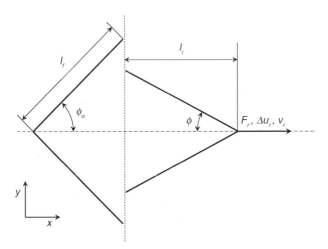

2.7 Schematic view on half of the picture frame: undeformed (left) and deformed shape (right).

$$\phi = \arccos\left(\cos\phi_o + \frac{\Delta u_r}{l_f}\right). \qquad 2.28$$

When a left shear deformation is applied, the crosshead displacement Δu_l results in a frame angle ϕ given by:

$$\phi = \arcsin\left(\sin\phi_o + \frac{\Delta u_l}{l_f}\right), \qquad 2.29$$

where Δu_l is in the positive vertical direction. The material shear angle θ can be found straightforwardly by substituting equation (2.28) or (2.29) into equation (2.5).

The fibre directions during testing are parallel to the sides of the frame,

$$\boldsymbol{a} = (\cos\phi, -\sin\phi, 0), \quad \boldsymbol{b} = (\cos\phi, \sin\phi, 0). \qquad 2.30$$

The rate of change of the frame angle for right shear deformation can be expressed in terms of the velocity:

$$\dot\phi = -\frac{v_r}{l_f \sin\phi}, \qquad 2.31$$

and the time derivative of the frame angle for left shear deformation is:

$$\dot\phi = \frac{v_l}{l_f \cos\phi}. \qquad 2.32$$

Finally, the rate of deformation tensor $\boldsymbol{D}$ is (Spencer, 2000):

$$\boldsymbol{D} = \begin{bmatrix} -\tan\phi & 0 & 0 \\ 0 & \cot\phi & 0 \\ 0 & 0 & -2\cot\phi \end{bmatrix} \dot\phi, \qquad 2.33$$

where the out-of-plane component follows from the assumption of incompressibility.

Considering that the fabric is in a state of plane stress without fibre elongation, the stress components can be found from equation (2.23):

$$\sigma_{xx} = \tfrac{1}{2} V_f m (e^{n\theta} - e^{-n\theta}) a_x b_x +$$
$$+ 2V_m D_{xx}(\eta + 2\eta_1 a_x^2 + 2\eta_s b_x^2 + 2(\eta_3 + \eta_4)a_x b_x) - 2V_m \eta D_{zz}$$

$$\sigma_{xy} = \tfrac{1}{4} V_f m \mathrm{left}(3^{n\theta} - e^{-n\theta})(a_x b_y + a_y b_x) +$$
$$+ 2V_m D_{xx}(\eta_1 a_x a_y + \eta_2 b_x b_y + \eta_3 a_y b_x + \eta_4 a_x b_y) + \qquad 2.34$$
$$+ 2V_m D_{yy}(\eta_1 a_x a_y + \eta_2 b_x b_y + \eta_3 a_x b_y + \eta_4 a_y b_x)$$

$$\sigma_{yy} = \tfrac{1}{2} V_f m (e^{n\theta} - e^{-n\theta}) a_y b_x +$$
$$+ 2V_m D_{yy}(\eta + 2\eta_1 a_y^2 + 2\eta_2 b_y^2 + 2(\eta_3 + \eta_4)a_y b_y) - 2V_m \eta D_{zz}$$

where all fabric deformation characteristics are defined in (2.5, 2.30 and 2.33).

Picture frame equilibrium

The shear fixture imposes a shear deformation on the fabric. When neglecting inertia effects, the stress and the strain distributions in the specimen are homogeneous as a result. The external loading as imposed by the tensile testing machine must be in equilibrium with these stresses, multiplied by the appropriate cross-sectional area. Volume changes can be assumed negligible during testing (McGuinness and Ó Brádaigh, 1997). The thickness of the composite thus increases during testing, inversely related to the reduction in surface area of the specimen. The thickness h during testing can be expressed in terms of the frame angle ϕ as:

$$h = \frac{h_o}{\sin 2\phi}, \qquad 2.35$$

where h_o is the initial thickness of the specimen.

Performing a static analysis (Lamers, 2004) provides the load response of the frame corresponding to a homogeneous fabric stress distribution. Analysing right shear deformation leads to:

$$F_r = \frac{h_o}{\sin 2\phi}(\sigma_{xx}l_f \sin \phi - \sigma_{yy}l_f \cot \phi \cos \phi) \qquad 2.36$$

and for left shear the load response is found as:

$$F_l = \frac{h_o}{\sin 2\phi}(\sigma_{xx}l_f \tan \phi \sin \phi - \sigma_{yy}l_f \cos \phi) \qquad 2.37$$

The load response of the fabric is found in terms of the material parameters by substituting (2.34) into (2.36) and (2.37). A non-linear fitting procedure is required to fit the material parameters m, n, η, η_1, η_2, η_3 and η_4 on the experimental data.

Picture frame experiments

Picture frame tests were performed on pre-consolidated laminates. Four layered glass fibre reinforced PPS 8H satin weave (Ten Cate CetexSS303) laminates were pressed. The PPS has a melting temperature of 285°C. The specimens were prepared to fit into the frame. The dimensions of the specimens are shown in Fig. 2.8. Some fibres were removed in the zone next to the central area which delays inhomogeneous deformation due to wrinkling. Five experiments were performed at 300°C and three at 350°C for right shear deformation. Also, two tests were performed for left shear deformation at 300°C. The velocity of the traverse of the tensile testing rig was set at 1000 mm/min for all

40 Composites forming technologies

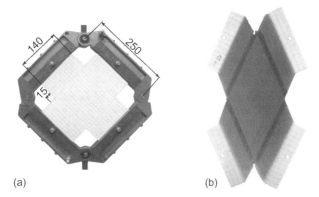

(a) (b)

2.8 (a) Composite clamped in the picture frame prior to testing. (b): Deformed shape of the specimen after the shear test. All dimensions are in mm.

tests. The deformed shape of the specimens is depicted in Fig. 2.8b. The corresponding load versus material deformation angle response of the tests is given in Fig. 2.9. This load versus angle response is typical for the picture frame tests. The load increases with increasing deformation angle in a non-linear manner. The load response of the specimens tested at 350°C is significantly lower than the load response for the specimens tested at 300°C. The lower load can be attributed to the lower viscosity of the PPS matrix material at higher temperatures, resulting in a lower shear resistance of the viscous part of the composite. The load versus angle response of the impregnated fabric was independent of the shearing direction. The composite intra-ply shear behaviour is symmetrical.

In-plane symmetrical behaviour implies that the composite viscosities are mutually related (Spencer, 2000). The viscosities η_1 and η_2 are equal even functions of cos 2ϕ, while η_3 and η_4 are equal odd functions of cos 2ϕ. Fitting

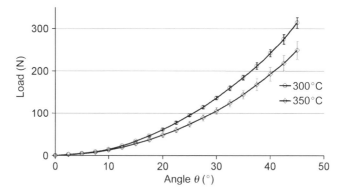

2.9 Load versus material deformation angle response during the picture frame test; test velocity 1000 mm/min, temperatures 300°C and 350°C.

Table 2.1 Fitted intra-ply material data for 8H glass PPS (Cetex SS303)

Property		$2\phi \leq \frac{1}{2}\pi$	$2\phi > \frac{1}{2}\pi$	$2\phi \leq \frac{1}{2}\pi$	$2\phi > \frac{1}{2}\pi$
m	MPa	0.218	0.218	0.182	0.182
n	–	3.290	3.290	3.156	3.156
η	MPa s	0.281	0.281	0.205	0.205
η_1	Pa s	16.503	16.503	30.315	30.315
η_2	Pa s	16.503	16.503	30.315	30.315
η_3	MPa s	−0.394	0.394	−0.335	0.335
η_4	MPa s	−0.394	0.394	−0.335	0.335

the material property data on the experimental results gives the parameter values listed in Table 2.1.

2.5.2 Bias extension

Alternatively, the intralaminar shear response of the composite can be characterised with a bias extension experiment. A strip of the composite reinforced with a ±45° woven fabric is extended in the bias direction as illustrated in Fig. 2.10. Three zones can be distinguished in the deformed specimen. The central area (*I*) shears uniformly, with an enclosed angle 2ϕ. Neglecting fibre extension, this angle is related to the clamp displacement Δu by

$$\phi = \arccos\left(\frac{1}{\sqrt{2}} \cdot \left(1 + \frac{\Delta u}{L_o - w_o}\right)\right), \qquad 2.38$$

with L_o and w_o as the original length and width of the sample respectively. The clamps induce an area with zero shear (*III*), whereas the intermediate zone (*II*) has a material shear angle θ_{II} which is half of the value in the central zone,

$$\theta_{II} = \tfrac{1}{2}\theta = \phi_o - \phi, \qquad 2.39$$

with ϕ_o as the half original enclosed angle (45°).

The stress state is uniaxial in the central region, with

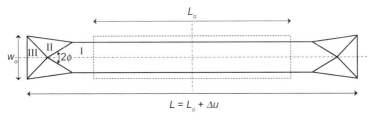

2.10 Schematic view of the bias extension specimen. The undeformed geometry is indicated with dashed lines.

$$\sigma_{xx} = \frac{F}{wh} = \frac{F\cos\phi}{w_o h_o \sqrt{2}}, \qquad 2.40$$

where F is the external force.

Combining equations (2.38), (2.40) with the relations (2.30–2.33) presented earlier for the picture frame analysis and the constitutive equation (2.23) leads to a nonlinear equation which can be fitted to the experimental force-displacement results. Note, however, that these data are obtained for non-zero fibre stresses, contrary to the picture frame results. This biaxial stress state is likely to affect the shear response of the fabric reinforced composite. This phenomenon is beyond the scope of the current chapter.

2.5.3 Pull-out

The individual plies slide with respect to each other during draping. This interlaminar behaviour is accounted for in the multi-layer material model. A viscous traction was assumed to model this phenomenon. This traction can be measured using fabric pull-out tests. Four layers of fabric are placed in a tensile testing machine in a temperature controlled environment (Murtagh and Mallon, 1997). The middle layers are then pulled out of the other two (see Fig. 2.11), while measuring the load as a function of the crosshead velocity. After a certain time, the average surface traction (the shear load divided over the current contact area) will attain a constant value. The average interlaminar slip factor β can be determined by elaborating (2.25) to

$$\beta = \frac{v_{rel}}{\gamma} = \frac{2wL(t)V}{F(t)}, \qquad 2.41$$

with V as the crosshead velocity, F as the tensile load, w as the width of the specimen and L as the momentary contact length.

Alternatively, a resin rich layer can be assumed of a certain thickness as found from microscopy (Fig. 2.5), similarly to McEntee and Ó Brádaigh (1998). The slip factor can be determined from (2.26), using this interface layer thickness and the resin viscosity data.

The actual tribology of these composite systems is complicated, however. The response is time dependent while the steady state values vary with fibre

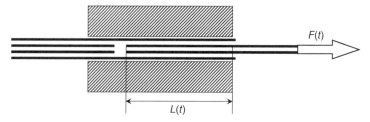

2.11 Schematic view of the pull-out experiment on a four layered laminate.

orientation, pressure, temperature and velocity. The simple viscous law introduced here is only a first approximation which should be adapted according to further findings.

Similar arguments hold for the friction between the composite and the tools, which can have a major effect on the draping process. The same pull-out set-up can be used to characterise ply-tool friction. The tools are placed on both sides of the composite and pressed with a pre-set load on the composite surface. The full laminate is now pulled through or out of the tooling blocks. The load displacement data can now be translated to a ply-tool slip factor or friction coefficient.

2.6   Future trends

Realistic process simulations of the composite forming process are within reach for product development purposes. Simulations can be made with acceptable computational efforts. The presented constitutive relations provide a basic framework which captures the most important phenomena: intra-ply shear of a fabric reinforced fluid, elastic fibres, shear locking of the fabric and inter-ply shear in multi-layered composite laminates. The required material property data can be determined by conceptually simple experiments in combination with nonlinear regression.

The various elements can be elaborated further, as the computer performance is still increasing rapidly and further complexities can be introduced. The description of interaction between the yarns in the fabric needs to be improved. Biaxial stresses have been shown to have a distinct effect on the fabric response (Boisse *et al.*, 2001). This, however, also holds for intralaminar shear, where little experimental data are available at present. The constitutive equations need to be established and appropriate characterisation methods need to be developed.

This argument holds even more for the interlaminar shear and the friction of the laminate with the forming tools. It is evident that the governing parameters are pressure, temperature, velocity, time, fabric structure and fibre orientation. However, the exact physics are poorly understood to date. The underlying mechanisms need to be identified, modelled and quantified in order include a more accurate description in the forming simulations.

Accurate process simulations further require the material property data over the full temperature range in the process. This implies a huge experimental effort, considering the vast number of combinations of fibres, matrices and fabric architectures. Micromechanical models can be used as an alternative to predict the composite properties with only the constituent properties and the reinforcement structure as an input. It will be very difficult to keep the characterisation efforts within reasonable limits without this type of model.

Finally, the success of forming simulations depends strongly on the ability of designers to work with them. On one hand, the developers need to make their software user friendly to make it accessible to the designer. On the other hand, the composite designer has to be aware of the complex material behaviour and the resulting effects on the forming process. This can only be achieved with thorough education and experience. Using these simulations as a black box is an approach which is doomed to fail.

2.7 References

Bergsma O, *Three-dimensional Simulation of Fabric Draping*, PhD thesis Delft University of Technology, 1995.

Boisse P, Borr M, Buet K and Cherouat, A, 'Finite element simulation of textile composite forming including the biaxial fabric behaviour', *Composites Part B*, 1997, **28**, 453–464.

Boisse P, Gasser A and Hivet G, 'Analyses of fabric tensile behaviour: determination of the biaxial tension-strain surface and their use in forming simulations', *Composites Part A*, 2001, **32**, 1395–1414.

Breen DE, House DH and Wozny MJ, 'A particle-based model for simulating the draping behavior of woven cloth', *Textile Research Journal*, 1994, **64**, 663–685.

Chen B, and Govindaraj M, 'A physically based model of fabric drape using flexible shell theory', *Textile Research Journal*, 1995, **65**, 324–330.

Cherouat A and Billoët JL, 'Mechanical and numerical modelling of composite manufacturing processes in deep-drawing and laying-up of thin pre-impregnated woven fabrics', *Materials Processing Technology*, 2001, **118**, 460–471.

Colombo G, Prati M and Rizzi C, 'Design and evaluation of a car soft top' In *XII ADM International Conference*, Rimini, Italy, 2001.

Cordier F and Magnenat-Thalmann N, 'Real-time animation of dressed virtual humans' *Computer Graphics Forum (Eurographics 2002)*, **21**, 327–336.

De Luca P, Lefébure P and Pickett AK, 'Numerical and experimental investigation of some press forming parameters of two fibre reinforced thermoplastics: APC2-AS4 and PEI-CETEX', *Composites Part A*, 1998, **29**, 101–110.

Heisey FL and Haller KD, 'Fitting woven fabric to surfaces in three dimensions', *Journal of the Textile Institute*, 1988, **2**, 250–263.

Ivanov I and Tabiei A, 'Flexible woven fabric micromechanical material model with fiber reorientation', *Mechanics of Advanced Materials and Structures*, 2002, **9**, 37–51.

Kang TJ and Yu WR, 'Drape simulation of woven fabric by using the finite- element method', *Journal of the Textile Institute*, 1995, **86**, 635–648.

Kato S, Yoshino T and Minami, H, 'Formulation of constitutive equations for fabric membranes based on the concept of fabric lattice model', *Engineering Structures*, 1999, **21**, 691–708.

Kawabata S, *The standardization and analysis of hand evaluation*, Osaka, Japan, Hand evaluation and standardization committee of the Textile Machinery Society of Japan, 1975.

Lamers EAD, *Shape distortions in fabric reinforced composite products due to processing induced fibre reorientation*, PhD thesis University of Twente, 2004.

Lim T-C and Ramakrishna S, 'Modelling of composite sheet forming: a review', *Composites Part A*, 2002, **33**, 515–537.

Long AC and Rudd CD 'A simulation of reinforcement deformation during the production of preforms for liquid moulding processes', *Journal of Engineering Manufacture*, 1994, **208**, 269–278.

Long AC, Souter B, Robitaille F and Rudd CD, 'Effects of fibre architecture on reinforcement deformation', *Plastics Rubber and Composites*, 2002, **31**, 87–97.

Mack C and Taylor H, 'The fitting of woven cloth to surfaces', *Journal of Textile Institute*, 1956, **47**, 477–487.

McEntee SP and Ó Brádaigh CM, 'Large deformation finite element modelling of single-curvature composite sheet forming with tool contact', *Composites Part A*, 1998, **29**, 207–213.

McGuinness GB and Ó Brádaigh CM, 'Development of rheological models for forming flows and picture-frame shear testing of fabric reinforced thermoplastic sheets', *Journal of Non-Newtonian Fluid Mechanics*, 1997, **73**, 1–28.

Mohammed M, Lekakou C, Dong L and Bader MG, 'Shear deformation and micromechanics of woven fabrics', *Composites Part A*, 2000, **31**, 299–308.

Murtagh A and Mallon P. 'Characterisation of shearing and frictional behaviour during sheet forming', In Bhattacharyya D and Pipes R, *Composite sheet forming*, Composite material series vol. 11, Amsterdam, Elsevier, 1997, 163–214.

Peng XQ, Xue P, Cao J, Lussier DS and Chen J, 'Normalization in the picture frame test of woven composites, length or area', In *Proceedings of the 5th ESAForm conference on material forming*, Krakow, Poland, 2002, Akapit Publishers.

Peters ST, *Handbook of composites*, London, Chapman & Hall, 1998.

Potter KD, 'The influence of accurate stretch data for reinforcements on the production of complex structural mouldings', *Composites*, 1979, **10**, 161–167.

Prodromou AG and Chen J, 'On the relationship between shear angle and wrinkling of textile composite preforms', *Composites Part A*, 1997, **28A**, 491–503.

Robertson RE, Hsiue ES and Yeh GSY, 'Fibre rearrangements during the moulding of continuous fibre composites II', *Polymer Composites*, 1984, **5**, 191–197.

Rogers TG, 'Rheological characterisation of anisotropic materials', *Composites*, 1989, **20**, 21–27.

Schowalter WR, *Mechanics of non-Newtonian fluids,* Oxford, Pergamon Press, 1978.

Sharma SB and Sutcliffe MPF, 'A simplified finite element approach to draping of woven fabric', In *Proceedings of the 6th ESAFORM conference on material forming*, Salerno, Italy, 2003, Nuova Ipsa Editore.

Sidhu RMJS, Averill RC, Riaz M and Pourboghrat F, 'Finite element analysis of textile composite preform stamping', *Composite Structures*, 2001, **52**, 483–497.

Smiley AJ and Pipes RB, 'Analysis of the diaphragm forming of continuous fiber reinforced thermoplastics', *Journal of Thermoplastic Composite Materials*, 1988, **1**, 298–321.

Spencer AJM, 'Theory of fabric-reinforced viscous fluids', *Composites Part A*, 2000, **31**, 1311–1321.

Spencer AJM, 'A theory of viscoplasticity for fabric-reinforced composites', *Journal of the Mechanics and Physics of Solids*, 2001, **49**, 2667–2687.

Tanov RR and Brueggert M, 'Finite element modelling of non-orthogonal loosely woven fabrics in advanced occupant restraint systems', *Finite Elements in Analysis and Design*, 2003, **39**, 357–367.

Trochu F, Hammami A and Benoit Y, 'Prediction of fibre orientation and net shape definition of complex composite parts', *Composites Part A*, 1996, **27**, 319–328.

Weën F van der, 'Algorithms for draping fabric on doubly-curved surfaces', *International journal for numerical methods in engineering*, 1991, **31**, 1415–1426.

3
Finite element analysis of composite forming

P BOISSE, INSA de Lyon, France

3.1 Introduction: finite element analyses of composite forming, why and where?

The finite element (FE) method is a relatively recent approach (Argyris, 1960) to finding numerical solutions to physical problems, especially in the field of solid and fluid mechanics. The method has been extended to some manufacturing process simulations, in particular to metal forming (Wagoner et al., 1996) and crash simulation (Halquist et al., 1985). Commercial software packages have been developed that allow fairly efficient analyses for industrial cases. The FE method often requires lengthy computation times, especially in manufacturing process simulations, but advances in computer efficiency make this drawback less important. The FE method is currently the most commonly used numerical approach, but several other numerical methods (such as the Natural Element Method (NEM) and Smooth Particle Hydrodynamics (SPH) (Sukumar et al., 1998; Gingold et al., 1977)) have appeared or are being developed. All these methods have the same goal, which is to give an approximated numerical solution to a global physical problem from its governing equations.

The most commonly used approach for analysing composite forming processes, and especially draping in woven composite reinforcements, is 'kinematical models' (Mark et al., 1956; Van Der Ween, 1991; Long et al., 1994; Borouchaki et al., 2003). Several packages are commercially available, such as FiberSIM and ESI-QuickFORM. This method is fairly efficient for hand draping in classical prepreg fabrics, but the models do not account for load boundary conditions, for possible sliding of the fabric in relation to the tools, or the mechanical behaviour of the woven reinforcement, although constitutive aspects have been introduced in some approaches (Long et al., 2001).

For a physical (or mechanical) analysis of a composite forming process, the complete model must include all the equations for the mechanics, especially equilibrium, constitutive equations, and boundary conditions. These equations must be solved numerically, with some approximations. Finite element analysis of the composite forming process includes modelling the tools, the contact and friction between the different parts, and above all, the mechanical behaviour of the composite during forming. As mentioned above, problems of computation time are reduced through improved processing capabilities. Better optimisation of methods has also improved efficiency. The main problem for the FE approach therefore lies in the requirement for accurate models of all the significant aspects of the forming process. When the models are defined and implemented, an FE analysis of a composite process gives detailed results concerning all the fields of unknowns, such as displacement, strain, stress, tensions in the yarns, and temperature, for any time during the process.

There are two main reasons for simulating the composite forming process. The first concerns the process itself and its feasibility. The simulation can give the conditions (for instance loads on the tools, initial orientation, type of material, etc.) that will make the forming possible, and also describe any possible defects after forming (wrinkles, porosities, yarn fractures, etc.). This is similar to simulating metal forming. Secondly, and unique to composite forming analysis, is the need to know at any point the directions and density of the fibres after forming. A woven composite reinforcement can be formed on a double curved shape because of the angle variations between warp and weft yarns, i.e. the in-plane shear strains. These angle variations can be very large, up to 50°, consequently the yarn directions depend significantly on the forming process. The directions and densities of fibres are also very important for analysing how the composite part will behave in use, with regards to stiffness, damage, fatigue, etc.

Fabric draping is not the only field in composite forming where FE analysis is used. Another important field for composite manufacturing simulations concerns mould filling in liquid composite moulding. In these processes, a liquid resin is injected through a fibrous reinforcement that has been shaped previously in the mould. Simulation of mould filling is usually based on the Darcy law (Darcy, 1856), associated with FE (or/and finite volume) approximations (Trochu *et al.*, 1993; Breard *et al.*, 2005; Comas-Cardona *et al.*, 2005). This flow analysis is mainly influenced by draping, when the part is double curved, because the permeability depends strongly on the angle between warp and weft yarns (Fournier *et al.*, 2005).

Although the above descriptions concern FE analysis of the entire composite forming process, because of the multiscale nature of composites, it is necessary to compute phenomena at lower scales. Analysing the deformations of a woven unit cell permits us to understand and identify the mechanical properties of the fabric as a whole (Boisse *et al.*, 2001; Hagège *et al.*, 2005; Lomov *et al.*, 2005), and resin flow simulations in this woven unit determine the permeability overall (Fournier *et al.*, 2005; Laine *et al.*, 2005). This chapter will focus on FE analysis of composite reinforcement forming.

3.2 The multiscale nature of composite materials and different approaches for composite forming simulations

3.2.1 Multiscale materials

Composite materials are made of fibres and matrix, and the fibre arrangements are numerous. Fibres can be short (SMC, BMC, GMT, etc.) or continuous, as is the case for composite structures that must withstand strong loads. Continuous fibres can be simply juxtaposed, or unidirectional (UD), woven, juxtaposed and stitched (non-crimp fabrics, NCF), braided, knitted or randomly oriented (mats). The arrangement of fibres is very important and the materials resulting from the various arrangements are very different. Composite materials are typically multiscale materials because the architecture and the properties at lower scales strongly influence the behaviour of the composite. This is true for composites during use, but it is even more the case in the forming process because the lack of matrix (or of matrix efficiency) renders the internal reinforcement architecture more essential.

Three scales can generally be distinguished in composite reinforcements. The macroscopic scale is that of the composite part for which the forming is being studied. The mesoscopic scale is an intermediate scale that concerns the yarn architecture. Typically, it is the scale of the unit woven cell, or knitted cell in the case of textile composites. Finally, the microscopic scale is that relating to the numerous fibres that make up the yarns (or tows or bundles), for instance a carbon yarn may contain 6000 fibres (in the case of the yarn in Fig. 3.14). The arrangement and density of the fibres within the yarn affect its behaviour. Fibre diameters for carbon, glass, aramid, polypropylene, etc., are usually between $5\,\mu$m and $50\,\mu$m.

3.2.2 Mechanism and specificity of composite forming

It is necessary to distinguish between composites with short fibres and composites with continuous fibres. Short fibres improve the mechanical properties of a material through the matrix, but this matrix remains the basic element. Injection is one of the manufacturing technologies used for materials with short fibres. The forming simulation in this case is a resin flow problem, but with fibres that are oriented by the forming (Dumont *et al.*, 2003; Chinesta *et al.*, 2005). In the case of continuous fibrous reinforcements, the reinforcements play a major role in the mechanical behaviour of the composite, the main function of the matrix being to prevent relative displacements of continuous filaments. This type of composite is used where load-bearing is important. The forming of composites with continuous reinforcements will be analysed below.

Forming of composite materials differs from that of materials such as metals or polymers because it uses the fibre-matrix composition of the composite. The forming process is performed in a state in which the matrix does not play a role.

In Liquid Composite Moulding (LCM) processes, the reinforcement is shaped before the resin is injected. Bending and in-plane shear are possible, thus double curved shapes can be obtained. In the case of thermoset prepreg draping, the matrix is present but is not solid because it is not yet polymerised, thus the reinforcement can be deformed as well. For Continuous Fibres Reinforced Thermoplastics (CFRTP) prepregs forming, the matrix is heated above the melting temperature, and this permits deformation of the reinforcement as well as the final forming. In these examples, the matrix is lacking or very weak and the internal architecture of the fibrous reinforcement permits the forming to take place. This is specific to composite materials and the forming modes are quite different to those of metals. There are usually very small extensions in the fibre directions, and the in-plane shear strains play a major role in forming. Codes for simulating the composite forming process and the approaches that are used are therefore specific to those materials.

3.2.3 Different approaches to FE analysis of composite forming

As described above, FE analysis of composite forming requires all the different aspects involved in the process to be modelled, in particular that of the fibrous reinforcement. This reinforcement can be with or without matrix, but as discussed in Section 3.2.2, this matrix is weak and allows the reinforcement to deform. The multiscale nature of the composite and of its fibrous reinforcement means that there are different possible approaches for FE analysis of the forming process.

The first approach considers the fibrous reinforcement as a continuum. The reinforcement is not continuous at lower scales, but that is the case for most materials under large strains (for instance a metal in plasticity), and a continuous material superimposed on the fibrous material can be postulated. This makes the assumption that there is no significant sliding between the fibres, i.e. that two neighbouring points in the initial state remain close after forming, which can be verified experimentally for most usual reinforcements. For instance, Fig. 3.1 (Boisse, 1994) shows a fibre woven fabric formed on a hemispherical, i.e. a double curve, shape. A set of lines following the warp and weft directions have been drawn on the fabric prior to forming. These lines become strongly curved after forming but remain continuous, which implies that, due to the weaving, there is no large sliding between warp and weft yarns (if there was, the lines would have become discontinuous), and consequently that the continuous approach is possible. The advantage of the continuous approach is that it can be used in standard finite elements, although the constitutive model of the continuum will have to convey the very specific mechanical behaviour of the fibrous reinforcement. Such behaviour depends mainly on the fibre directions, which change strongly during forming. Section 3.3 will demonstrate why the approaches normally used for anisotropic metal forming cannot be used here.

50 Composites forming technologies

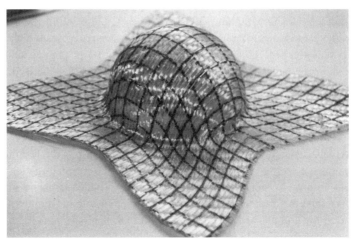

3.1 Deformation of straight lines drawn on the fabric prior to forming.

Other specific aspects of fibrous reinforcement mechanical behaviours, such as crimp change and shear locking, should also be taken into account in the model.

The opposite approach is to see the fibrous reinforcement as a set of elements at lower scales, such as the yarns, woven cells, fibres, etc. The FE analysis is then concerned with those elements that are in contact or are linked by springs. Although some approaches at the microscopic scale exist (Durville, 2002; Duhovic *et al.*, 2006), the large number of filaments means that it is more realistic to analyse a small number of elements at the mesoscale (woven or knitted cells, for instance), and it is not currently possible to analyse an entire forming process. The advantage over the continuous approach is that the description of the internal structure of the reinforcement naturally accounts for some aspects of the material, such as the directions of the fibres. Nevertheless, it is difficult to delineate models that are efficient enough at the mesoscale but simple enough to be able to analyse a forming process. Continuous and mesoscopic FE models are discussed in the next two sections.

3.3 The continuous approach for composite forming process analysis

This approach assumes that the composite can be considered as a continuum media during forming, consequently the mechanics of continuous media can be used. This assumption is questionable, especially for stress. What is the stress in a woven fabric? A stress tensor is defined as a function that gives in any point P the stress vector $d\vec{F}/dS$ for any normal vector. This is not clear if P is a point within a woven fabric. It is therefore necessary to consider a continuous media that is not exactly equal to the fibrous reinforcement, but that is homogenous and has the same global mechanical behaviour. This is possible if (as seen in Section

3.2.3 and Fig. 3.1) large sliding between the yarns is minimal. This method also makes an approximation, and so the local quantities in the fibrous reinforcement are only those of the continuous media on average.

The main difficulty in using the continuous approach is to capture the effects of the fibre architecture and its evolution. There are many models (see Chapter 2). Most of them assume that the fibrous reinforcement is elastic while forming. That is usually true for extensions in the fibre directions, but not usually true in the other directions, such as in-plane shear, bending, and transverse compression. Nevertheless, the forming process is a more or less constant operation and making this assumption doesn't change the result of the analysis greatly.

Continuous behaviour models that capture macro-level phenomena at lower scales generally concern homogenisation, although homogenisation typically refers to techniques applied to a two-scale periodic material, in which the analysis of a unit cell reveals the properties of the homogenised material (Hsiao *et al.*, 1999; Peng *et al.*, 2002). This approach is elegant but requires lengthy computational times. Furthermore, extending it to non-linear problems is difficult. Three continuous approaches used in FE analysis are described below.

3.3.1 Hypoelastic model for fibrous materials

In this method, fibres are considered to have a single direction. Approaches traditionally developed in finite element codes (such as ABAQUS, for instance) for anisotropic metal at large strains are based on Jaumann corotational formulation (Dafalias, 1983; Gilormini *et al.*, 1993) or the Green-Naghdi approach (Dienes, 1979; Gilormini *et al.*, 1993). In these models, a rotation is used both to define an objective derivative for the hypoelastic law and to update the othotropic frame. The rotations used in Green-Naghdi and Jaumann derivatives are average rotations of the material. The polar rotation $\mathbf{R}$ is used in Green-Naghdi:

$$\mathbf{R} = \mathbf{F}\mathbf{U}^{-1} \qquad 3.1$$

where $\mathbf{F}$ is the deformation gradient and $\mathbf{U}$ is a symmetrical strain tensor. In the Jaumann approach, the rotation $\mathbf{Q}$ of the corotational frame is used. These well-known approaches cannot be used for a fibrous material under large strains because the update for the strong direction must follow the fibre direction strictly. Figure 3.2 shows the evolution of the orthotropic direction in a finite element analysis of the extension of a knitted fibrous material (Hagège, 2004). The analysis, using ABAQUS, uses the hypoelastic Green-Naghdi approach and it can be seen that the orthotropic direction does not follow the fibre direction.

The following approach uses the rotation of the fibre denoted Δ (Hagège, 2004; Hagège *et al.*, 2005). The initial orientation of the orthotropic axes $\{\mathbf{\kappa}^0\}$ is defined by a rotation tensor field $\mathbf{O}$ that transforms the unitary vectors of the global basis $\{\mathbf{G}\}$:

52   Composites forming technologies

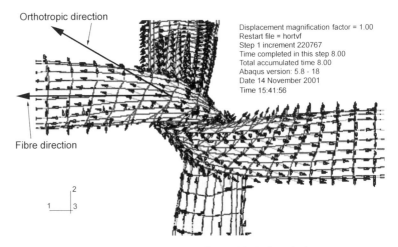

3.2 Orthotropic direction evolution in a Green-Naghdi analysis.

$$\kappa_i^0 = \mathbf{O} \cdot \mathbf{G}_i \qquad 3.2$$

The material rotation $\mathbf{\Delta}$ is then used to update the initial constitutive axes $\{\kappa^0\}$ to the current constitutive axes $\{\kappa^t\}$ (Fig. 3.3):

$$\kappa_i^t = \mathbf{\Delta} \cdot \kappa_i^0 \qquad 3.3$$

Some developments of equation (3.3) (Criesfield, 1991) lead to equations (3.4) that explicitly give the constitutive axes $\{\kappa^t\}$ as functions of the initial constitutive axes $\{\kappa^0\}$ and the deformation gradient $\mathbf{F}$:

$$\kappa_1^t = \frac{\mathbf{F} \cdot \kappa_1^0}{\|\mathbf{F} \cdot \kappa_1^0\|}$$

$$\kappa_2^t = \kappa_2^0 - \frac{b_2}{1+b_1}\left(\kappa_1^0 + \kappa_1^t\right)$$

$$\kappa_3^t = \kappa_3^0 - \frac{b_3}{1+b_1}\left(\kappa_1^0 + \kappa_1^t\right) \qquad 3.4$$

with $b_k = \kappa_1^t \cdot \kappa_k^0$ and $b_k \neq 1$. In this formulation, the fibre direction, i.e. the strong anisotropic direction, remains aligned with the first vector of $\{\kappa^t\}$. The constitutive behaviour is then fully defined at each time point. In fact, the initial constitutive tensor $^0\mathbf{C}$ has known components that can be computed with the traditional engineer's constants:

$$^0\mathbf{C} = {^0C_{ijkl}}\kappa_i^0 \otimes \kappa_j^0 \otimes \kappa_k^0 \otimes \kappa_l^0 \qquad 3.5$$

The current constitutive tensor $\mathbf{C}$ can be deduced from $^0\mathbf{C}$ by a rotational transport based on the fourth order rotation tensor Λ:

$$\mathbf{C} = \Lambda : {^0\mathbf{C}} : \Lambda^T \qquad 3.6$$

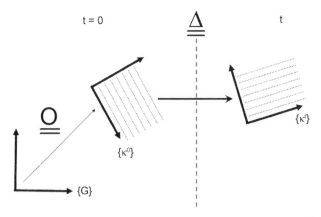

3.3 2D representation of the evolution of the constitutive axes $\{\kappa^t\}$.

Λ is the fourth order rotational tensor obtained from Δ:

$$\forall \mathbf{A} \quad \Lambda : \mathbf{A} = \Delta \cdot \mathbf{A} \cdot \Delta^T \qquad 3.7$$

Consequently $\mathbf{C}$ is given by:

$$\mathbf{C} = {}^0C_{ijkl}\kappa_i^t \otimes \kappa_j^t \otimes \kappa_k^t \otimes \kappa_l^t \qquad 3.8$$

The constitutive tensor $\mathbf{C}$ can be used in a hypo-elastic law written:

$$\sigma^\nabla = \mathbf{C} : \mathbf{D} \qquad 3.9$$

$\mathbf{D}$ is the strain rate and σ^∇ is the objective derivative of the Cauchy stress associated to the fibre rotation Δ.

$$\sigma^\nabla = \Delta \cdot \frac{d}{dt}\left(\Delta^T \cdot \sigma \cdot \Delta\right) \cdot \Delta^T \qquad 3.10$$

The cumulated tensorial strain tensor ε and stress tensor σ associated with such an objective derivative are given by:

$$\varepsilon = \Delta \cdot \left(\int_0^t \Delta^T \cdot \mathbf{D} \cdot \Delta \, dt\right) \cdot \Delta^T \qquad 3.11$$

$$\sigma = \Delta \cdot \left(\int_0^t \Delta^T \cdot \mathbf{C} \cdot \mathbf{D} \cdot \Delta \, dt\right) \cdot \Delta^T \qquad 3.12$$

It can be shown that equation (3.11) will always give a logarithmic strain in the strong anisotropic direction and that equation (3.12) ensures the summation of the stress increments along this direction.

Finally, the use of the material rotation tensor Δ for the objective derivative (3.10) and the evolution law (3.6) entails a consistent approach for fibrous media with one strong anisotropic direction. Figure 3.4 shows FE analysis of the same knitted material as Fig. 3.2, but uses the above approach implemented in

54   Composites forming technologies

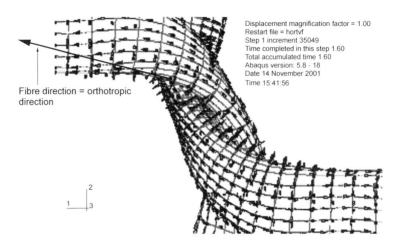

3.4 Orthotropic axes updated using the fibre rotation ≜.

ABAQUS via a VUMAT routine (User Material routine) (Hagège, 2004). In this case the orthotropic direction remains strictly in the fibre direction and it has been shown that the analysis is consistent with biaxial tension on knitted material (Hagège *et al.*, 2005).

The above approach is carried out for a single fibre direction, although many fibrous reinforcements, and especially woven fabrics, have two fibre directions. The behaviour here is no longer orthotropic because there are very large angle variations between the warp and weft directions due to in-plane shear. The above formulation can be used by superposition on the same point for two materials using their own fibre directions. Sliding between warp and weft directions can be ignored because the two materials (warp and weft) are in the same finite element. However, that does not take account of the interaction between fibres, such as crimp changes. In order to extend the approach to two fibre directions using rotation of the fibre, it is necessary to leave out the rotational derivatives that lead to orthotropic frames. A Lie derivative based on the deformation gradient **F** has been defined and implemented in ABAQUS. It is not exactly **F** that is used, but **F**$_N$, which gives the same directions but keeps the unit vector with a norm equal to one. Examples of forming fabric simulations have been analysed in Hagège (2004). Figure 3.5 presents the different updates for the material axis and the associated objective derivatives described in this section.

3.3.2 Non-orthogonal constitutive models

In this method, the stress and strain of a continuous material are related to fibrous reinforcement using the constitutive relation in a non-orthogonal frame directed by the fibre directions. Models have been developed by Yu *et al.* (2002)

Finite element analysis of composite forming

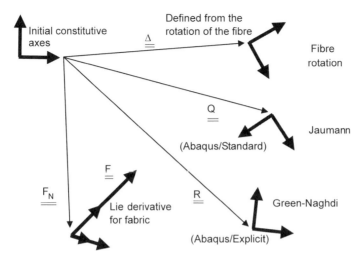

3.5 Material direction updates and objective derivatives (Hagège, 2004).

and Xue *et al.* (2003). They consider two yarn directions and use them to define the non-orthogonal frame. The second model, Xue *et al.* (2003) is briefly described here.

Figure 3.6 shows a local element of the continuum and the coordinates ξ and η along the warp and weft directions. x', y' are orthogonal coordinates with $x' = \xi$. The stress components in theses two frames can be related:

$$\begin{bmatrix} \sigma_{x'} \\ \sigma_{y'} \\ \sigma_{x'y'} \end{bmatrix} = \begin{bmatrix} 1 & 0 & 0 & 0 \\ 0 & 1 & -\cot\theta & -\cot\theta \\ -\cot\theta & \cot\theta & 0 & 1 \end{bmatrix} \begin{bmatrix} \sigma_{\xi} \\ \sigma_{\eta} \\ \tau_{\xi\eta} \\ \tau_{\eta\xi} \end{bmatrix} = T_2 \begin{bmatrix} \sigma_{\xi} \\ \sigma_{\eta} \\ \tau_{\xi\eta} \\ \tau_{\eta\xi} \end{bmatrix} \quad 3.13$$

$\tau_{\xi\eta}$ is not equal to $\tau_{\eta\xi}$ because ξ, η is non-orthogonal. The strains in the two frames are also related:

$$\begin{bmatrix} \epsilon_{\xi} \\ \epsilon_{\eta} \\ \gamma_{\xi\eta} \\ \gamma_{\eta\xi} \end{bmatrix} = \begin{bmatrix} 1 & 0 & 0 \\ \cos^2\theta & \sin^2\theta & \cos\theta\sin\theta \\ 0 & 0 & 1 \\ 0 & 0 & 1 \end{bmatrix} \begin{bmatrix} \epsilon_{x'} \\ \epsilon_{y'} \\ \epsilon_{x'y'} \end{bmatrix} = T_3 \begin{bmatrix} \epsilon_{x'} \\ \epsilon_{y'} \\ \epsilon_{x'y'} \end{bmatrix} \quad 3.14$$

ϵ_{ξ} and ϵ_{η} are the normal strains along warp and weft directions, $\gamma_{\xi\eta}$ and $\gamma_{\eta\xi}$ are the angular changes from the initial right-angle position. It can be assumed that the biaxial tensile properties and in-plane shear properties are independent. It has been shown experimentally that the biaxial tensions are affected only weakly by in-plane shear (Buet-Gautier *et al.*, 2001). The relation between stress and strain components in the non-orthogonal coordinates can be supposed in the form:

56 Composites forming technologies

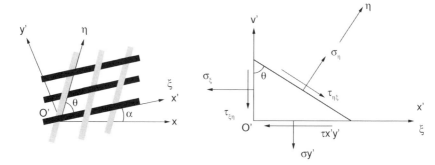

3.6 Orthogonal and non-orthogonal local frame and stress components on an isolated element (Xue *et al.*, 2003).

$$\begin{bmatrix} \sigma_\xi \\ \sigma_\eta \\ \tau_{\xi\eta} \\ \tau_{\eta\xi} \end{bmatrix} = \begin{bmatrix} D_{11} & D_{12} & 0 & 0 \\ D_{21} & D_{22} & 0 & 0 \\ 0 & 0 & \beta D_{33} & 0 \\ 0 & 0 & 0 & (2-\beta)D_{33} \end{bmatrix} \begin{bmatrix} \epsilon_\xi \\ \epsilon_\eta \\ \gamma_{\xi\eta} \\ \gamma_{\eta\xi} \end{bmatrix} = D \begin{bmatrix} \epsilon_\xi \\ \epsilon_\eta \\ \gamma_{\xi\eta} \\ \gamma_{\eta\xi} \end{bmatrix} \qquad 3.15$$

D_{11}, D_{22}, D_{12}, D_{21} are tension coefficients. D_{33} denotes the shear property. β is a coefficient that stands for contributions from each stress to shear. The constitutive equation can be expressed in the local orthogonal coordinates, and then in the global orthogonal coordinates.

$$[\sigma]_{x'y'} = \mathbf{T}_2 \mathbf{DT}_3 [\varepsilon]_{x'y'} \qquad [\sigma]_{xy} = \mathbf{RT}_2 \mathbf{DT}_3 \mathbf{R}^T [\varepsilon]_{xy} \qquad 3.16$$

R is the rotation matrix (3×3) that gives the components of an in-plane second order tensor in x, y from its components in x', y'. Equation (3.16) is a constitutive relation in the global frame using the properties in the fibre directions. The components D_{ij} of (3.15) can be identified from biaxial tests and in-plane shear tests (Xue *et al.*, 2003).

3.3.3 A macro-mechanical model used in a commercial FE code

The initial research and development work presented in this section was undertaken within a European Brite Euram project (BE 5092). Developments made under ESI PAM-STAMP led to the commercial FE code PAM-FORM, dedicated to composite forming and especially to reinforced thermoplastics (De Luca *et al.*, 1998; Pickett *et al.*, 2005).

An important point, frequently present in composite forming simulations, concerns the interface between plies. Several plies are often shaped together, especially in the case of reinforced thermoplastic composites. Each ply is kept discrete using shell elements. The friction loads are the summation of dry friction and viscous friction, see Fig. 3.7 (Pickett *et al.*, 2005).

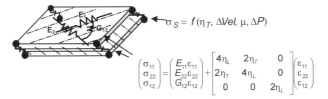

$$\begin{pmatrix} \sigma_{11} \\ \sigma_{22} \\ \sigma_{12} \end{pmatrix} = \begin{pmatrix} E_{11}\varepsilon_{11} \\ E_{22}\varepsilon_{22} \\ G_{12}\varepsilon_{12} \end{pmatrix} + \begin{bmatrix} 4\eta_L & 2\eta_T & 0 \\ 2\eta_T & 4\eta_L & 0 \\ 0 & 0 & 2\eta_L \end{bmatrix} \begin{pmatrix} \varepsilon_{11} \\ \varepsilon_{22} \\ \varepsilon_{12} \end{pmatrix}$$

3.7 Two stacked plies and the constitutive laws for composite sheet forming (Pickett *et al.*, 2005).

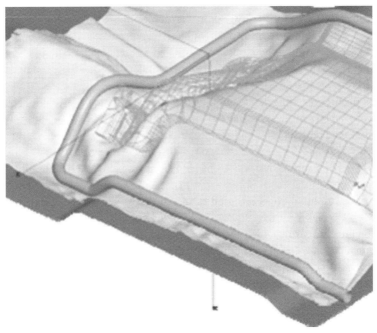

3.8 Simulation of the forming of a non-crimp fabric on the floor pan of a car (PAM-FORM ESI Group).

The mechanical behaviour of the ply is likened to elastic fibres embedded in a viscous resin (Ó Brádaigh *et al.*, 1993). Both resin longitudinal viscosity η_L and transverse viscosity η_T can be included. For dry fabrics, viscosity is ignored. The model can also be extended to NCF by modelling the stitching using spring elements (Pickett *et al.*, 2005). The simulation of the forming of a non-crimp fabric on the floor pan of a car is presented in Fig. 3.8, conducted by the ESI-Group using PAM-FORM within the European project TECABS.

3.4 Discrete or mesoscopic approach

The discrete approach is the opposite of the continuous approach, and considers and models the components of fibrous reinforcement at low scale. These components can be yarns, woven cells or stitching, and also sometimes fibres. Because these elements are usually at the mesoscale (as defined in Section 3.2.1), the approach is also known as meso-mechanical modelling. Some analyses have been proposed where all fibres are modelled (microscale modelling) (Durville, 2002; Duhovic *et al.*, 2006; Pickett, 2002), but the number of fibres in a composite structure limits these computations to small subdomains, for instance a woven cell or a few braided or knitted loops. Work in the field of discrete or mesoscopic analyses includes studies by Ben Boubaker *et al.* (2002, 2005), Cherouat *et al.* (2001), Ramgulam *et al.* (2005) and Pickett *et al.* (2005). A major difficulty lies in the description of the components at mesoscopic scale, usually the woven yarns. A compromise must be found between a precise description (which will be expensive from the computation time point of view) and a simple description, where it is possible to compute the entire forming process.

Beam and truss element are the more common descriptions for yarns (Cherouat *et al.*, 2001; Sharma *et al.*, 2003; Skordos *et al.*, 2005). Figure 3.9 shows the unit cell of a material made of four two-node trusses that represent the tows, and one (or two) truss elements to model shear stiffness (Sharma *et al.*, 2003). This approach has been used to simulate hemispherical draping (Skordos *et al.*, 2005).

Masses and springs are used to model the woven reinforcement in Fig. 3.10 (Ben Boubaker *et al.*, 2005). Springs are used to describe stretching, shear, bending and also elastic foundations. Drape deformation of a square fabric on a square surface has been performed using this model. The same authors seek to account for the crimp variations and yarn interaction, but the model is restricted to a plane cross-section and concerns a small number of woven cells.

Based on the above descriptions, a parallel can be drawn between fabric FE analysis and truss or beam structures such as those in civil engineering. Beam or truss elements can be used for FE analysis, but equivalent continuous models

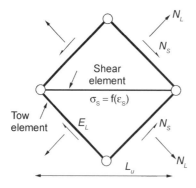

3.9 Unit cell made of four trusses for tows and one truss for shear stiffness (Sharma *et al.*, 2003).

Finite element analysis of composite forming 59

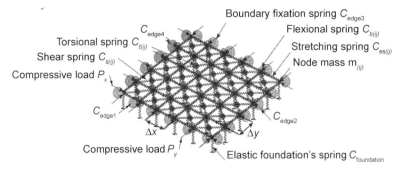

3.10 Discrete fabric model based on masses and springs (Ben Boubaker *et al.*, 2005).

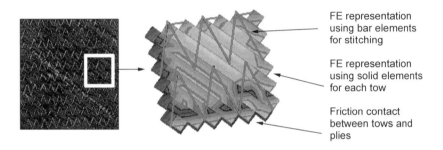

3.11 A 'representative cell' of the meso-mechanical model for NCF (Pickett *et al.*, 2005).

have been defined in order to compute structures with very large numbers of periodic beam patterns (Tollenaere *et al.*, 1998). Nevertheless, the complexity of the behaviour of a woven reinforcement cell is greater than that of a lattice cell.

In Pickett *et al.* (2005), meso-mechanical FE modelling of NCF uses 3D elements for each yarn and bar element for stitching (Fig. 3.11). Friction contact between tows and plies is taken into account. Although the complete model probably needs extensive computational time, a forming process has been simulated in this way (Pickett, 2005).

A point in favour of analysis at the mesoscopic scale lies in the strong increase in computer efficiency. Some discrete models that only work on small fabric parts will probably be used on a whole forming process in the near future.

3.5 Semi-discrete approach

The semi-discrete approach associates the FE method and a mesoscopic analysis of the woven unit cell. Specific finite elements are defined that are made of a discrete number of woven unit cells. The description of the fabric by finite

60   Composites forming technologies

elements assumes that two points of weft and warp yarns initially superimposed remain superimposed after forming, i.e. there is no sliding between the yarns. This has been experimentally shown in most forming cases, as shown in Section 3.2.3 (Fig. 3.1).

3.5.1   Simplified dynamic equation for fabrics

Fabrics classically used as textile composite reinforcements have yarns made of thousands of small fibres, such as glass, carbon or aramid (Fig. 3.12a, b). Warp and weft yarns are woven following classical weaves, plain, satin or twill (Fig. 3.12c). This constitution leads to a fabric with very specific mechanical behaviour. Most of the rigidities are small or very small in comparison to tensile stiffness in the yarn directions. The mechanical behaviour of the unit woven cell for the model must be kept as simple as possible, i.e. must only account for significant mechanical quantities. The model will have to describe the specificities of textile reinforcement mechanical behaviour, especially:

- the non-linear tensile behaviour due to crimp interchange; and
- shear locking and the very different in-plane shear behaviour before and after this locking angle.

The diameter of the glass, carbon or aramid fibres is very small (a few μm) with regard to length. Consequently, the fibres can only be submitted to tension in their longitudinal direction $\mathbf{h}_1$ (Fig. 3.12a). The yarns usually used for composite reinforcements are composed of juxtaposed fibres in the same direction $\mathbf{h}_1$ (roving). Because relative sliding of the fibres is possible, the Cauchy stress state in the yarn, as well as in the fibres, is in the form (Fig. 3.12a,b):

$$\boldsymbol{\sigma} = \sigma^{11} \mathbf{h}_1 \otimes \mathbf{h}_1 \qquad 3.17$$

The tension in the yarn can be defined as:

$$T^{11} = \int_A \sigma^{11} \, dS \qquad \mathbf{T} = T^{11} \mathbf{h}_1 \otimes \mathbf{h}_1 \qquad 3.18$$

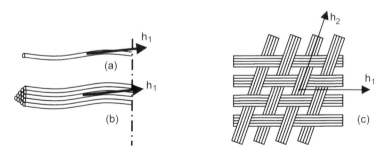

3.12 (a) Single fibre, (b) yarn made of juxtaposed fibres, (c) woven yarns.

A is the section of the yarn. This tension is a better defined quantity than the stress and is easier to measure. If a woven domain is considered (Fig. 3.12c), the tension tensor is in the form:

$$\mathbf{T} = T^{11} \mathbf{h}_1 \otimes \mathbf{h}_1 + T^{22} \mathbf{h}_2 \otimes \mathbf{h}_2 \quad \quad 3.19$$

where $\mathbf{h}_1$ and $\mathbf{h}_2$ are the vectors in warp and weft directions. In order to account for warp-weft interactions due to in-plane shear, i.e. warp-weft angle variations, a couple C, normal to the fabric, is considered at each crossover, or set of crossovers in the case of a more complex woven unit cell. Consequently, the global simplified dynamic equation is:

$$\sum_{p=1}^{\text{ncell}} \left({}^P\epsilon_{11}(\boldsymbol{\eta})^P T^{11} L_1 + {}^P\epsilon_{22}(\boldsymbol{\eta})^P T^{22} L_2 \right) + \sum_{p=1}^{\text{ncell}} ({}^P C \, {}^P\gamma(\boldsymbol{\eta}) - W_{\text{ext}}(\boldsymbol{\eta})$$

$$= \int_\Omega \rho \, \ddot{\mathbf{u}} \cdot \boldsymbol{\eta} \, dV \quad \quad 3.20$$

where $\varepsilon(\boldsymbol{\eta}) = \nabla^s \boldsymbol{\eta} = \epsilon_{\alpha\beta}(\boldsymbol{\eta}) \mathbf{h}^\alpha \otimes \mathbf{h}^\beta$ is the symmetrical gradient in the virtual displacement $\boldsymbol{\eta}$ (α and β are indexes taking value 1 or 2). $\mathbf{h}^1$, $\mathbf{h}^2$ are the contravariant vectors related to $\mathbf{h}_1$, $\mathbf{h}_2$, i.e. $\mathbf{h}_\alpha \cdot \mathbf{h}^\beta = \delta_\alpha{}^\beta$. L_1 and L_2 are the lengths of the warp and weft yarns in the midplane of the fabric. $\gamma(\boldsymbol{\eta})$ is the virtual relative rotation between warp and weft fibres (or virtual shear angle). ncell is the number of woven unit cells of the textile structure, ${}^P Q$, which means that the quantity Q is considered for the woven unit cell number p. $\ddot{\mathbf{u}}$ is the acceleration, ρ is the mass per volume of the fabric Ω. $W_{\text{ext}}(\boldsymbol{\eta})$ is the virtual work of the exterior prescribed loads.

To make an FE simulation of composite woven reinforcement forming based on the above approach, it is necessary to be able to calculate the tensions T^{11} and T^{22} and the shear couple C for a given strain field in the woven unit cell. It is assumed that the tension does not depend on the shear angle and that the shear couple does not depend on the axial strain, i.e. $T^{11}(\epsilon_{11}, \epsilon_{22})$, $T^{22}(\epsilon_{11}, \epsilon_{22})$ and $C(\gamma)$. In Buet-Gauthier et al. (2001), biaxial tensile tests performed for different angles between warp and weft yarns showed that the influence of this angle is small and can be neglected. The second assumption (C only depending on γ) is probably less true (Dumont, 2003; Lomov et al., 2004), nevertheless, all the currently available experimental results give the shear load as a function of the shear angle without any information on the tensions, so the assumption $C(\gamma)$ will be made by default. Bending stiffness is not taken into account. For single- or few-layer textiles obtained by weaving yarns made of very small glass or carbon fibres and used as composite reinforcements, this is a justified assumption in most fabric forming processes.

3.5.2 Experimental and virtual test for tensile and in-plane shear behaviour

The weak form (3.20) requires knowledge of the two 'tension surfaces' $T^{11}(\epsilon_{11}, \epsilon_{22})$ and $T^{22}(\epsilon_{11}, \epsilon_{22})$. These surfaces can be obtained from a biaxial tensile test. Due to the weaving, the biaxial behavior of a fabric is complex and generally non-linear. Some experiments have been described in Kawabatta *et al.* (1973) and Buet-Gauthier *et al.* (2001). Biaxial tensile strains are prescribed to a cross-shaped woven specimen (Fig. 3.13) with different strain ratios (denoted k) in warp and weft directions. The tension measurements give the curve tensions as a function of axial strain for different warp-weft ratios. From this set of curves, the surface $T^{\alpha\alpha}(\epsilon_{11}, \epsilon_{22})$ for the given fabric can be extrapolated (Fig. 3.14). The tensile surfaces can also be obtained from three-dimensional FE computations on a unit cell as shown in Fig. 3.14 (Gasser *et al.*, 2000; Boisse *et al.*, 2001). These analyses are not classical, as the yarn is made of thousands of fibres and the mechanical behaviour of this assembly is very specific. The objective derivative based on the fibre rotation, as described in Section 3.3.1, must be used in order to follow the fibre direction exactly (Hagège *et al.*, 2005).

As shown in equation (3.17), a single fibre, and consequently an isolated yarn, can only be submitted to a tensile stress along its direction. When the yarns are woven, the interactions between warp and weft can create other stresses in the fabric. In-plane shear rigidity is of particular interest. It is very weak in most cases and sometimes insignificant in comparison with tension stiffness (Boisse *et al.*, 2001). The in-plane shear behaviour of fabric has been studied extensively, probably because it is the main deformation mode of fabrics (McGuinness *et al.*, 1997, 1998; Prodomou *et al.*, 1997; McBride *et al.*, 1997; Cao *et al.*, 2004).

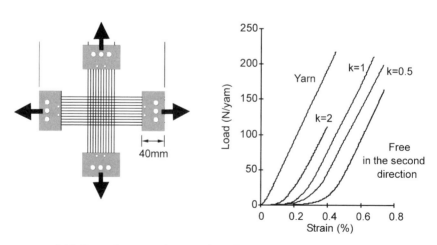

3.13 Cross shape specimen and tensile curves for different warp weft strain ratios k.

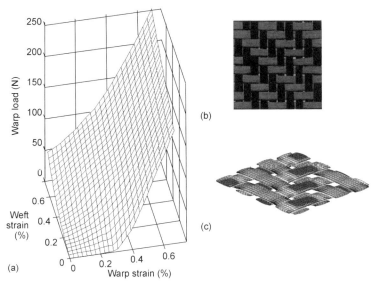

3.14 Tensile surface for a 2 × 2 twill of carbon.

The experimental in-plane shear behaviour of woven reinforcements can be analysed using a classical picture frame device (Fig. 3.15). Optical strain measurements can be taken at the macroscopic (pictures of the whole specimen), mesoscopic (a few woven cells) and microscopic levels within a yarn (Dumont, 2003). They permit both strain measures independent of the device and help us to understand the internal behaviour of the fabric during shearing. The strain fields are computed using an image correlation method (Raffel *et al.*, 1998; Vacher *et al.*, 1999). The macroscopic measures give the shear field versus the load on the picture frame and allow the homogeneity of the shear in the specimen to be checked. Figure 3.16 shows the load on the picture frame versus the shear strain in the case of a glass plain weave. In zone 1, the load is weak and the displacement field within the yarn (microscopic scale) shows that the yarn is submitted to rotation without local strain. The global shear of the fabric is entirely due to the relative motions of the yarns. The beginning of zone 2 corresponds to the shear-limit angle or shear-locking angle. The yarns start to be in contact with their neighbour and are laterally compressed, partially (zone 2), then totally (zone 3).

For a given shear angle γ the couple C can be deduced from the tension load on the picture frame. The power provided by the tension machine is assumed to be equally distributed on all woven unit cells.

$$\text{ncell } C(\gamma) = F(\gamma)V$$

and consequently

64   Composites forming technologies

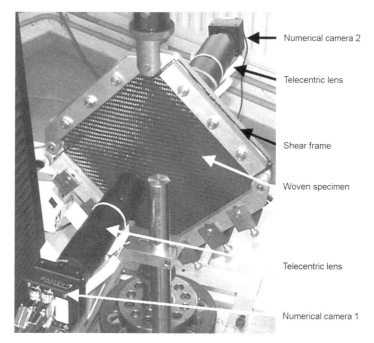

3.15 Shear frame device equipped with an optical system.

$$C(\gamma) = \frac{a}{\text{ncell}} \frac{\sqrt{2}}{2} \left(\cos \frac{\gamma}{2} - \sin \frac{\gamma}{2} \right) F(\gamma) \qquad 3.21$$

where a is the length of the side of the frame (between two pin-joints), F is the load on the tension machine for a shear angle γ and V is the speed of the tension machine.

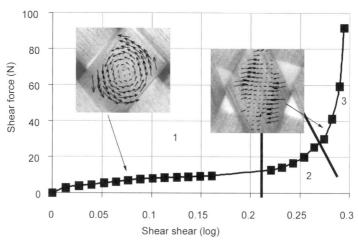

3.16 Shear curve and optical analysis (Dumont, 2003).

Finite element analysis of composite forming

This shear curve can also be obtained from three-dimensional FE analysis on a unit cell submitted to in-plane shear (Boisse et al., 2005b), although this analysis is difficult due to the contact localisation beyond the shear-locking angle.

3.5.3 A four-node finite element made of woven cells

From our knowledge of the tension surfaces ($T^{11}(\epsilon_{11}, \epsilon_{22})$ and $T^{11}(\epsilon_{11}, \epsilon_{22})$) and of the shear curve ($C(\gamma)$), the simplified dynamic equation (3.5) permits us to construct specific finite elements for fabric forming (Boisse et al., 2005a; Zouari et al., 2006).

The four-node element is presented in Fig. 3.17. It is made of ncell^e woven cells. The directions of the yarns are those of the natural coordinates in the reference element ξ_1, ξ_2, i.e. the directions of the sides of the element. There are two main reasons for this. First, the numerical efficiency is improved because the expressions of the interior load components are much simpler. Secondly, it has been shown that, in the case of a material with two directions that are very stiff in comparison with others (especially woven materials), the finite element analyses can lead to locking if these directions are not those of the element sides (Yu et al., 2004).

Because the computations are made using an explicit approach, the only quantity needed is the elementary interior nodal load $\mathbf{F}^e_{int}$ that is related to the interior elementary work W^e_{int}:

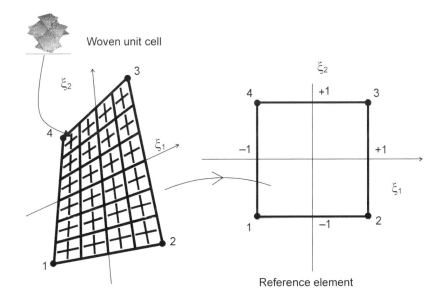

3.17 Four node finite element made of woven cells.

$$W^e_{\text{int}}(\boldsymbol{\eta}) = \sum_{p=1}^{\text{ncell}^e} \left({}^P\epsilon_{11}(\boldsymbol{\eta}) \, {}^PT^{11} \, {}^PL_1 \right) + {}^P\epsilon_{22}(\boldsymbol{\eta}) \, {}^PT^{22} \, {}^PL_2 + \sum_{p=1}^{\text{ncell}^e} {}^P\gamma(\boldsymbol{\eta}) {}^PC$$

$$= \eta_s (F^e_{\text{int}})_s \qquad 3.22$$

The nodal index s varies from 1 to 12 in the case of the four-node quadrilateral.

The symmetrical gradient of the virtual displacement is expressed in $\mathbf{h}^1$, $\mathbf{h}^2$ and $\mathbf{g}^1$, $\mathbf{g}^2$:

$$\nabla^s \boldsymbol{\eta} = \epsilon_{\alpha\beta} \mathbf{h}^\alpha \otimes \mathbf{h}^\beta = \bar{\epsilon}_{\alpha\beta} \mathbf{g}^\alpha \otimes \mathbf{g}^\beta \qquad 3.23$$

$\mathbf{g}_1$, $\mathbf{g}_2$ is the covariant material base such as $\mathbf{g}_\alpha = \partial x / \partial \xi_\alpha$ and $\mathbf{g}^\alpha$ the related contravariant vectors.

The strain interpolation components $B_{\alpha\alpha s}$ are defined from the virtual strain components $\bar{\epsilon}_{\alpha\alpha}$:

$$\bar{\epsilon}_{\alpha\alpha} = \frac{\partial \boldsymbol{\eta}}{\partial \xi_\alpha} \cdot \mathbf{g}_\alpha = \frac{\partial N^k}{\partial \xi_\alpha} (\mathbf{g}_\alpha)_m \eta_s = B_{\alpha\alpha s} \eta_s \qquad 3.24$$

where k = integer part of $(s+2)/3$ and $m = s - 3(k-1)$. To define the interpolation of the angle ${}^P\gamma(\boldsymbol{\eta})$, a first order development is required. The rotation between warp and weft yarns corresponding to the virtual displacement η is:

$$\gamma(\boldsymbol{\eta}) = \arccos\left(\frac{\mathbf{g}_{1\eta} \cdot \mathbf{g}_{2\eta}}{\|\mathbf{g}_{1\eta}\| \|\mathbf{g}_{2\eta}\|} \right) - \arccos\left(\frac{\mathbf{g}_1 \cdot \mathbf{g}_2}{\|\mathbf{g}_1\| \|\mathbf{g}_2\|} \right) \qquad 3.25$$

where $\mathbf{g}_\alpha = \partial \mathbf{x}/\partial \xi_\alpha$ and $\mathbf{g}_{\alpha\eta} = [\partial (\mathbf{x} + \boldsymbol{\eta})]/\partial \xi_\alpha$ are the material covariant vectors respectively in the current and virtual configuration.

Denoting $\theta = (\mathbf{g}_1, \mathbf{g}_2)$ and $\theta_\eta = (\mathbf{g}_{1\eta}, \mathbf{g}_{2\eta}) = \theta + \gamma(\boldsymbol{\eta})$ and neglecting the second order terms, equation (3.12) can be approximated by:

$$\gamma(\boldsymbol{\eta}) = \frac{\partial \boldsymbol{\eta}}{\partial \xi_1} \cdot \left[\cotg \theta \frac{\mathbf{g}_1}{\|\mathbf{g}_1\|^2} - \frac{\mathbf{g}_2}{\sin \theta \|\mathbf{g}_1\| \|\mathbf{g}_2\|} \right]$$

$$+ \frac{\partial \boldsymbol{\eta}}{\partial \xi_2} \cdot \left[\cotg \theta \frac{\mathbf{g}_2}{\|\mathbf{g}_2\|^2} - \frac{\mathbf{g}_1}{\sin \theta \|\mathbf{g}_1\| \|\mathbf{g}_2\|} \right] \qquad 3.26$$

This gives the shear strain interpolation:

$$\gamma(\boldsymbol{\eta}) = B_{\gamma s} \eta_s \qquad 3.27$$

With:

$$B_{\gamma s} = \frac{\partial N^k}{\partial \xi_1} \left[\cotg \theta \frac{(g_1)_m}{\|\mathbf{g}_1\|^2} - \frac{(g_2)_m}{\sin \theta \|\mathbf{g}_1\| \|\mathbf{g}_2\|} \right]$$

$$+ \frac{\partial N^k}{\partial \xi_2} \left[\cotg \theta \frac{(g_2)_m}{\|\mathbf{g}_2\|^2} - \frac{(g_1)_m}{\sin \theta \|\mathbf{g}_1\| \|\mathbf{g}_2\|} \right] \qquad 3.28$$

From the strain interpolation coefficients $B_{\alpha\alpha s}$ and $B_{\gamma s}$ obtained in (3.11) and (3.15), the virtual elementary interior load work is related to virtual displacements:

$$W^e_{int}(\mathbf{\eta}) = \eta_s \left(\sum_{p=1}^{ncell^e} \|\mathbf{g}_1\|^{-2} \, {}^pB_{11s} \, {}^pT^{11} + \|\mathbf{g}_2\|^{-2} \, {}^pB_{22s} \, {}^pT^{22} + {}^pC^p B_{\gamma s} \right)$$

$$= \eta_s (\mathbf{F}^e_{int})_s \qquad 3.29$$

As shown in Boisse (1994), computation of the nodal interior loads does not require the summation on all the woven cells of the element. Accounting for the bilinear interpolation, the summation can be done on only four cross-overs, the positions of which depend on the number of warp and weft yarns. If n_c and n_t are the number of yarns in warp and weft directions, the position of the four cross-overs are:

$$\lambda_1 = (-1)^\gamma \left((n_c^2 - 1)(3n_c^2)^{-1} \right)^{1/2} \quad \lambda_2 = (-1)^\delta \left((n_t^2 - 1)(3n_t^2)^{-1} \right)^{1/2} \qquad 3.30$$

and δ are equal to 1 or 2. Consequently, the nodal interior load components can be calculated by:

$$(\mathbf{F}^e_{int})_s = \sum_{\gamma=1}^{2} \sum_{\delta=1}^{2} \frac{n_c n_t}{4} \begin{pmatrix} B_{11s}(\lambda_1, \lambda_2) L_1 T^{11}(\lambda_1, \lambda_2) \|\mathbf{g}_1(\lambda_1, \lambda_2)\|^{-2} \\ + B_{22s}(\lambda_1, \lambda_2) L_2 T^{22}(\lambda_1, \lambda_2) \|\mathbf{g}_2(\lambda_1.\lambda_2)\|^{-2} \\ + B_{\gamma s}(\lambda_1, \lambda_2) C(\lambda_1, \lambda_2) \end{pmatrix} \qquad 3.31$$

This expression is explicit, there is no matrix multiplication, no computation of terms equal to zero (corresponding for instance to non-existent stiffness in fabrics). Consequently, the numerical efficiency of the element is good.

3.5.4 Hemispherical forming of an unbalanced fabric

The hemispherical forming of a 2×2 nylon twill is analysed in this section. This type of fabric is used in the automotive industry (Dumont, 2003). It exhibits a very unbalanced tensile behaviour in warp and weft directions (Fig. 3.18). The shear behaviour of this fabric has been analysed experimentally using the picture frame test. Its rigidity has been investigated using two straight segments, whose slopes are $k_1 = 0.03$ mmN/rd and $k_2 = 0.095$ mmN/rd with a critical shear angle $\gamma_c = 0.5$rd. Tests of hemispheric sheet-forming have been carried out by F. Dumont in the S3MFM composites laboratory of the University of Nottingham (Daniel et al., 2003). The forming process was simulated using two approaches, tension only, and tension plus shear. The results of these two simulations as well as the experimental final shape are shown in Fig. 3.19. The experimental deformed shape is very different in warp and weft directions. The warp direction (vertical in Fig. 3.19 and corresponding to the most rigid yarns), shows significant sliding of fabric in the matrix. In the weft direction, there is no visible

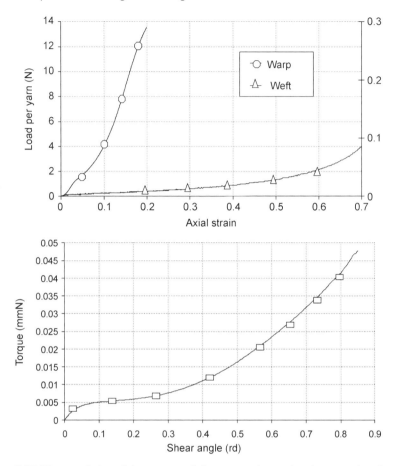

3.18 Characteristics of the woven reinforcement, in tension (warp and weft direction), and in shear.

sliding. The yarns are strongly stretched. At the summit of the hemisphere, an initially square quadrilateral becomes a rectangle with a side ratio equal to 1.8 (Fig. 3.19a). Both simulations give a value of this ratio close to the experimental value (Figs 3.19b and 3.19c). The deformation of the hemispherical part is computed well by both approaches.

The asymmetry of the deformed shape in warp and weft directions is obtained by both approaches. The main difference concerns the wrinkles. In the tension plus shear approach, the shear strain energy leads to the appearance of wrinkles in the plane region of the preform, whereas for tension only, there are no wrinkles. The shapes are in good agreement with those of the experimental preform (Figs 3.19a and 3.19c). There are no wrinkles in the hemispherical zone and the two approaches give close results in this region. But there are regions where the shear angle is higher than the shear-limit angle. When the shear

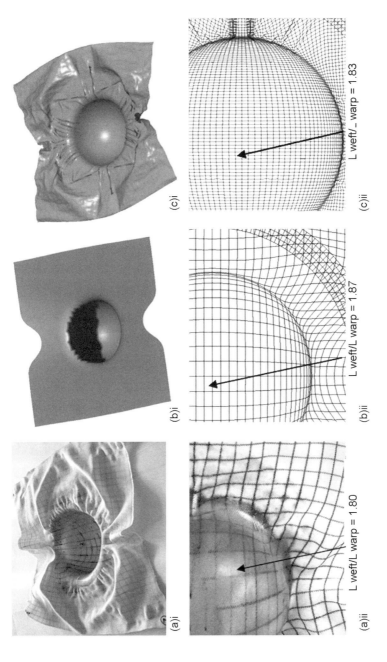

3.19 Deformed shape after the hemispheric forming in the experimental case (a), simulation in tension only (b) and in tension plus shear (c).

energy is taken into account, the minimisation of the total deformation energy leads to an out-of-plane solution, i.e. the wrinkles (Fig. 3.19c). The contribution of shear behaviour is mainly in describing the state after the appearance of wrinkles. At this stage, the relative rotations between warp and weft yarns are reduced. For instance, the maximum shear angle is 38° in Fig. 3.19c as opposed to 50° in Fig. 3.19b. This value (38° in the case of tension plus shear analysis) is in good agreement with angles measured experimentally (37°). It should be pointed out that the mesh used in Fig. 3.19c (tension plus shear) is finer than that used in Fig. 3.19b (tension only). These meshes can be used because their thickness does not change the solution significantly. The mesh size can be much finer in the case of tension plus shear because the wrinkles that appear in this case need much smaller elements in order to describe the wrinkles correctly, without using any perturbation methods or initial imperfections.

3.6 Multi-ply forming and re-consolidation simulations

There are several composite forming processes in which the transverse behaviour, i.e. the strain and stress across the thickness, is a major issue. That is the case for Continuous Fibres and ThermoPlastic (CFRTP) matrix forming processes (Maison *et al.*, 1998). After heating at a temperature higher than the melting point (Fig. 3.20a), forming is carried out using a punch and die process, normally using a rubber on the die (Fig. 3.20b). Re-consolidation is obtained by applying pressure on the punch (Fig. 3.20c). The objective of this last stage is to remove any residual pores at the interface of the plies, which is critical for products such as load-bearing aeronautical parts.

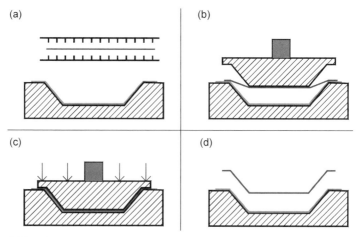

3.20 Different stages of forming. (a) Heating of the CFRTP. (b) Forming with punch and die. (c) Reconsolidation phase. (d) Final part.

Finite element analysis of composite forming   71

3.6.1 Simulation of CFRTP forming processes

One of the main forming modes is the relative sliding between the plies. In order to permit the necessary sliding, each ply is modelled as a set of shell elements. An example is shown in Fig. 3.21 for a Z reinforcement made of ten plies that slide during bending of the initially flat plate. The relative sliding between the plies (Fig. 3.22) agrees with experimental results (Cheruet *et al.*, 2002).

An important quality issue for the final part is the absence of pores in the thickness of the composite. Any pores could be the source of a fracture during the service life of the composite and must be avoided, particularly for aeronautical applications. This is the reason behind the re-consolidation stage of the forming process (Fig. 3.20). It has been shown that many gaps appear in the material after the heating stage (Fig. 3.23) and these are removed by the pressure applied during the re-consolidation phase (Fig. 3.24). This stage is essential to

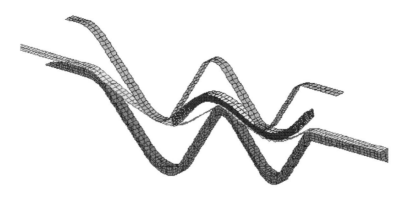

3.21 Forming of a Z reinforcement. Forming stage (Cheruet *et al.*, 2002).

3.22 Interply sliding during the forming stage (Cheruet *et al.*, 2002).

72   Composites forming technologies

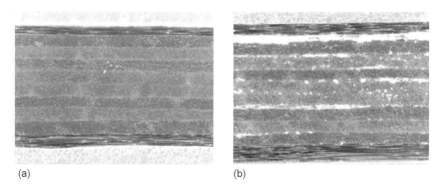

3.23 Material before and after the heating stage (Cheruet *et al.*, 2002).

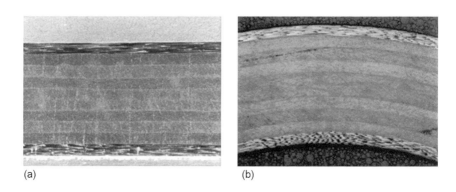

3.24 Material state after the re-consolidation stage (Cheruet *et al.*, 2002).

the final quality of the part. It can be analysed by an FE simulation to determine if the stress through the thickness is sufficient to remove any pores.

3.6.2   Shell element with pinching

Re-consolidation has been studied by Lee *et al.* (1987) and some models for local consolidation have been proposed. These studies have shown that re-consolidation depends on the stress state in the laminate, and mainly on the normal stress in the re-consolidation stage. This stress component is not present in classical shell theory. Some finite elements with stress/strain through the thickness have been proposed (Simo *et al.*, 1990; Butcher *et al.*, 1994; Bletzinger *et al.*, 2000). A shell element is used where a degree of freedom through the thickness strain is introduced (Coquery, 1999; Cheruet *et al.*, 2002). The thickness stress is not equal to zero but is related to the thickness variation β by the constitutive law. Assuming small rotations between two computation steps, the displacement expression is (Fig. 3.25):

Finite element analysis of composite forming

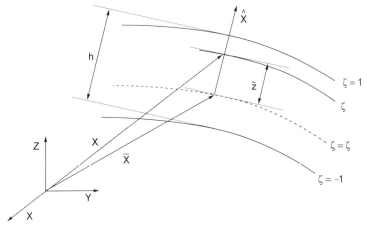

3.25 Kinematics of the shell with pinching.

$$u = \bar{u} - \bar{z}^0 \hat{X}^t \times \theta + z\beta\hat{X}^t \qquad 3.32$$

If $\beta = 0$ then equation (3.32) leads to classical shell kinematics without pinching. β is a additional pinching degree of freedom. The strain tensor $\epsilon(\mathbf{u}) = \frac{1}{2}[\nabla(\mathbf{u}) + \nabla^T(\mathbf{u})]$ can be derived from to the displacement (3.9).

In an orthogonal frame $(\hat{\mathbf{e}}_1, \hat{\mathbf{e}}_2, \hat{\mathbf{e}}_3 = \hat{\mathbf{X}})$ the membrane bending strain components are:

$$\begin{Bmatrix} \epsilon_{11} \\ \epsilon_{22} \\ 2\epsilon_{12} \end{Bmatrix} = \begin{Bmatrix} u^m_{x,1} \\ u^m_{y,2} \\ u^m_{y,1} + u^m_{x,2} \end{Bmatrix} + z \begin{Bmatrix} \theta_{y,1} \\ -\theta_{x,2} \\ -\theta_{x,2} + \theta_{y,1} \end{Bmatrix} \qquad 3.33$$

The transverse shears are:

$$\begin{Bmatrix} 2\epsilon_{13} \\ 2\epsilon_{23} \end{Bmatrix} = \begin{Bmatrix} u^m_{z,1} + \theta_y \\ u^m_{z,2} - \theta_x \end{Bmatrix} + z \begin{Bmatrix} \beta_{,1} \\ \beta_{,2} \end{Bmatrix} \qquad 3.34$$

and the strain through the thickness (pinching) is:

$$\{\epsilon_{33}\} = \beta \qquad 3.35$$

The transverse shear strains are modified by pinching. In contrast to the classical shell, the normal stress through the thickness is not zero. It is deduced from $\epsilon_{33} = \beta$ using the compaction behavior law. It has been shown that this element exhibits a 'pinching' locking. To avoid this locking, it is necessary to modify the constitutive relation in order to remove the coupling between pinching and bending (Coquery, 1999; Cheruet et al., 2002). The example presented in Fig. 3.26 (cantilever plate) shows the pinching locking obtained with a complete behaviour law and the accurate result if pinching and bending are not coupled. Analysis of the reason for this locking can be found in Soulat et al. (2006).

74   Composites forming technologies

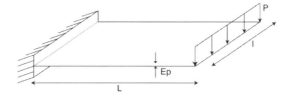

Thickness (m)	Transverse displacement (m)		
	Traditional shell element	Shell with pinching element, including bending uncoupling	Shell with pinching element with 3D behaviour
0.01	-0.044	-0.044	-0.038
0.001	-0.044	-0.044	-0.037
0.0001	-0.044	-0.044	-0.031

3.26 Effectiveness of uncoupling bending-pinching.

3.6.3 Simulation of the forming and re-consolidation stage of a Z profile

Using the shell element described above, the stress through the thickness is computed for an initially flat ply which is formed into a Z shape (Fig. 3.21) and then compacted.

During and at the end of the forming stage (Fig. 3.27), the stress through the thickness is equal to zero in most areas except in the radius of the tools. That confirms micrographic observations which have shown that consolidation only occurs near the radius during the forming stage (Cheruet *et al.*, 2002).

After the compaction phase, the entire part is in compression (Fig. 3.28). The value of the stress through the thickness is related to the angle of the curved part of the Z profile. In order to obtain an accurate value of the thickness stress, it is important to used an efficient non-linear compaction law (Gutowski, 1985; Baoxing *et al.*, 1999).

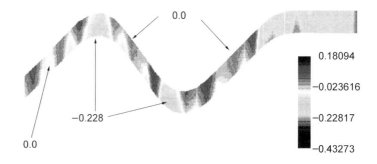

3.27 Stress component through the thickness at the end of forming.

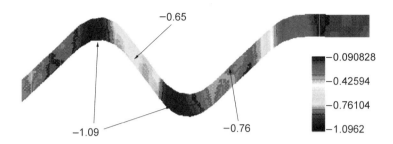

3.28 Stress component through the thickness after the re-consolidation stage.

3.7 Conclusions

Finite element analysis is probably the approach that will be used to simulate composite forming processes in the future. These simulations are especially important for composite materials because not only do they help determine the feasibility of the forming process itself, but they also give fibre directions and densities, which are essential for further analysis of the composite part in service. The deformation modes of the composite during forming are related mainly to the internal fibrous structure of the reinforcement, consequently the forming modes are specific to these materials. Fibre extensions are usually small, but large in-plane shear can occur.

At present, two approaches are used. The continuous approach involves defining an equivalent continuous mechanical behaviour for the fibrous reinforcement at the macroscopic level, and the discrete or mesoscopic approach models the components of the reinforcement at the mesoscopic level. While most industrial analysis currently uses the continuous approach, the mesoscopic approach will become more common as analysts take advantage of computing improvements.

One point must be underlined at the end of this chapter. It concerns the necessity for accurate tests (experimental or virtual) to determine the mechanical properties of the materials, including friction. The FE method is a mechanical approach and understanding the behaviour of the material during forming is essential for the computations. Tests for composite reinforcements are not as well established as those for metallic materials, for instance a cooperative benchmark performed by several labs on in-plane shear properties has shown a large variation in results for a glass fibre fabric used in automotive applications. This area must be improved if FE analysis for composite forming is to be accurate.

3.8 References

ABAQUS FEA software, www.abaqus.com
Argyris J (1960), *Energy Theorems and Structural Analysis*, Butterworths, London.

Baoxing C, Chou T W (1999), Compaction of woven-fabric preforms in liquid composite molding processes: single layer deformation, *Composite Science and Technology*, 59, 1519–1526.

Ben Boubaker B, Haussy B, Ganghoffer J F (2002), 'Discrete models of woven structures, draping and stability analysis', *CRAS Paris, Série Mécanique*, 330, 871–877.

Ben Boubaker B, Haussy B, Ganghoffer J F (2005), 'Discrete models of fabrics accounting for yarn interactions', *European Journal of Computational Mechanics*, 14 (6–7), 653–676.

Bletzinger K U, Bischoff M, Ramm E (2000), 'A unified approach for shear-locking free triangular shell finite elements', *Computer and Structures*, 75, 321–334.

Boisse P, (1994), *Modèles mécaniques et numériques pour l'analyse non-linéaire des structures minces*, Thesis for PhD direction enabling, University of Besançon, France.

Boisse P, Gasser A, Hivet G (2001), 'Analyses of fabric tensile behaviour: determination of the biaxial tension-strain surfaces and their use in forming simulations', *Composites A*, 32, 1395–1414.

Boisse P, Zouari B, Gasser A (2005a), 'A mesoscopic approach for the simulation of woven fibre composite forming', *Composites Science and Technology*, 65, 429–436.

Boisse P, Gasser A, Hagège B, Billoet J L (2005b), 'Analysis of the mechanical behaviour of woven fibrous material using virtual tests at the unit cell level', *Int. Journal of Material Science*, 40, 5955–5962.

Borouchaki H, Cherouat A (2003), 'Geometrical draping of composites', *Comptes Rendus de l'Académie des Sciences*, Paris, Série II B, 331, 437–442.

Bréard J, Saouab A (2005), 'Numerical simulation of liquid composite molding processes', *European Journal of Computational Mechanics*, 14 (6–7), 841–865.

Buet-Gautier K, Boisse P (2001), 'Experimental analysis and modeling of biaxial mechanical behavior of woven composite reinforcements', *Experimental Mechanics*, 41 (3), 260–269.

Butcher N, Ramm E, Roehl D (1994), 'Three dimensional extension of non-linear shell formulation based on the enhanced assumed strain concept'. *Int. J. for Num. In Engng.*, 37, 2551–2568.

Cao J, Cheng H S, Yu T X, Zhu B, Tao X M, Lomov S V, Stoilova Tz, Verpoest I, Boisse P, Launay J, Hivet G, Liu L, Chen J, De Graaf E F, Akkerman R (2004), 'A cooperative benchmark effort on testing of woven composites', *Proceedings of the 7th Int. ESAFORM Conference on Material Forming*, Trondheim (Norway), 305–308.

Cherouat A, Billoët J L (2001), 'Mechanical and numerical modelling of composite manufacturing processes deep-drawing and laying-up of thin pre-impregnated woven fabrics', *J. Mat. Proc. Technology*, 118, 460–471.

Cheruet A, Soulat D, Boisse P, Soccard E, Maison-Le Poec S (2002), 'Analysis of the interply porosities in thermoplastic composites forming processes', *International Journal of Forming Processes*, 5, (2–4), 247–258.

Chinesta F, Cueto E, Ryckelynck D, Ammar A (2005), 'Numerical simulation of liquid composite molding processes', *European Journal of Computational Mechanics*, 14, (6–7), 903–923.

Comas-Cardona S, Groenenboom P H L, Binétruy C, Krawczak P (2005), 'Simulation of liquid composite molding processes using a generic mixed FE-SPH method', *European Journal of Computational Mechanics*, 14 (6–7), 867–883.

Coquery M H (1999), *Modélisation d'un joint de culasse multifeuille*, PhD Thesis, ENSAM Paris.

Criesfield M A (1991), *Non linear finite element analysis of solids and structures, II: Advanced topics*, John Wiley & Sons.

Dafalias Y F (1983), 'Corotational rates for kinematic hardening at large plastic deformations', *Trans. of the ASME, J. of Ap. Mech.*, 50, 561–565.

Daniel J L, Soulat D, Dumont, F, Zouari B, Boisse P, Long A C (2003), 'Forming of a very unbalanced fabric. Experiment and simulation', *International Journal of Forming Processes*, 6 (3–4), 465–480.

Darcy H (1856), 'Les Fontaines Publiques de La Ville De Dijon: Distribution d'Eau et Filtrage Des Eaux', *Appendice – Note D*, Victor Dalmont, Paris.

De Luca P, Pickett A K (1998), 'Numerical and experimental investigation of some press forming parameters of two fibre reinforced thermoplastics: APC2-AS4 and PEI-CETEX', *Composites Part A*, 29, 101–110.

Dienes J K (1979), 'On the analysis of rotation and stress rate in deforming bodies', *Acta Mechanica*, 32, 217–232.

Duhovic M, Bhattacharyya D (2006), 'Simulating the deformation mechanisms of knitted fabric composites', *Composites A*, 37 (11) 1897–1915.

Dumont F (2003), *Contribution à l'expérimentation et à la modélisation du comportement mécanique de renforts de composites tissés*, PhD Thesis, Université Paris 6.

Dumont P, Orgéas L, Corre S L, Favier D (2003), 'Anisotropic viscous behavior of sheet molding compounds (SMC) during compression molding', *Int . J. Plasticity*, 19, 625–646.

Durville D (2002), 'Modélisation par éléments finis des propriétés mécaniques de structures textiles: de la fibre au tissu', *European Journal of Computational Mechanics*, 11 (2–4), 463–477.

ESI-QuickFORM, PAM-FORM, PAM-STAMP, *www.esi-groupe.com*

FiberSIM, www.vistagy.com

Fournier R, Coupez T, Vincent M (2005), 'Numerical determination of the permeability of fibre reinforcement for the RTM process', *European Journal of Computational Mechanics*, 14 (6–7), 803–818.

Gasser A, Boisse P, Hanklar S (2000), 'Analysis of the mechanical behaviour of dry fabric reinforcements. 3D simulations versus biaxial tests'. *Computational Material Science*, 17, 7–20.

Gilormini P, Roudier P, Rougee P (1993), 'Cumulated tensorial deformation measures', *Comptes-rendus de l'Académie des Sciences de Paris* II, 316, 1499–1504.

Gingold R A, Monaghan J J (1977), 'Smoothed particle hydrodynamics: Theory and application to non-spherical stars', *Mon. Not. R. Astr. Soc.*, 181, 375–389.

Gutowski T G (1985), 'A resin flow/fibre deformation model for composites', *SAMPE Quart*, 16 (4), 58–64.

Hagège B (2004), *Simulation du comportement mécanique des milieux fibreux en grandes transformations: application aux renforts tricotés*, PhD Thesis, ENSAM Paris, France.

Hagège B, Boisse P and Billoët J-L (2005), 'Finite element analyses of knitted composite reinforcement at large strain', *European Journal of Computational Mechanics*, 14 (6–7), 767–776.

Halquist J O, Goudreau G L, Benson D J (1985), 'Sliding interfaces with contact impact in large scale Lagrangian computations', *Comp. Meth. Appl. Engng.*, 51, 107-137.

Hsiao S W, Kikuchi N (1999), 'Numerical analysis and optimal design of composite thermoforming process', *Comp. Meth. Appl. Mech. Engrg.*, 177, 1–34.

Kawabata S, Niwa M, Kawai H (1973), 'The finite deformation theory of plain weave fabrics part I: The biaxial deformation theory', *J. Textile Inst.*, 64 (1), 21–46.

Laine B, Hivet G, Boisse P, Boust F, Lomov S (2005), 'Permeability of the woven fabrics: A parametric study', *Proceedings of the 8th Int.Conf. ESAFORM on Material Forming*, Cluj-Napoca (Roumanie), 2, 995–998.

Lee W, Springer G (1987), 'A model of the manufacturing process of thermoplastic matrix composites', *Journal of Composite Materials*, 21, 1017–1055.

Lomov S V, Stoilova T, Verpoest I (2004), 'Shear of woven fabrics: theoretical model, numerical experiments and full strain measurements', *Proceedings of the Int. Conf ESAFORM 7*, Trondheim, 345–348.

Lomov S V, Bernal E, Ivanov D S, Kondratiev S V, Verpoest I (2005), 'Homogenisation of a sheared unit cell of textile composites: FEA and approximate inclusion model', *European Journal of Computational Mechanics*, 14 (6–7), 709–729.

Long A C, Rudd C D (1994), 'A simulation of reinforcement deformation during the production of preform for liquid moulding processes', *I. Mech. E. J. Eng. Manuf.*, 208, 269–278.

Long A C, Souter B J, Robitaille F (2001), 'Mechanical modelling of in-plane shear and draping for woven and non-crimp reinforcements', *J. of Thermoplastic Composite Materials*, 14, 316–326.

Maison S, Thibout C, Garrigues C, Garcin J L, Payen H, Sibois H (1998), 'Technical developments in thermoplastic composites fuselages', *SAMPLE Journal*, 34 (5), 33–39.

Mark C, Taylor H M (1956), 'The fitting of woven cloth to surfaces', *Journal of Textile Institute*, 47, 477–488.

McBride T M, Chen J (1997), 'Unit-cell geometry in plain-weave fabrics during shear deformations', *Composites Science and Technology*, 57, 3, 345–351.

McGuinness G B, Bradaigh C M O (1997), 'Development of rheological models for forming flows and picture-frame shear testing of fabric reinforced thermoplastic sheets', *Journal of Non-Newtonian Fluid Mechanics*, 73, 1–2, 1–28.

McGuinness G B, Bradaigh C M O (1998), 'Characterisation of thermoplastic composite melts in rhombus-shear: the picture-frame experiment', *Composites Part A*, 29, 1–2, 115–132.

Ó Brádaigh C M, McGuinness G B, Pipes R B (1993), 'Numerical analysis of stresses and deformations in composite materials sheet forming: Central indentation of a circular sheet', *Composites Manufacturing*, 4, (2), 67–83.

Peng X, Cao J (2002), 'A dual homogenization and finite element approach for material characterization of textile composites', *Composites B*, 33, 45–56.

Pickett A K (2002), 'Review of finite element methods applied to manufacturing and failure prediction in composite structures', *Applied Composite Material*, 9, 43–58.

Pickett A K, Creech G, de Luca P (2005), 'Simplified and advanced simulation methods for prediction of fabric draping', *European Journal of Computational Mechanics*, 14 (6–7), 677–691.

Pickett T (2005), 'Modelling drape and impact of textile composites: meso and macro approaches', *Conference Advances in Multi-Scale Modelling of Composite Material Systems and Components*, Monterey, USA, 65–66.

Prodromou A G, Chen J (1997), 'On the relationship between shear angle and wrinkling of textile composite preforms', *Composites Part A*, 28, 491–503.

Ramgulam R, Potluri P (2005), 'Mechanics of woven fabrics using cruciform elements', *European Journal of Computational Mechanics*, 14 (6–7), 653–676.

Raffel M, Willert C, Kompenhaus J (1998), *Particle Image Velocimetry. A practical guide*, Experimental Fluid Mechanics, Springer Berlin.

Sharma S B, Sutcliffe M P F (2003), 'A simplified finite element approach to draping of woven fabric', *Proceedings of the 6th Int.Conf. ESAFORM on Material Forming*, Salerno, Italy, 887–890.

Simo J C, Fox D D, Rifai M S (1990), 'On a stress resultant geometrically exact shell model. Part 4: Variable thickness shells with trough-the-thickness stretching', *Comp. Meth. in App. Mech and Engng*, 79, 21–70.

Skordos A, Monroy Aceves C, Sutcliffe M (2005), 'Development of a simplified finite element model for draping and wrinkling of woven material', *Proceedings of the 8th Int.Conf. ESAFORM on Material Forming*, Cluj-Napoca (Romania).

Soulat D, Cheruet A, Youssef M, Boisse P (2006), 'Simulation of continuous fibre reinforced thermoplastic forming using a shell finite element with transverse stress', *Computers and Structures*, 84, 13–14, 888–903.

Sukumar N, Moran B, Belytschko T (1998), 'The natural element method in solid mechanics', *International Journal for Numerical Methods in Engineering*, 43 (5), 839–887.

Tollenaere H, Caillerie D (1998), 'Continuous modelling of lattice structures by homogenization', *Advances in Engineering Software*, 29 (7), 600–705.

Trochu F, Gauvin R, Gao D-M (1993), 'Numerical analysis of the resin transfer molding process by the hinite element method', *Advances in Polymer Technology*, 12 (4), 329–342.

Vacher P, Dumoulin S, Arrieux R (1999), 'Determination of the forming limit diagram from local measurement using digital image analysis', *International Journal of Forming Processes*, 2–4, 395–408.

Van Der Ween F (1991), 'Algorithms for draping fabrics on doubly curved surfaces', *International Journal of Numerical Method in Engineering*, 31, 1414–1426.

Wagoner R, Chenot J L (1996), *Fundamentals of metal forming analysis*, Wiley.

Xue P, Peng X, Cao J (2003), 'A non-orthogonal constitutive model for characterizing woven composites', *Composites Part A*, 34, 183–193.

Yu W R, Pourboghrat F, Chung K, Zamploni M, Kang T J (2002), 'Non-orthogonal constitutive equation for woven fabric reinforced thermoplastic composites', *Composites Part A*, 33, 1095–1105.

Yu X, Ye L, Mai Y-W (2004), 'Finite element spurious wrinkles on the thermoforming simulation of woven fabric reinforced composites', *Proceedings of the Int. Conf ESAFORM 7*, Trondheim, 325–328.

Zouari B, Daniel J L, Boisse P (2006), 'A woven reinforcement forming simulation method influence of the shear stiffness', *Computers and Structures*, 84, 5–6, 351–363.

4
Virtual testing for material formability

S V LOMOV, Katholieke Universiteit Leuven, Belgium

4.1 Introduction

Deformability of textile preforms plays a key role in the quality of a composite part formed into a 3D shape and processed using liquid moulding techniques such as resin transfer moulding (RTM). Ill-chosen placement of the preform, disregarding its behaviour under heavy strain in complex deformation may result in the preform wrinkling or even damage, deteriorating the performance of the composite part. This explains the importance of predictive modelling of deformability of textile preforms. The deformation modes of primary importance are in-plane deformation (tension and shear) and compression of the preform. Deformability of woven preforms in these modes is the subject of the present chapter. Out-of-plane bending may also be considered, as it can affect internal geometry of the preform, especially for small bending radii; a model of the woven fabric bending can be found in refs 1 and 2. Naturally the deformability of woven fabrics is also important for apparel textiles, and has attracted attention of textile materials researchers. Works of Kawabata, Niva and Kawai,[3–5] de Jong and Postle,[6] Hearle and Shanahan[7] have established an approach to mathematical modelling of deformation of woven fabrics, which can be summarised by three principles.

First, the model uses deformations, rather than loads, as input for in-plane deformation (for compression model the applied pressure is the input). An overall deformation pattern is imposed over the woven fabric repeat (unit cell) to change the spacing of the yarns in tension and the angle between them in shear. As formulated by Komori and Ito,[8] the unit cell is subject to transformation of coordinates defined by the given deformation. The contacts between yarns stay unchanged in tension, and experience rotation (not sliding) of the contacting yarns in shear.

Second, the principle of minimum energy is applied to compute the internal geometry of the deformed fabric. With the spacing and orientation of the warp and weft given, the yarn paths are defined using one of available geometrical models (Peirce's, elastica, splines, etc.), with crimp heights and dimensions of the cross-sections of the yarns (which can change under transversal force caused by tension) as parameters. These parameters are calculated via the principle of minimum total energy, associated with the yarn tension, bending and compression. Tension of the yarn is computed using the experimental tension diagram. Elongation of the yarn is estimated by the difference between yarn length in the repeat before and after the deformation. Experimental bending and compression diagrams are used to compute bending energy and resistance to compression. All these experimental diagrams are non-linear. Application of the principle of minimum energy to fibrous assemblies must be considered as heuristic, as it can be applied rigorously to conservative systems only. This means that frictional effects are not taken into account in the solution of the minimisation problem. The internal friction between fibres in the yarns enters the calculations via non-linear bending and compression diagrams. The interyarn friction is absent in tension, and manifests itself in the rotation of the contacting yarns in shear.

Third, the applied loads are computed via the balance between, on one hand, the mechanical work done by the loads on deformations of the unit cell, and, on the other hand, the sum of the change of the total energy of the deformed yarns and the work of friction (if any).

In the papers cited above,[3–7] and some others,[9–14] the approach has been successfully applied to plain woven fabrics. It has also been reported to work for twills and satins.[15] We will use this scheme for all three types of deformation under consideration: compression, bi-axial tension and shear.

Returning to deformability of woven reinforcements for composites, one finds quite a number of publications on modelling. The model complexities range from simple empirical models to elaborate finite element descriptions. Certain important points have been investigated, which were not covered by earlier 'apparel-oriented' models.

In studies of compression the attention was given to compressibility of the reinforcement at high loads, which are characteristic of composite processing. The compression curve is broken into three regions (low, medium and high loads), each dominated by different phenomena.[16–19] The nesting of layers of the reinforcement is taken into account.[17,20,21] Models of shear of woven reinforcements[22–29] have to consider very high shear angles (up to 60–70°) occurring in forming of complex 3D parts. This is dealt with by an introduction of models of lateral compression of the yarns, which come into contact when the shear angle reaches and exceeds the locking angle of the fabric 'trellis'. The simple, but not true-to-life concept of preserving the volume of the unit cell to calculate the change of the thickness of fabric in shear, has been advanced to more correct considerations of yarn compression.

Recently, finite element descriptions of deformability of textiles have been introduced for bi-axial tension,[30–37] shear[38,39] and compression[20] of woven reinforcements (see Chapter 3). Based on the increasing power of computers, these approaches aim to describe in detail the 3D behaviour of the fabric constituents including contact and friction and to obtain result fields at a local level.

Finite element modelling encounters two difficulties. First, the geometry of a textile unit cell is very complex, and creating a solid model manually is not an easy task. The solution is provided by the use of a textile geometry modeller as a preprocessor, capable of creating a finite element model automatically.[40–48] Second, the description of the material behaviour used in the finite element model must realistically represent the actual behaviour of the fibrous assemblies – yarns. Development of such a library of material models for textiles, available in finite element packages, presents a serious challenge to researchers.

Models of textile deformability developed so far have built a solid foundation for their generalisation, encapsulating the achievements of textile material science in a modelling software tool, allowing wide variability of textile structure and yarn parameters, instrumented with visualisation features and able to transfer the models of textile geometry into specialised micro-mechanical and flow modelling software as well as into general purpose finite element packages. Such a tool can be considered as a preprocessor for calculation of homogenised properties – permeability tensor and stiffness matrix – of deformed textile reinforcement. These properties, in their turn, are used as input to provide local parameters in modelling of Darcy flow through the deformed preform and structural finite element analysis of a 3D shaped composite part. This work is in progress in Composite Materials Group in the Department MTM, K.U. Leuven.[21,49–57] It has resulted in the development of textile modelling software *WiseTex* (http://www.mtm.kuleuven.be/Research/C2/poly/software.html). The chapter describes models of deformability of woven fabrics implemented in *WiseTex*. First versions of the models described here were developed in the early 1990s.[58–63]

4.2 Mechanical model of the internal geometry of the relaxed state of a woven fabric

The comprehensive description of the model of the relaxed state of a woven fabric can be found elsewhere.[49,51,53,64] Here we state its main components used in the simulation of the fabric deformation.

4.2.1 Weave pattern and elementary crimp intervals

A weave pattern (for one- and multilayered fabrics) is coded with matrix coding.[21,49,51,53] It allows separation of the crimped shape of the warp and weft yarns into elementary bent intervals (Fig. 4.1), representing sections of the yarn between interlacing sites. The shape of the yarn on an elementary interval is described using a parameterised function $z(x; h/p)$, where z and x are coordinates of the yarn middle line, h is the crimp height and p is the distance

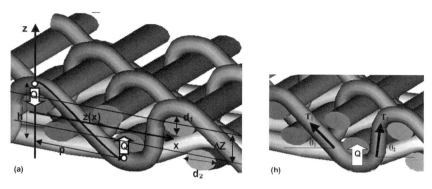

4.1 Model of internal geometry (a) and tension (b) of woven fabric.

between the interval ends (spacing of the yarns). The shape $z(x; h/p)$ for a given relative crimp height h/p is computed using the principle of minimum bending energy of the yarn on the interval and has the form:

$$z(x)h[1/2 - 3(x/p)^2 + 4(x/p)^3 + A(h/p)(x/p)^2((x/p) - 1)^2((x/p) - 1/2)] \quad 4.1$$

where the first term is a spline function, corresponding to the solution of the linearised minimum energy problem, and the second term represents a correction for a non-linear formulation. The function $A(h/p)$ is calculated from the solution of the minimum energy problem and is tabulated.

With this function known, the *characteristic function F* of the crimp interval is computed, representing the bending energy of the yarn:

$$w = \frac{1}{2} \int_0^p B(\kappa) \frac{(z'')^2}{(1 + (z')^2)^{5/2}} dx = \frac{B(\bar{\kappa})}{p} F(h/p) \quad 4.2$$

where $B(\kappa)$ is the (measured experimentally) bending rigidity of the yarn, which depends (non-lineary) on the local curvature $\kappa(x)$, or, after the integration, on an average curvature over the interval

$$\bar{\kappa} = \sqrt{\frac{1}{p} \int_0^p \frac{(z'')^2}{(1 + (z')^2)^{5/2}} dx} \quad 4.3$$

Function $F(h/p)$ is also tabulated. With the function F known, the transversal forces acting on the interval ends can be estimated as

$$Q = \frac{2w}{h} = \frac{2B(\bar{\kappa})p}{p^2} \frac{p}{h} F(h/p) \quad 4.4$$

4.2.2 Compression of the yarns in the relaxed fabric

Warp and weft yarns in the relaxed fabric are compressed by the transversal forces Q [4.4] according to experimental diagrams, measured on 'virgin' yarns

$$d_1 = d_{10}\eta_1(Q), \quad d_2 = d_{20}\eta_2(Q) \qquad 4.5$$

where subscript '0' refers to the uncompressed state of the yarn, d_1 and d_2 are dimensions of the yarn cross-section (Fig. 4.1). These dimensions and crimp heights of the yarns are interconnected:

$$h^{Wa} = \Delta Z + (d^{Wa} + d^{We}) - (h_1^{Wa} + h_2^{We})/2 \qquad 4.6$$

where superscripts refer to the warp and weft yarns, subscripts '1' and '2' refer to two weft yarns in different layers, ΔZ is the distance between fabric layers (Fig. 4.1).

With crimp heights of weft yarns given, equations [4.4–4.6] provide a closed system of non-linear equations for calculation of the transversal forces Q and yarn dimensions d_1 and d_2.

4.2.3 Minimum energy problem – calculation of the weft crimp heights

The weft crimp heights are found using the principle of minimum bending energy of the yarns inside the unit cell. It is written as

$$W_\Sigma = \sum_{i=1}^{N_{Wa}} \sum_{k=1}^{K_i^{Wa}} w_{ik}^{Wa} + \sum_{j=1}^{N_{We}} \sum_{k=1}^{K_j^{We}} w_{jk}^{We} \to \min \qquad 4.7$$

where subscripts i,j refer to different warp and weft yarns, k to the elementary crimp interval of the warp/weft yarn, and energies of the elementary intervals are calculated using (2). The minimum problem [4.7] is solved for the weft crimp heights, with all other parameters defined inside the minimisation algorithm obtained via solution of the system [4.4–4.6] for the given current crimp heights. It takes about 1 s on 1 GHz PC to compute parameters for a 3D fabric with 20 yarns in the repeat, and about 0.05 s – for a plain weave fabric.

4.3 Model of compression of woven fabric

4.3.1 Outline of the algorithm

When a fabric is compressed, the following changes in geometry take place:

- warp and weft yarns are compressed;
- the less crimped yarn system increases its crimp, and the more crimped system reduces in crimp.

These two processes are treated in the model separately. This follows from the assumption of an even distribution of the compressive force over warp/weft intersections, because this assumption implies that force per intersection, which compresses the yarn cross-sections and bends the yarns, is independent of any changes of warp and weft crimp or cross-section dimensions.

To compute compression of the yarns, the compression force per intersection is evaluated:

$$Q_c = F/(N_{Wa}N_{We}) \qquad 4.8$$

where F is the pressure force on fabric repeat. This value is added to all the Q_{ij} – transversal forces acting on the intersections and computed with [4.4], to evaluate the dimensions of the yarns with [4.5]. Hence, both the compression due to yarn bending and the compression due to external force are accounted for. The algorithm presented above is then applied to yield the compressed dimensions of the yarn cross-sections and the new values of the yarn crimp.

The change of crimp in compression (increasing for warp and decreasing for weft, or vice versa) leads to a decrease of the fabric thickness. Therefore the basic mechanical equation governing this process is

work of compressive force Q on change of thickness db
$\quad =$ change of bending energy of yarns dW $\qquad 4.9$

The compression of yarns has been accounted for before and the resulting changed cross-section dimensions are 'frozen', that is why the work of yarn compression does not enter the balance [4.9].

Changes of the fabric thickness db and bending energy of the yarns dW depend on the change of the *set* of weft crimp heights $\{dh_j^{We}\}$ and therefore [4.9] has a *set* of unknown variables. A reasonable assumption to cope with this difficulty is: 'The crimp changes in such a way as to provide the maximum possible change of thickness'. This means that if the function $b(\{h_j^{We}\})$ is considered (b being the fabric thickness), then changes of crimp will follow the direction of the maximum slope (in the opposite direction):

$$\{dh'_{ij}\} = -x \; \mathrm{grad} \; b(\{h_j^{We}\})$$

where x is computed to satisfy [4.9]; dashed values refer to changed crimp. Equation [4.9] is then written as follows:

$$F \cdot \lfloor b(\{h_j^{We}\}) - b(\{h_j'^{We}\}) \rfloor = \sum_{i,k} W^{Wa}(\{h_j'^{We}\}) + \sum_{j,k} W^{We}(\{h_j'^{We}\})$$

$$- \sum_{i,k} W^{Wa}(\{h_j^{We}\}) - \sum_{j,k} W^{We}(\{h_j^{We}\}) \qquad 4.10a$$

$$h_j'^{We} = h_j^{We} - x \cdot \mathrm{grad} \; b(\{h_j^{We}\}) \qquad 4.10b$$

to be solved numerically for x, which should also satisfy

$$0 \le h_j^{We} \le h_j^{We,max} \quad \text{and} \quad b(\{h_j^{We}\}) - b(\{h_j'^{We}\}) \ge 0 \qquad 4.11$$

4.3.2 Compression of 2D laminates

Woven preforms are usually compressed in a mould as a stack of fabric layers, which are not precisely positioned one against another, causing a geometric and mechanical phenomenon of *nesting*. Nesting plays an important role in determining the permeability of the laminate and the mechanical properties of the composite. Nesting causes a statistical distribution of the laminate properties, both at different positions within a composite sample and between different samples in a set of otherwise identical parts.

To calculate compression of a laminate, consider first compression of one layer of the fabric. After the calculation described above, the dimensions and placement of the yarns inside the unit cell of compressed fabric are known. The placement hence defines the *surface profile functions* of the face and back surface of the laminate (Fig. 4.2a):

$$h_f(x,y) = \frac{Z}{2} - z_f(x,y)$$
$$h_b(x,y) = z_b(x,y) + \frac{Z}{2}$$

4.12

where x, y, z are Cartesian coordinates, with the centre of the coordinate system in the centre of the unit cell, Z is the fabric thickness, z_f and z_b are coordinates of the face and back surface of the fabric. Equation [4.12] applies if there is a point (x, y, z) inside the yarns or fibrous plies for the given (x, y). If no such point exists, then

$$h_f(x,y) = h_b(x,y) = Z$$

With the surface profile functions defined, it is easy to calculate the nesting of the layers. Consider two identical layers of the laminate, with one layer shifted relative to another by dx and dy in x and y directions (Fig. 4.2b). To define the nested position, we must calculate the distance h^* between centre planes of the layers, when the yarns in the layers are just touching one another and there is no inter-penetration of the yarns. When nesting is zero, $h^* = Z$. Consider a certain distance between centre planes h. The distance (depending on the (x, y) position) between the back surface of the upper layer and the face surface of the lower layer is

$$\delta(x, y; h) = h - Z + h_f(x, y) + h_b(x - dx, y - dy)$$

Defining

$$\delta^*(h) = \min_{x,y} \delta(x, y; h)$$

we can compute the nesting distance h^* as a solution of the equation

$$\delta^*(h) = 0$$

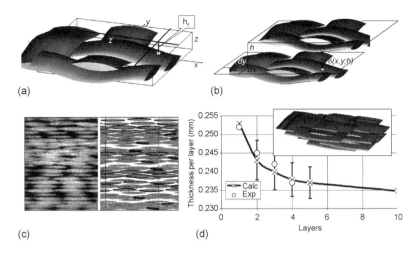

4.2 Nesting model: (a) Surface depth function h_f; (b) Shifted layers; (c) Captured with X-ray microCT and computed cross-section of 25 ply laminate; (d) thickness of laminate per layer vs number of layers.

If the surface depth functions are defined on a finite mesh, these calculations are easily performed if shifts dx, dy are done in integer mesh units. In this case the calculations involve $O(Nx * Ny)$ comparisons, Nx and Ny being the mesh size in x and y directions.

For a laminate of L layers, the set of shifts dx_l, dy_l, $l = 1 \ldots L$ is defined, and the algorithm is applied to one layer after another. When the nested positions of layers are defined, the descriptions of the yarns and fibrous plies of the original one-layer fabric are copied and positioned according to the in-plane shifts and vertical placement of the layers (Fig. 4.2c). Note that the configuration shown in Fig. 4.2 is subject to the translational symmetry transformation in the x–y plane, and the apparent voids in the unit cell volume are actually filled by the yarns belonging to the adjacent repeatable units.

Figure 4.2c compares the simulations with experiment for a glass fabric with 3.34 warp yarns/cm and 3.62 weft yarns/cm, 600 tex yarns, areal density 420 g/m². The compressibility of the fabric has been studied by Lomov and Verpoest,[21] where the precision of the *WiseTex* compression model has been validated. Consider laminates made of this fabric under a pressure of 1 MPa. Comparison between an experimental cross-section of the laminate and a result of the random simulation (Fig. 4.2c) reveals similarity between the qualitative characteristics of the placement of the layers. In both cases, there exist regions with high packing density of the yarns, where the local fibre volume fraction is very high, and regions with high porosity, effectively creating channels in the fabric. Figure 4.2c shows also the experimental data on the thickness of the laminate with one, two, three and four layers,[21] in comparison with the results of Monte-Carlo simulations of the random stacking of the fabric layers (size of the sampling 500). The experimental and computed data correspond quite well.

4.3.3 Compression of 3D fabric

The algorithm described above does not preserve the length of the yarns after compression, hence introducing an error on the final fabric geometry. In the compaction of a 2D fabric this error is small and can be considered to be negligible. When a 3D fabric is compacted, its z-yarns (going through the thickness), deviate considerably from their paths, as the length of the yarn must be preserved when the thickness of the fabric is reduced. In a 3D fabric with z-yarns initially almost vertical, they will acquire S/Z shapes. If the interlacing yarns are oblique, their initial slope is increased.

To describe such behaviour rigorously, one would have to account for all kinds of contacts between yarns occurring during the compaction. This task may be an interesting challenge for finite element modelling. In the present model, implemented in *WiseTex*, a simple geometrical approach is followed. A spline correction is added to the yarn path, with spline coefficients chosen as to preserve the yarn length after the deformation. The result of the correction is illustrated in Fig. 4.3, which shows the results of measurement and simulation of compression of 3D glass fabric (yarns $4 \times 4 \times 33$ tex, 35 yarns/cm in warp and weft, areal density $3900 \, g/m^2$, weave is shown in the figure). The fabric is proposed in Parnas *et al.*[65] as a benchmarking case for study of RTM composite processing. Compression of the fabric has been measured on the KES-F (Kawabata Evaluation System) textile compression tester for the low load range and on an Instron for higher loads. The input data on compressibility of the yarns and their bending rigidity has been measured on KES-F. Other fabric data was

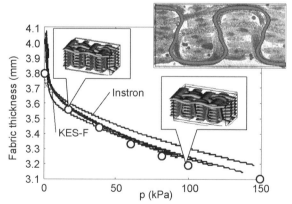

4.3 Compression of 3D fabric. Above: Compression diagram, measured (lines) and computed (circle). Inset: Computed and observed shape of z-yarn at 300 kPa.

taken as specified in Parnas et al.[65] We see that the described algorithm provides a reasonable prediction of the compression diagram as well as the fabric internal structure after compression.

4.4 Model of uniaxial and biaxial tension of woven fabric

4.4.1 Outline of the algorithm

The algorithm described here is a generalisation of the method, proposed by Kawabata et al.[3,4] and Olofsson[66] and widely used by other researchers for plain weave fabrics (with possible extension to other 2D structures).[15] The computational scheme was also implemented into FEA.[30,31] We apply the method to a generalised description of an arbitrary weave, including 3D architectures.

Consider first a woven fabric under biaxial tension characterised by deformations in warp (x-axis) and weft (y-axis) directions $e_x = Y/Y_0 - 1$, $e_y = X/X_0 - 1$, where X and Y are sizes of the fabric repeat, subscript '0' designates the undeformed state. As discussed above, the internal structure of the fabric is described based on weft crimp heights h_j^{We} and weft and warp cross-section dimensions at the intersections d_{ij}^{Wa} and d_{ji}^{We} (subscripts designate different yarns in the fabric repeat). These values change after the deformation. Tension of the yarns induces transversal forces, which compress the yarns, changing the d values. The same transversal forces change the equilibrium conditions between warp and weft, which leads to a redistribution of crimp and change of crimp heights. When the mentioned values in the deformed configuration are computed, the internal geometry of the deformed fabric is built as explained above. Change of length of the yarns determines their average (in the repeat) deformations, which, through the tension-deformation diagrams of the yarns allow yarn tensions to be computed. When summed, with yarn inclinations due to the crimp accounted for, the yarn tensions are transformed into loads, causing the fabric deformations.

The key problem in biaxial modelling is the computation of the crimp heights and transversal forces in the deformed structure. Assuming that the spacing of the yarns in the fabric is changed proportionally to the change of the repeat size, we compute the x and y positions of intersections of warp and weft in the deformed structure. The configuration of the yarns in the crimp intervals between the intersections is determined by these positions and (unknown) crimp heights. Consider some values of the crimp heights. Then the geometrical model determines the positions of the ends of crimp intervals (warp/weft intersections) and bent shape of the yarns in the intervals.

The transversal forces are computed using the following formula (Fig. 4.1b):

$$Q = Q_{bend} + T_1 \sin \theta_1 + T_2 \sin \theta_2 \qquad 4.13$$

where Q_{bend} is the transversal force due to yarn bending [4.4], $T_{1,2}$ are the yarn tensions on two crimp intervals adjacent to the point of application of the transversal force, $\theta_{1,2}$ are the angles of inclination of the yarn on these crimp intervals. We assume that the tension of the yarn can be computed based on the average deformation ε of the yarns (therefore $T_1 = T_2$):

$$T_1 = T_2 = T(\epsilon); \epsilon = \frac{l - l_0}{l_0} \qquad 4.14$$

where l is the yarn length. Note that T depends on yarn length after the deformation, which in its turn depends on crimp heights and yarn dimensions.

The transversal forces compress the yarns according to an experimental compression law [4.5]. When the yarn dimensions are computed and 'frozen', crimp heights are determined using the minimum energy condition:

$$W = W_{bend} + W_{tens} \to \min$$

where W_{bend} and W_{tens} are the bending and tension energy of the yarns. The former is computed summing up bending energies of the yarns in crimp intervals between yarn intersections [4.7], the latter is the sum of tension energies of all the yarns, which are computed using their (linear or non-linear) tension diagrams and yarn deformations.

The computations described above determine one step in the iteration process: starting from current values of the crimp heights we compute yarn lengths, yarn tensions, transversal forces, yarn compressed dimensions and then new values of the crimp heights. The full algorithm is depicted in Fig. 4.4.

If one side of the fabric is kept free (uni-axial tension, say, along the warp), then the described algorithm has another, outer iteration loop, searching for $X < X_0$ (negative e_y) which would lead to zero loads along the weft (y) direction. This allows computing the Poisson coefficient for the fabric.

4.4.2 Comparison with finite element simulation and experiment

Experimental biaxial tensile properties of balanced glass woven fabric (plain weave, 2.2 yarns/cm, 1220 tex) and results of finite element (FE) analysis have been reported in refs 32, 33 and 55. The main feature of FE calculations is the specific mechanical behaviour of the single yarn, which is composed of thousands of fibres with very small sections, which are very flexible and can slide relative to others. The behaviour of the yarn is assumed to be orthotropic. Shear moduli are very small. Young's moduli in the direction perpendicular to the yarn are all very small in comparison to the modulus in the direction of the yarn. A hypoelastic orthotropic model is used and the rotation of the orthotropic frame (as well as the rotational objective derivative) is based on the rotation of the yarn. The compression of the yarn is very important in biaxial tension,

Step 1. Set initial deformations and tensions. Step 2. Compute dimensions of yarns and transversal forces. Step 3. Compute length of the yarns. Step 4. Compute deformations and tensions. Step 5. Check convergence for the deformations. If not, go to Step 2.	Step 1. Compute spacing $p_{Wa0}(1+\epsilon_y); p_{We} = p_{We0}(1+\epsilon_x)$. Step 2. Set changes of weft crimp heights $\Delta h_{ij} = 0$. Step 3. Compute fabric **internal structure** for $h_{ij} = h_{ij0} + \Delta h_{ij}$. Step 4. Compute average **yarns strains** $\epsilon = l/l_0 - 1$. Step 5. Compute yarns tensions $F = F(\epsilon)$. Step 6. Compute **transversal forces** Q (due to bending and tension). Step 7. Compute **compression** of the yarns under the forces Q. Check convergence of Q. Step 8. Compute Δh_{ij} using the condition of **minimum of total (bending plus tension) energy** of the yarns in the repeat. Step 9. Check convergence of Δh_{ij}; if not, go to Step 3. Step 10. Compute applied forces summing up the yarns' tensions.
Step 1. Set an approximation $\{h_{ij}^{We}\}$. Step 2. Compute $\mathrm{grad}(W\{h_{ij}^{We}\})$. Step 3. Solve the minimization problem in the gradient direction. Step 4. Check the convergence of $\{h_{ij}^{We}\}$; if not, go to Step 2.	

4.4 Biaxial tension algorithm.

because the undulation variations directly depend on these thickness changes. The transverse Young's modulus is assumed to be of the form.

$$E_3 = E_\epsilon + E_0 |\epsilon_{33}^n| \epsilon_{11}^m \qquad 4.15$$

E_ϵ is the transverse Young's modulus of the unloaded state. It is very weak (nearly equal to zero) for the single yarns of the studied fabrics. The parameters E_0, m, n are determined using an inverse method from the biaxial test for equal forces in the warp and weft direction (Fig. 4.5a). The results of the finite element analysis are in good agreement with the experimental data.[32–34]

The current formulation of the approximate model (Fig. 4.4) does not envisage the dependency of the yarn compression diagrams on the applied tension. Therefore comparison of the experimental data and results of finite element simulations can answer two questions:

- Does the iterative algorithm of Fig. 4.4, simplified vis-à-vis finite element modelling, provide results close to the latter (and to experiment)?
- What are the errors introduced by using compression diagrams obtained without the yarn tension?

To answer these questions, *WiseTex* modelling of the tension of the fabrics, described above, have been performed. The yarn tension resistance has been taken from the experiment (Fig. 4.5). Bending rigidity of the yarns has been measured on KES-F to provide the value of $0.5\,\text{Nmm}^2$. The compression law was derived from equation [4.15] using different constant levels of tension (ϵ_1). These diagrams are referred as 'Compression $\epsilon_1 = \ldots \%$'.

Figure 4.5 shows the results. The calculations describe well the qualitative difference between deformation regimes of uniaxial and biaxial tension. For the uniaxial tension the error of calculations is small whatever compression diagram is used. The standard textile compression tester can be used to gather input data for simulation of this tension regime. For the biaxial tension the difference between calculations with compression laws corresponding to different tensions can be as large as 30%. A reasonable correlation is found when the compression

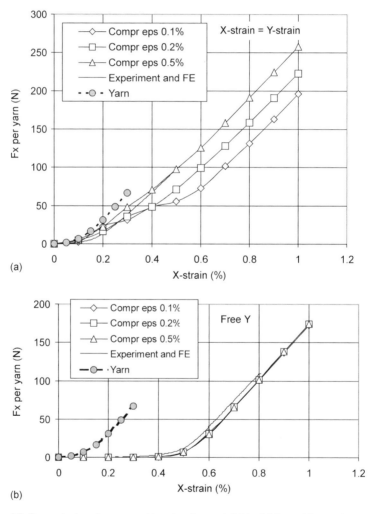

4.5 Computed and measured tension force: (a) biaxial (equal forces in warp and weft direction) and (b) uniaxial tension.

diagram used corresponds to the highest level of strain of the yarns. For the low strain region the experimental curve corresponds better to the calculations with compression diagrams for low tension, and vice versa for higher strain.

4.5 Model of shear of woven fabric

In formulating the model we again follow the approach outlined by S. Kawabata in the 1970s,[5] which is also used in more recent publications.[22–25,67,68]

Our aim is, given a value of the shear angle, γ, to compute the shear force, T, in the presence of (pre)tension of the fabric. The tension is dealt with according to the algorithms of the previous section, resulting in values of the tension of yarns and transversal forces Q, associated with it [4.13]. Therefore we introduce an important assumption: tension of the yarns and transversal forces are computed for the non-sheared configuration and do not change during the shear deformation. This may lead to errors of two types:

- Neglecting the forces in the direction of the yarns, developed during shear. Using general equations, describing the plane stress-strain state of an equivalent continuous membrane, these forces can be calculated as $F^* = T \sin 2\gamma$. With pretension force in the order of 1 N/mm, and shear forces of 0.010.1 N/mm (see the examples in Section 4.6.3 below), the corresponding error is a few percent.
- Change of the yarn tension due to the change of the angle of intersection of the yarns, yarn dimensions, crimp heights during shear. These factors could affect yarn tension (for the same overall deformation of the fabric in the direction of the yarns) also by only a few percent.

Introducing an appropriate calculation loop could eliminate both errors. This complication of the algorithms has not been deemed necessary, as overall precision of the calculation of the shear force, connected with uncertainties of the input data, is also not better than 10 to 20%.

Consider a sheared unit cell of a woven fabric (Fig. 4.6). For simplicity the illustrations below show a plain weave unit cell. The equations are, however, applied to elementary crimp intervals and the forces are summed for the actual fabric repeat. An account is taken therefore of the differences imposed by the weave pattern, as shown in comparison with experimental data below.

Consider a change of shear angle $\Delta\gamma$. The mechanical work ΔA of the shear deformation is (Fig. 4.6):

$$\Delta A = TXY \cos \gamma \Delta \gamma \qquad 4.16$$

where X and Y are the dimensions of the unit cell; units of T are N/mm, corresponding to the 2D nature ('membrane') of the problem. Usage of shear force per unit length also has been proven experimentally to be a correct normalisation procedure for the picture frame test.[69] We will take into account the following mechanisms of the yarns deformation, determining the shear

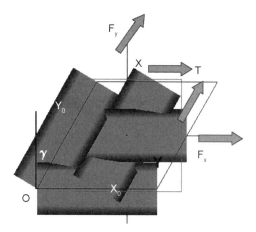

4.6 Sheared unit cell of woven fabric.

resistance: friction; (un)bending; lateral compression; torsion; vertical displacement. Accordingly the mechanical work A is subdivided:

$$\Delta A = \Delta A_{friction} + \Delta A_{disp} + \Delta A_{bending} + \Delta A_{torsion} \qquad 4.17$$

The lateral compression of the yarns is introduced via the transversal forces. We do not consider intra-yarn friction here, as suggested in ref. 70. It is felt that this factor is accounted for by the lateral compression calculations, but the question needs more careful examination in future work, especially in the light of experimental evidence of intra-yarn shear.[71]

The transversal forces, acting on the yarns and determining the friction between them, are caused by tension and bending of the yarns [4.13]. During the shear, yarns are subject to the lateral compression, which creates a pressure inside yarns, which results in an additional component of the transversal force acting on the yarns of the interlacing system. Therefore the transversal forces at yarn intersections will be

$$Q = Q_{bending} + Q_{tension} + Q_{compression} \qquad 4.18$$

where the first two terms are computed with [4.4] and [4.13]. The last term is calculated by the change of fibre volume fraction inside the yarn (Fig. 4.7a) and experimental compression diagram of the yarn:

$$Q_{compression} = (P_{Wa} + P_{We}) \cdot d_{Wa2} d_{We2}$$

where P_{Wa} and P_{We} are pressures inside warp and weft yarn, calculated as

$$P = P(Vf(d_1, d_2))$$

where the fibre volume fraction Vf is inversely proportional to the area of the compressed cross-section with dimensions d_1 and d_2, and the dependency $P(Vf)$ is measured in the compression experiment.

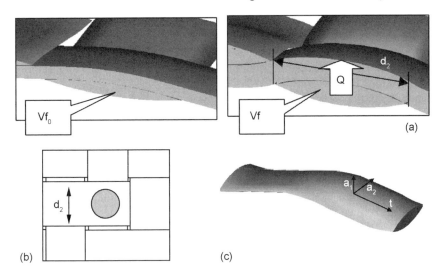

4.7 Components of shear resistance: (a) lateral compression; (b) friction; (c) torsion.

The components of the shear resistance are computed as follows.

Friction moment (Fig. 4.7b)

$$M_{friction} = fQr; \quad r = \frac{2R}{3}; \quad R = \tfrac{1}{2}\sqrt{d_{Wa2}d_{We2}} \qquad 4.19$$

$$\Delta A_{friction} = M_{friction}\Delta\gamma$$

where f is the coefficient of friction, r is the effective radius of the zone of friction for the normal force Q evenly distributed over a circle with radius R, d_2 is the width of the intersecting warp and weft yarns.

The definition of R is an assumption, valid for flat, low crimp rovings, when the contact zone of the yarns can be expected to cover all the yarn width. Leaf and Sheta[72] use a correction accounting for a possible lesser contact zone for more rounded yarns. The accurate definition of the contact zone is out of scope of the present model, hence introduction of the correction does not seem to be justified. Note that assumption of wide contact [4.19] may lead to overestimation of $M_{friction}$, whilst neglect of the curvature of the contacting surfaces (more important for less flat yarns) leads to underestimation of it. Comparisons[73] of estimations by [4.19] with direct measurements of the friction moment between two intersecting yarns by Kawabata,[5] place the experimental values within the range of the predictions, given uncertainties in the friction coefficient values, and do not show systematic errors of equations [4.19].

Mechanical work of torsion (Fig. 4.7c)

$$\Delta A_{torsion} = C\tau\Delta\tau; \tau = \int_0^S \mathbf{t} \cdot \left(a \times \frac{d\mathbf{a}}{ds}\right) ds$$

where C is the torsional rigidity of the yarn, τ is the full angle of torsion, computed by integration of rotation of vector $\mathbf{a}$, determining orientations of the yarn cross-section axis, about the tangent to the yarn middle line $\mathbf{t}$, over the yarn length. Measurement of the torsional rigidity of yarns requires non-standardised equipment,[74] which is normally not suitable for heavy yarns used in composite reinforcements. It the absence of direct measurement, it can be estimated as $C = B/(d_2/2)$, where B is the bending rigidity of the yarn, and d_2 its width.[75]

Mechanical work of (un)bending of the yarns

$$\Delta A_{bending} = \frac{1}{2} B \int_0^S \left(\frac{d^2 \delta z}{ds^2}\right)^2 ds$$

where δz is the difference between the z-coordinate of the centre line of the yarn before and after the deformation (at a given coordinate along the yarn s).

Mechanical work of vertical displacement of the yarns (this displacement goes against the transversal forces Q)

$$\Delta A_{disp} = Q \cdot \frac{1}{l} \int_{contact} \delta z \cdot ds$$

where l is the yarn length, and the integration is performed over the contact zones between warp and weft yarns, determined by the geometrical model.[51,76]

With the components of mechanical work [4.16] known, equation [4.15] is integrated to yield the dependency $T(\gamma)$. Note that the result of the calculation depends on the applied tension (via $Q_{tension}$). This dependency and comparison of the model with experiment is discussed in detail in the next section.

4.6 Parametric description of fabric behaviour under simultaneous shear and tension

Using the model outlined above, a designer can get input data for forming simulations. However, getting input data for the model is also difficult, as this involves testing of the yarns on specialised textile testing equipment. For some classes of reinforcement materials it is possible to make a generic characterisation of them, depending on a few parameters, and then, using the theoretical model, produce analytical expressions for shear behaviour.

4.6.1 Properties of glass rovings

Woven glass fabrics are a common reinforcement for composite materials. The raw materials, glass rovings, are produced by different manufacturers (terms *rovings, yarns* and *tows* are used below as synonyms). The results shown here[21,55,77] cover the full range of linear density from 150 to 5000 tex. All the measurements were done on 'virgin' (before weaving) rovings.

Width and thickness

The width of a roving was measured at 20 points along the yarn length on the scanned (1200 dpi) images of rovings lying flat. The thickness of a roving was determined by the extrapolation of the compression curve (see below) to zero pressure. Measurements of the thickness with a calliper or a textile thickness meter is done under a certain pressure (uncontrollable in the former and controlled in the latter case) and therefore does not provide correct values for a free yarn. Figure 4.8a,b shows a summary of the results. The data are reasonably approximated by a linear dependency of the roving width and thickness on the linear density.

Bending

Bending of rovings was also measured on KES-F. Single yarns were tested. In the first stage of bending, the fibres tend to buckle, breaking the sizing and resulting in a sharp increase of stiffness.[21] Therefore the first loading cycle was discarded from data processing. Sometimes the same phenomena occur when reversing the direction of bending. In these cases the corresponding loading part of the curve was also discarded. For each type of yarn three samples were tested; for each sample three full bending cycles were performed and the second cycle was used in the data processing using the standard KES-F routine: bending rigidity is determined by the slope of the diagram between curvatures of 0.5 and $1.5\,\text{cm}^{-1}$. Figure 4.8c shows dependency of the bending rigidity on the linear density of the yarns. The tested rovings have fibres of different diameter d (from 16 to 21 m). The scatter of 30% in the diameter gives a scatter of 300% in the bending rigidity of the fibres (proportional to d^4) and 70% in the bending rigidity of fibre bundles with the same linear density, but with fibres of the different diameter (proportional to d^2 for the case of non-interacting fibres). This explains the significant deviation of some data points from the average 'master' curve for yarns with linear density less then 1000 tex. However, the general agreement of the quadratic approximation of the 'master' curve with the data is sufficiently high.

Compression

Compression of rovings was measured on KES-F apparatus, following the routine prescribed by the KES-F manual. The maximum force used was 100 cN,

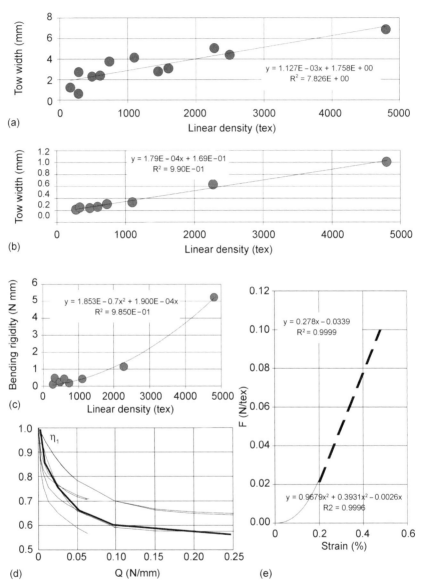

4.8 Properties of glass rovings: (a) width; (b) thickness; and (c) bending rigidity: points – measurements, line – linear/polynomial approximation; (d) compression diagram: thin lines – measurements for different rovings, thick line – least square approximation with [4.20]; (e) tension: solid line – non-linear, dashed line – linear region.

which corresponds to the compression force per unit length (4 mm under the machine's head) of 25 cN/mm. For the thicker rovings a larger head was used, which gives lower maximum force. For each type of yarn three samples were tested, with three compression cycles for each yarn. Following Lomov and Verpoest,[21] the second cycle of compression was used for final results.

For simulation of the compression behaviour of the yarns in the model the following data are needed: (1) dimensions of the yarn cross-section in free state d_{10} and d_{20} (thickness and width of the roving, see above); (2) functions $\eta_1 = \eta_1(Q)$, $\eta_2 = \eta_2(Q)$, $\eta_1 = d_1/d_{10}$; $\eta_2 = d_2/d_{20}$ [4.4]. The KES-F equipment does not provide data for the yarn flattening. Therefore an empirical relationship between η_1 and η_2 has been used:[73]

$$\eta_2 = 1/\eta_1^{0.33}.$$

The compression curves for the different rovings (Fig. 4.8d) demonstrate a significant scatter, without any recognisable trends relating to linear density or the tow thickness. The average 'master' diagram is built by the least square approximation by the formula

$$\eta_1 = \frac{1 + \left(\dfrac{Q}{Q_0}\right)^a \eta_{1min}}{1 + \left(\dfrac{Q}{Q_0}\right)^a}, \quad \eta_{1min} = 0.5583, Q_0 = 0.0277, a = 1.7987 \quad 4.20$$

where Q_0, a and η_{1min} are the fitting parameters. The error of this approximation is about 15% ($R^2 = 0.76$).

Friction

The friction coefficient between roving and steel was measured on KES-F. The obtained value $f = 0.24$ was retained as the value for roving-roving friction, as it does not contradict with the literature data.[22]

Tension

Tension diagrams of the rovings were taken from the measurements in Boisse et al.[30] (Fig. 4.8e). We assume that the tension resistance of the rovings is proportional to the linear density. The tension diagram has two regions: non-linear, up to approximately 0.2% of strain and linear. The former represents straightening of the fibres in the tow. The tow resistance to tension is a combination of the (un) bending and tension resistance of the fibres. Proportionality of the tension force to the linear density of the tow is equivalent to the assumption that the waviness of the fibres is roughly the same in different rovings. The linear part of the diagram corresponds to tension of the straight fibres. The experimentally measured tensile modulus for this part of the diagram is slightly less than the tensile modulus of glass fibres 28.3 N/tex (72 GPa).

100   Composites forming technologies

Summary of the 'master' descriptions of the properties of glass rovings

The result of the parametrisation of the properties of glass rovings is shown in Table 4.1. All the data necessary for the calculation of the shear diagram of a roving, are given as functions of the roving linear density.

4.6.2 Parameterisation of the shear diagram

A woven fabric is characterised by weave pattern, weaving density, and warp and weft yarn descriptions.

For the weave patterns, the three most widely used types were chosen for the numerical experiments: plain, twill 2/2 and satin 5/2. These patterns represent the range of the interaction intensity of the yarns in the weave: maximum possible for plain weave, average for twill 2/2 and weak for satin 5/2. One can expect a monotonic decrease of the shear resistance with weakening of the interaction of the yarns.

For the weaving density, only square (= identical warp and weft parameters) fabrics have been considered, being the most important practical case. Shear behaviour of non-balanced fabrics can be simulated using the same model, if needed. To characterise the weaving density, a *looseness factor s* has been introduced:

Table 4.1 Averaged dependencies of properties of glass rovings on their linear density t, tex

Property	Formula
Thickness d_1, mm	$d_1 = 1.79\text{E-04} * t + 1.69\text{E-01}$
Width d_2, mm	$d_2 = 1.127\text{e-03} * t + 1.758\text{E} + 00$
Coefficient of compression, η_1, as function of compressive force per roving length $[Q] = \text{N/mm}$	$\eta_1 = \dfrac{d_1}{d_{10}} = \dfrac{1 + \left(\dfrac{Q}{Q_0}\right)^a \eta_{min}}{1 + \left(\dfrac{Q}{Q_0}\right)^a} = const\,(t)$, $\eta_{min} = 0.5583, Q_0 = 0.0277, a = 1.7987$
Coefficient of flattening, η_2, as function of compressive force per roving length $[Q] = \text{N/mm}$	$\eta_2(Q) = \dfrac{d_2}{d_{20}} = (\eta_1(Q))^{-0.33} = const\,(t)$
Bending rigidity B, N mm^2	$B = 1.85\text{e-07} * t^2 + 1.90\text{E=04} * t$
Coefficient of friction f	$f = 0.24$
Tension diagram 'force F vs strain ϵ', $[F] = \text{N/tex}$, $[\epsilon] = \%$	$F = \begin{cases} 0.9679 \cdot \epsilon^3 + 0.3931 \cdot \epsilon^2 - 0.0026 \cdot \epsilon, & \epsilon \leq 0.2\% \\ 0.278 \cdot \epsilon - 0.0339, & \epsilon > 0.2\% \end{cases}$

$$s = \frac{p - d_2}{d_2} \qquad 4.21$$

where p is the spacing of the yarns (inverse to the ends/picks count), d_2 is the yarn width. The looseness factor represents ratio of the width of the pores in the fabric to the yarn width. The range $s = (0.01, 0.02, 0.05, 0.075, 0.1, 0.2, 0.5, 1.0)$ has been considered, where value $s = 0$ represents an extremely tight, and $s = 1$ an extremely loose fabric. The value of d_2 in [4.16] can be calculated for a given linear density of the yarns using the formula of Table 4.1.

Properties of glass rovings of a given linear density t are fully described by the formulae of Table 4.1. Hence t is the only parameter needed to characterise the warp and weft yarns (identical because of the assumption of square fabric construction). The range of t for the numerical experiments was $t = (100, 200, 500, 1000, 2000, 5000)$ tex.

Finally, the shear resistance (shear diagram $T(\gamma)$) depends on the fabric (pre)tension. Only equal tension of the warp and weft has been considered (the case of non-symmetrical tension can be simulated with the same model if needed). The pretension is characterised by the tensile strain of the fabric $\epsilon = 0$ to 1%. The maximum value of 1% corresponds to the start of a considerable extension of the glass fibre (as opposed to the decrimping of the yarns)[30,55] and is not likely to be exceeded in the real forming processes.

The calculations followed the following steps for all combinations of parameters:

- For a given linear density of the yarns t, calculate yarn width d_1;
- Using a given looseness factor s, calculate spacing of the yarns p;
- Assign properties of the warp and weft yarns using the value of t and formulae of Table 4.1;
- Build a geometrical model of non-sheared fabric of a given weave pattern, calculated yarn spacing and calculated properties of the yarns;
- Apply the model of coupled biaxial tension and shear of the fabric for a given pretension of the warp and the weft and calculate the shear diagram $T(\gamma)$.

The numerical experiments resulted in a vast set of shear diagrams $T(\gamma; weave, t, s, \epsilon)$. The diagrams for the different weaves were further processed separately. The processing has been done in the following steps.

Shape of the shear diagrams

Consider a diagram $T(\gamma)$ for a given set of parameters (t, s, ϵ), a typical example of which is shown in Fig. 4.9. The diagram has been approximated with an analytical expression

$$T(\gamma) = T_0 + T_1 \tan^a \gamma \qquad 4.22$$

This formula [4.22] represents the main features of the shear behaviour:

Composites forming technologies

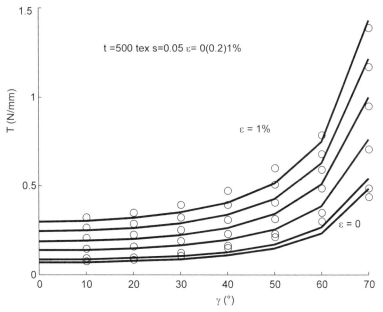

4.9 Simulated shear diagrams (points) and their approximation with equation [4.23] – lines. Plain weave, $t = 500\,\text{tex}$, $s = 0.05$, $\epsilon = 0(0.2)1\%$ (shear resistance monotonically increases with the increase of pre-strain s).

- Non-zero resistance from the very beginning of the shear ($\gamma = 0$), caused by the friction between the yarns
- Low shear modulus for low shear angles

$$G(\gamma) = \frac{dT}{d\gamma} = aT_1 \tan^{a-1}\gamma(1 + \tan^2\gamma) \xrightarrow[\gamma \to 0]{} 0 \quad (a > 1)$$

- Locking behaviour for high shear:

$$T(\gamma) \xrightarrow[\gamma \to \pi/2]{} \infty, G(\gamma) \xrightarrow[\gamma \to \pi/2]{} \infty$$

After fitting the calculated diagrams with [4.22], the value of a was found in all cases to lie inside the interval (1 ... 3), without any clear trend in dependency on (t, s, ϵ). It was decided then to try to use the approximation [4.22] with $a = 2$ in all the cases. After refitting the data, the agreement between the approximation

$$T(\gamma) = T_0 + T_1 \tan^2\gamma \qquad 4.23$$

was found to be quite good. An example of the approximation is shown in Fig. 4.9. A certain systematic underestimation of the shear force by [4.19] in this figure is explained by the fact that it refers to a certain combination of the parameters t and s; for other combinations the constant value $a = 2$ leads to a certain overestimation.

The value $a = 2$ and analytical expression [4.23] were used in all the subsequent calculations. The coefficients in the equation [4.23] have a clear mechanical meaning: T_0 is the shear resistance at $\gamma \to 0$, caused by friction between the yarns, T_1 determines the shear modulus

$$G(\gamma) = \frac{dT}{d\gamma} = 2T_1 \tan\gamma (1 + \tan^2\gamma)$$

Coefficients T_0 and T_1

Coefficients of the formula [4.23] depend on parameters (t, s, ϵ). To establish these dependencies in a closed form, the calculated values of the coefficients were tabulated:

$$T_0 = T_0(t, s, \epsilon), \quad T_1 = T_1(t, s, \epsilon) \qquad 4.24$$

and a regression analysis was performed over the whole set of the data. After analysis of different representations of the regression formulae, the following equations were chosen to fit the data:

$$\ln T_0 = b_1 + b_2 \cdot \ln t + b_3 \cdot \ln s + b_4 \cdot \ln \epsilon + b_5 \cdot (\ln t)^2 + b_6 \cdot (\ln t \cdot \ln s)$$
$$+ b_7 \cdot (\ln t \cdot \ln \epsilon) + b_8 \cdot (\ln s)^2 + b_9 \cdot (\ln s \cdot \ln \epsilon) + b_{10} \cdot (\ln \epsilon)^2 \qquad 4.25$$

$$\ln T_1 = c_1 + c_2 \cdot \ln t + c_3 \cdot \ln s + c_4 \cdot \ln \epsilon + c_5 \cdot (\ln t)^2 + c_6 \cdot (\ln t \cdot \ln s)$$
$$+ c_7 \cdot (\ln t \cdot \ln \epsilon) + c_8 \cdot (\ln s)^2 + c_9 \cdot (\ln s \cdot \ln \epsilon) + c_{10} \cdot (\ln \epsilon)^2 \qquad 4.26$$

Regression coefficients of the formulae [4.25] and [4.26] are given in Table 4.2. Figures 4.10 and 4.11 illustrate the quality of the fitting. The simulated values (dashed curves) sometimes show significant variability for smaller pretensions and looser fabrics, which can violate expected monotonic trends

Table 4.2 Coefficients of the regression equations [4.25, 4.26]

	Plain		Twill 2/2		Satin 5/2	
	b	c	b	c	b	c
1	−9.87	−23.1	−12.1	−17.57	−11.4	−21.0
2	1.86	4.17	2.35	2.96	2.14	3.36
3	0.102	−2.12	−0.381	−0.884	−0.042	−1.72
4	3.01	2.05	2.64	1.68	2.12	6.91
5	−0.10	−0.207	−0.131	−0.192	−0.119	−0.176
6	−0.085	0.111	−0.021	−0.090	−0.062	0.021
7	0.017	−0.100	0.001	0.256	−0.027	−0.150
8	0.028	−0.094	−0.030	−0.113	−0.015	−0.133
9	0.436	0.215	0.367	0.463	0.277	0.557
10	0.435	−0.021	0.263	−0.623	−0.057	−2.52

104   Composites forming technologies

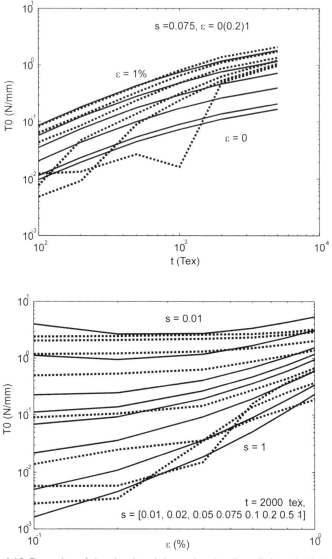

4.10 Examples of the simulated dependencies $T_0 = T_0(t, s, \epsilon)$ (dashed lines) and regression curves [4.25] – solid lines. Plain fabric.

(shear resistance increase with increase of the pre-strain and fabric tightness). These variations are due to the approximate nature of the model and certain instabilities in the iteration loop shown in Fig. 4.4, as well as errors in fitting the diagrams with [4.23]. However, after fitting (therefore smoothening) of the curves they become monotonic and exhibit the expected trends.

Parameterised analytical expressions [4.23–4.26] are ready to use in a user subroutine of forming simulations for woven glass reinforcements.

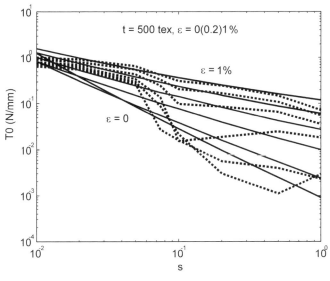

4.10 Continued

4.6.3 Comparison with experiments

Finally, the derived model [4.23–4.26] has to be compared with experimental data. Obtaining input data of the fabric for such a comparison is straightforward: the model requires parameters of the fabric normally provided by the manufacturer: linear density of the yarns, weave structure and ends/pick count (easily transformed into the looseness parameter s). Any publication of shear test results contains these data, which allows comparison of the model prediction with published results, without necessary thorough investigation of the fabric internal geometry.

However, the remaining parameter of the model, the pretension of the yarns, is not controlled in most cases of the picture frame test (see Chapter 1). To solve this problem, in cases where the pretension is not given by the experimentalists, an expected range of pretension strain values will be used in the comparison.

Table 4.3 summarises the test cases[22,78] used for the comparison. Apart from the data on the fabric construction, calculations require a value of pretension strain. These data are not readily available. The values given in Table 4.3 were obtained as follows.

In our own measurements[78] we tried to estimate the strain of the fabric in the picture frame measuring the distance between marks on the fabric, made before it had been mounted in the frame and tensed by the wavy grips. This measurement is not precise. The errors, caused by different sources (the main ones being small displacements to be measured and fibrous surface of the yarns, which cause the marks to widen), can be as high as 0.1 to 0.2% of the strain.

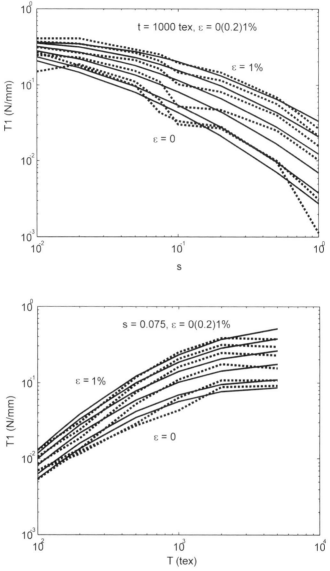

4.11 Examples of the simulated dependencies $T_1 = T_1(t, s, \epsilon)$ (dashed lines) and regression curves [4.26] – solid lines. Twill 2/2 fabric.

Long[22] provides the value of pretension force 1.1 N/mm for fabrics Long-1 and Long-2, and 0.69 N/mm for the fabric Long-3. Calculating the biaxial tensile diagram of the fabrics, using the model described above and input data of Tables 4.2 and 4.3, the pretension was estimated as 0.2% for the fabrics Long-1 (plain weave) and Long-2 (satin), and 0.1% for the twill fabric Long-3.

Virtual testing for material formability 107

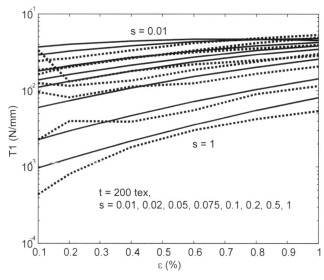

4.11 Continued

The model gives good predictions for the initial stage of shear, up to the locking of the fabric structure (lower estimation of the locking angle is given by a simple geometrical formula, Table 4.3): the set of experimental diagrams for the estimated range of pre-strain agrees well with the calculations (Figs 4.12 and 4.13).

Table 4.3 Input data for comparison with experiments. Data on fabrics 'Lomov' is from ref. 78, 'Long' is from ref. 22

Fabric ID	Weave	Linear density of warp and weft, t, tex	Ends/picks count, yarns/cm	Looseness factors s^*	Geometrical locking angle, 0^{***}	Pre-strain, %***
Lomov-1	twill 2/2	280	4.4	0.11	27	first cycle 0.3...0.5, second cycle 0.1...0.3
Lomov-2	plain	480	3.5	0.20	37	first cycle 0.4...0.6, second cycle 0.2...0.4
Lomov-3	plain	1200	2.3	0.29	45	first cycle 0.5...0.7, second cycle 0.3...0.5
Long-1	plain	1220	2.5	0.30	41	0.2
Long-2	satin	1450	2.7	0.32	41	0.2
Long-3	twill 2/2	2500	1.6	0.31	41	0.1

* Value of s is calculated based on t value and averaged ends/picks count, using formulae Table 4.1
** Geometrical locking angle $\gamma^* = \arccos \dfrac{d_2}{p} = \arccos(1-s)$
*** See text for the explanation of estimation of the pre-strain

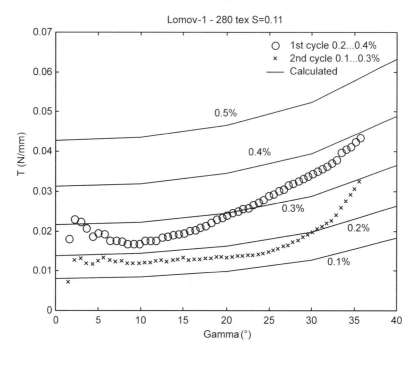
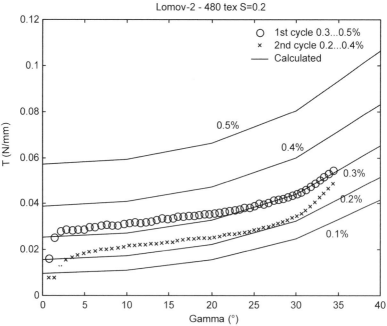

4.12 Comparison between calculated (curves) and experimental (points) data. Fabrics Lomov-1...3.[78]

Virtual testing for material formability 109

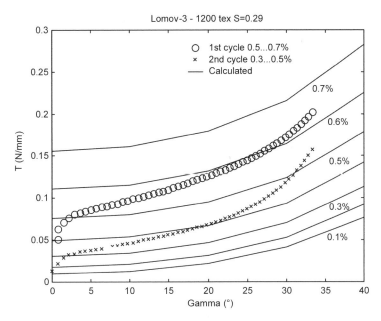

4.12 Continued

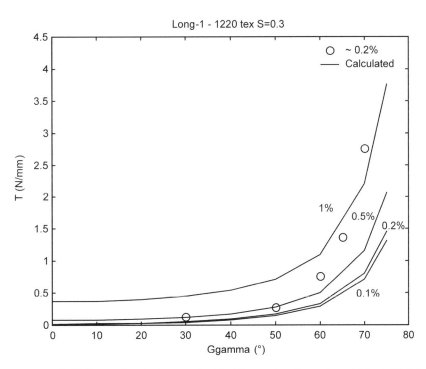

4.13 Comparison between calculated (curves) and experimental (points) data. Fabrics Long-1...3.[22]

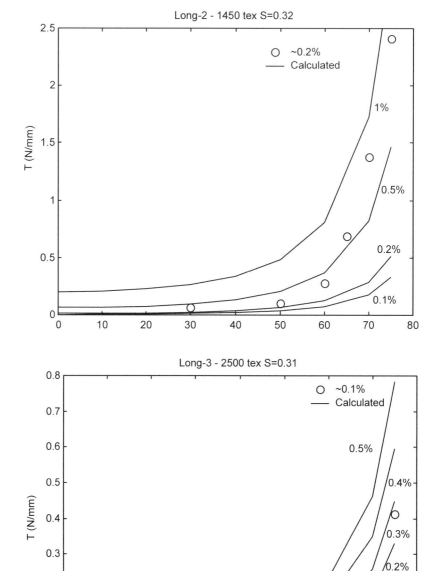

4.13 Continued

After the locking of the structure (distance between the yarns in the sheared fabric equal to the width of the yarns) the sharp increase of the shear resistance is caused by the lateral compression of the yarns. The model accounts for this phenomenon; however, the experimental diagram increases more steeply than the one calculated according to the estimated pre-strain. The discrepancy can be explained by two reasons. First, the 'master' compression curve for the glass rovings (Fig. 4.8d) is defined within a significant scatter of the experimental diagrams; the flattening coefficient is estimated by an approximate formula. These factors determine an approximate nature of the calculation of the lateral compression, employed in the model. Second, for the higher shear one can expect an increase of the tension of the fabric on the frame, which would increase the shear resistance. The experimental diagram intersects calculated iso-strain curves in the direction of increase of the pre-strain.

4.7 Conclusion: creating input data for forming simulations

The theoretical methods described in this chapter and in Chapter 3 provide the possibility to make predictions of the properties of textile reinforcements to be used in forming simulations, producing constitutive descriptions of the materials, as outlined in Chapter 2. The simulations of compressibility can give estimations for the reinforcement thickness (hence fibre volume fraction in the composite) in pressure-controlled resin infusion processes. Coupled shear–biaxial tension simulation provides complex material response surfaces, proposed in refs 32 and 33. Understanding of the scale of the tension-shear coupling effects may help to understand the level of blank-holding forces necessary to eliminate wrinkling.

Being approximate, such predictions are likely to be used for screening of *a priori* unacceptable variants and for qualitative analysis of the manufacturing process. For the latter, several challenges present themselves:

- Tension diagrams of textiles are definitely non-linear. Can a few percent of low-tension strain make a difference for determination of blank-holding conditions?
- Tension in different directions is coupled. Material models available in the existing software packages do not account for this – should they?
- Should the description of shear resistance in forming simulation account for low shear force in the initial stage of shear? For real non-linear behaviour after the geometrical locking angle, or is a simple 'zero resistance – locking – infinite resistance' model sufficient?
- The calculations shown in this chapter demonstrate a strong dependence of the shear resistance on tension. This is not accounted for in the existing forming software – is it necessary to include this effect?

- In general: Does a designer need to spend money, time and resources for complex measurements of non-linear coupled compression-tension-shear behaviour, or it is sufficient to use simpler material models in forming simulations, with the accuracy of predictions enough for practical purposes?

Finally, the 'virtual testing' can help to understand difficulties in non-standardised measurements of biaxial tensile and shear resistance of textiles, and inconsistencies of results obtained under different conditions, as described in Chapter 13, opening the way for standardising the test apparatus and procedures.

4.8 References

1. Lomov, S.V., A.V. Truevtzev and C. Cassidy, A predictive model for the fabric-to-yarn bending stiffness ratio of a plain-woven set fabric. *Textile Research Journal*, 2000. **70**(12) 1088–1096.
2. Sagar, T.V. and P. Potluri, Computation of bending behavior of woven structures using optimization techniques. *Textile Research Journal*, 2004. **74**(10) 879–886.
3. Kawabata, S., M. Niwa and H. Kawai, The finite-deformation theory of plain weave fabrics. Part I. The biaxial-deformation theory. *Journal of the Textile Institute*, 1973. **64**(1) 21–46.
4. Kawabata, S., M. Niwa and H. Kawai, The finite-deformation theory of plain weave fabrics. Part II. The uniaxial-deformation theory. *Journal of the Textile Institute*, 1973. **64**(2) 47–61.
5. Kawabata, S., M. Niwa and H. Kawai, The finite-deformation theory of plain weave fabrics. Part III. The shear-deformation theory. *Journal of the Textile Institute*, 1973. **64**(2) 42–85.
6. de Jong, S. and R. Postle, A general energy analysis in fabric mechanics using optimal control theory. *Textile Research Journal*, 1978. **48**(3) 127–135.
7. Hearle, J.W.S. and W.J. Shanahan, An energy method for calculations in fabric mechanics. *Journal of the Textile Institute*, 1978. **69**(4) 81–110.
8. Komori, T. and M. Itoh, Theory of general deformation of fibre assemblies. *Textile Research Journal*, 1991. **61**(10) 588–594.
9. Anandjiwala, R.D. and G.A.V. Leaf, Large-scale extension and recovery of plain woven fabrics. Part I. Theoretical. *Textile Research Journal*, 1991. **61**(11) 619–634.
10. Anandjiwala, R.D. and G.A.V. Leaf, Large-scale extension and recovery of plain woven fabrics. Part II. Experimental and discussion. *Textile Research Journal*, 1991. **61**(12) 743–755.
11. Huang, N.C., Finite biaxial extension of completely set plain woven fabrics. *Journal of the Applied Mechanics*, 1979. **46**(9) 651–655.
12. Dastoor, P.H., S.P. Hersh, S.K. Batra and W.J. Rasdorf, Computer-assisted structural design of industrial woven fabrics, Part III Modelling of fabric uniaxial/biaxial load deformation. *Journal of the Textile Institute*, 1994. **85**(2) 135–157.
13. Pastore, C.M., A.B. Birger and E. Clyburn, 'Geometrical modelling of textile reinforcements', in *Mechanics of Textile Composites Conference*, 1995, NASA Hampton, Virginia. 597–623.
14. Christoffersen, J., Fabrics orthotropic materials with a stress-free shear mode. *Journal of the Applied Mechanics*, 1980. **47**(1) 71–74.

15. Reumann, R.-D., Neuartiges Berechnungs-verfahren für das flächenstructurabhängige Kraft-Dehnungs-Verhalten textiler Flächengebilde. *Wissen. Z. Techn. Univ. Dresden*, 1988. **37**(6) 163–169.
16. Chen, B. and T.-W. Chou, Compaction of woven-fabric preforms in liquid composite molding processes single-layer deformation. *Composites Science and Technology*, 1999. **59** 1519–1526.
17. Chen, B. and T.-W. Chou, Compaction of woven-fanric preforms nesting and multi-layer deformation. *Composites Science and Technology*, 2000. **60** 2223–2231.
18. Chen, B., A.H.-D. Cheng and T.-W. Chou, A nonlinear compaction model for fibrous preforms. *Composites Part A*, 2001. **32** 701–707.
19. Chen, B., E.J. Lang and T.-W. Chou, Experimental and theoretical studies of fabric compaction behaviour in resin transfer moulding. *Materials Science and Engineering*, 2001. **A317** 188–196.
20. Kurashiki, T., M. Zako and I. Verpoest, 'Damage development of woven fabric composites considering an effect of mismatch of lay-up', in *Composites for the Future, Proceedings 10th European Conference on Composite Materials (ECCM-10)*, 2002 Brugge. CD edition.
21. Lomov, S.V. and I. Verpoest, Compression of woven reinforcements a mathematical model. *Journal of Reinforced Plastics and Composites*, 2000. **19**(16) 1329–1350.
22. Long, A., 'Process modelling for textile composites', in *International Conference on Virtual Prototiping EUROPAM 2000*. 2000 Nantes. 1–17.
23. Long, A.C., M.J. Clifford, P. Harrison and C.D. Rudd, 'Modelling of draping and deformation for textile composites', in *ICMAC – International Conference for Manufacturing of Advanced Composites*. 2001, IOM Communications Belfast. 66–76.
24. Long, A.C., F. Robitaille, B.J. Souter and C.D. Rudd, 'Permeability Prediction for Sheared, Compacted Textiles During Liquid Composite Modelling', in *13th International Conference on Composite Materials (ICCM-13)*. 2001. Beijing, China.
25. Crookston, J.J., A.C. Long and I.A. Jones, Modelling effects of reinforcement deformation during manufacture on elastic properties of textile composites. *Plastics, Rubber and Composites*, 2002. **31**(2) 58–65.
26. Harrison, P., J. Wiggers, A.C. Long and C.D. Rudd, 'Constitutive modelling based on meso and micro kinematics for woven and stitched fabrics', in *Proceedings ICCM-14*. 2003 San Diego. CD edition.
27. Harrison, P., M.J. Clifford, A. Long and C.D. Rudd, A constituent-based predictive approach to modelling the rheology of viscous textile composites. *Composites Part A*, 2004. **35** 915–931.
28. Harrison, P., M.J. Clifford and A.C. Long, Shear cheracterisation of viscous woven textile composites A comparison between picture frame and bias extension experiments. *Composites Science and Technology*, 2004. **64** 1453–1465.
29. Liu, L., J. Chen and J.A. Sherwood, Two-dimensional macro-mechanics shear models of woven fabrics. *Composites Part A*, 2004. **36** 105–114.
30. Boisse, P., M. Borr, K. Buet and A. Cherouat, Finite element simulations of textile composite forming including the biaxial fabric behaviour. *Composites Part B*, 1999. **28B** 453–464.
31. Boisse, P., A. Cherouat, J.C. Gelin and H. Sabhi, Experimenal study and finite element simulation of a glass fibre fabric shaping process. *Polymer Composites*, 1999. **16**(1) 83–95.

32. Boisse, P., K. Buet, A. Gasser and J. Launay, Meso/macro-mechanical behaviour of textile reinforcements for thin composites. *Composites Science and Technology*, 2001. **61** 395–401.
33. Boisse, P., A. Gasser and G. Hivet, Analyses of fabric tensile behaviour determination of the biaxial tension-strain surfaces and their use in forming simulations. *Composites Part A*, 2001. **32**(10) 1395–1414.
34. Gasser, A., P. Boisse and S. Hanklar, Analysis of the mechanical behaviour of dry fabric reinforcements. 3D simulations versus biaxial tests. *Computational Materials Science*, 2000. **17** 7–20.
35. Launay, J., K. Buet-Gautier, G. Hivet and P. Boisse, Analyse experimentale et modeles pour le comportement mechanique biaxial des renforts tisses de composites. *Revue des composites et des materiaux avances*, 1999. **9**(1) 27–55.
36. Kuwazuru, O. and N. Yoshikawa, 'Non-constitutive numerical modeling for plain-weave fabrics', in *Proceedings of 7th Japan International SAMPE Symposium & Exhibition, November 13–16*. 2001 Tokyo. 729–732.
37. Sakakibara, K., A. Yokoyama and H. Hamada, 'Deformation mechanism of textile under uni- and biaxial tensile loading', in *Proceedings of 7th Japan International SAMPE Symposium & Exhibition, November 13–16*. 2001 Tokyo. 705–708.
38. Zouari, B., F. Dumont, J.L. Daniel and P. Boisse, 'Analyses of woven fabric shearing by optical method and implementation in a finite element program', in *Proceedings of the 6th ESAFORM Conference on material Forming*. 2003 Salerno. 875–887.
39. Hivet, G., B. Laine and P. Boisse, 'Consistent preprocessor for the unit woven cell for meso-macro analyses of fabric forming', in *Procedings of the 8th ESAFORM Conference on Material Forming*. 2005 Cluj-Napoca. 947–950.
40. Kondratiev, S., *Finite element modelling of the spatial stress-strain state of textiles and textile composites*. Masters thesis, 2001, State Technical University St.-Petersburg.
41. Van Genechten, B., *Finite element modelling of textile composites*. Masters thesis, 2002, Vrije Universiteit Brussel – Katholieke Universiteit Leuven.
42. Robitaille, F., A.C. Long, I.A. Jones and C.D. Rudd, Authomatically generated geometric descriptions of textile and composite unit cells. *Composites Part A*, 2003. **34**(4) 303–312.
43. Robitaille, F., A. Long, M. Sherburn, C.C. Wong and C. Rudd, 'Predictive modelling of processing and performance properties of textile composite unit cells current status and perspectives', in *Proceedings ECCM-11*. 2004. CD Edition.
44. Lomov, S.V., I. Verpoest, E. Bernal, F. Boust, V. Carvelli, J.-F. Delerue, P. De Luka, L. Dufort, S. Hirosawa, G. Huysmans, S. Kondratiev, B. Laine, T. Mikolanda, H. Nakai, C. Poggi, D. Roose, F. Tumer, B. van den Broucke, B. Verleye and M. Zako, 'Virtual textile composites software Wisetex integration with micro-mechanical, permeability and structural analysis', in *Proceedings of the 15th International Conference on Composite Materials (ICCM-15)*. 2005 Durban. CD edition.
45. Lomov, S.V., X. Ding, S. Hirosawa, S.V. Kondratiev, J. Molimard, H. Nakai, A. Vautrin, I. Verpoest and M. Zako, 'FE simulations of textile composites on unit cell level validation with full-field strain measurements', in *Proceedings 26th SAMPE-Europe Conference*. 2005 Paris. 28–33.
46. Verpoest, I. and S.V. Lomov, Virtual textile composites software Wisetex integration with micro-mechanical, permeability and structural analysis. *Composites*

Science and Technology, 2005, **65** (15–16) 2563–2574.
47. Mikolanda, T., S.V. Lomov, M. Kosek and I. Verpoest, *Simple use of virtual reality for effective visualization of textile material structures*, CODATA Prague Workshop Information Visualization, Presentation, and Design, 2004, Prague. CD edition.
48. Lomov, S.V., E. Bernal, D.S. Ivanov, S.V. Kondratiev and I. Verpoest, Homogenisation of a sheared unit cell of textile composites FEA and approximate inclusion model. *Revue européenne des éléments finis*, **14**(6–7) 709–728.
49. Lomov, S.V., G. Huysmans, Y. Luo, R. Parnas, A. Prodromou, I. Verpoest and F.R. Phelan, Textile Composites Modelling Strategies. *Composites Part A*, 2001. **32**(10) 1379–1394.
50. Belov, E.B., S.V. Lomov, I. Verpoest, T. Peeters, D. Roose, R.S. Parnas, K. Hoes and V. Sol, Modelling of permeability of textile reinforcements Lattice Boltzmann method. *Composites Science and Technology*, 2004. **64** 1069–1080.
51. Lomov, S.V., A.V. Gusakov, G. Huysmans, A. Prodromou and I. Verpoest, Textile geometry preprocessor for meso-mechanical models of woven composites. *Composites Science and Technology*, 2000. **60** 2083–2095.
52. Lomov, S.V., I. Verpoest, T. Peeters, D. Roose and M. Zako, Nesting in textile laminates geometrical modelling of the laminate. *Composites Science and Technology*, 2002. **63**(7) 993–1007.
53. Lomov, S.V., G. Huysmans and I. Verpoest, Hierarchy of textile structures and architecture of fabric geometric models. *Textile Research Journal*, 2001. **71**(6) 534–543.
54. Lomov, S.V., A. Nakai, R.S. Parnas, S. Bandyopadhyay Ghosh and I. Verpoest, Experimental and theoretical characterisation of the geometry of flat two- and three-axial braids. *Textile Research Journal*, 2002. **72**(8) 706–712.
55. Lomov, S.V., T. Truong Chi, I. Verpoest, T. Peeters, V. Roose, P. Boisse and A. Gasser, Mathematical modelling of internal geometry and deformability of woven preforms. *International Journal of Forming Processes*, 2003. **6**(3-4) 413–442.
56. Lomov, S.V., B. Van den Broucke, F. Tumer, I. Verpoest, P. De Luka and L. Dufort, 'Micro-macro structural analysis of textile composite parts', in *Proceedings ECCM-11*. 2004 Rodos. CD Edition.
57. Van den Broucke, B., F. Tumer, S.V. Lomov, I. Verpoest, P. De Luka and L. Dufort, 'Micro-macro structural analysis of textile composite parts case study', in *Proceedings of the 25th International SAMPE Europe Conference, March 30th–April 1st*. 2004 Paris. 194–199.
58. Lomov, S.V. and B.M. Primachenko, Mathematical modelling of two-layered woven fabric under tension. *Technologia Tekstilnoy Promyshlennosty*, 1992(1) 49–53.
59. Lomov, S.V., Computer aided design of multilayered woven structures, part 1. *Technologia Tekstilnoy Promyshlennosty*, 1993(1) 40–45.
60. Lomov, S.V., Computer aided design of multilayered woven structures, part 2. *Technologia Tekstilnoy Promyshlennosty*, 1993(2) 47–50.
61. Primachenko, B.M., S.V. Lomov, V.V. Lemeshkov, O.P. Petrova and D.P. Pizvanova, Computer aided design of multilayered woven structures, part 3. *Technologia Tekstilnoy Promyshlennosty*, 1993(3) 42–45.
62. Lomov, S.V. and A.V. Gusakov, Modellirung von drei-dimensionalen gewebe Strukturen. *Technische Textilen*, 1995. **38** 20–21.
63. Lomov, S.V. and N.N. Truevtzev, A software package for the prediction of woven fabrics geometrical and mechanical properties. *Fibres & Textiles in Eastern Europe*,

1995. **3**(2) 49–52.
64. Lomov, S.V., I. Verpoest and F. Robitaille, 'Manufacturing and internal geometry of textiles', in *Design and manufacture of textile composites*, A. Long, Editor. 2005, Woodhead Publishing Ltd. 1–60.
65. Parnas, R.S., J.G. Howard, T.L. Luce and S.G. Adwani, Permeability characterisation. Part 1 A proposed standard reference fabric for permeability. *Polymer Composites*, 1995. **16**(6) 429–445.
66. Olofsson, B., A general model of a fabric as a geometric-mechanical structure. *Journal of the Textile Institute*, 1964. **55**(11) T541–T557.
67. Daniel, J.L., D. Soulat and P. Boisse, 'Shear and tension stiffness influence in composites forming modelling', in *Proceedings ESAFORM-2004*. 2004 Trondheim. 301–304.
68. Lomov, S.V. and I. Verpoest, Model of shear of woven fabric and parametric description of shear resistance of glass woven reinforcements. *Composites Science and Technology*, 2006. **66** 919–933.
69. Peng, X.Q., J. Cao, J. Chen, P. Xue, D.S. Lussier and L. Liu, Experimental and numerical analysis on normalisation of picture frame tests for composite materials. *Composites Science and Technology*, 2004. **64**(1) 11–21.
70. Harrison, P., M.J. Clifford, A.C. Long and C.D. Rudd, 'A micro-mechanical approach to stress-prediction during shear for woven continuous fibre-reinforced impregnated composites', in *Proceedings of the 5th International ESAFORM Conference on Material Forming*, 2002 Krakow. 275–278.
71. Lomov, S.V., A. Willems, M. Barburski, T. Stoilova and I. Verpoest, 'Strain field in the picture frame test large and small scale optical measurements', in *Proceedings of the 8th ESAFORM Conference on Material Forming*. 2005 Cluj-Napoca. 935–938.
72. Leaf, G.A.V. and A.M.F. Sheta, The initial shear modulus of plain-woven fabrics. *Journal of the Textile Institute*, 1984. **75**(3) 157–183.
73. Lomov, S.V., 'Prediction of geometry and mechanical properties of woven technical fabrics with mathematical modelling', in *Dept. Mechanical Technology of Fibrous Materials*. 1995, SPbSUTD St. Petersburg.
74. Belov, E.B., S.V. Lomov, N.N. Truevtsev, M.S. Bradshaw and R.J. Harwood, Study of yarn snarling. Part I Critical parameters of snarling. *Journal of the Textile Institute, Part 1 Fibre Science and Textile Technology*, 2002. **93**(4) 341–365.
75. Morton, W.E. and D.W.S. Hearle, *Physical properties of textile fibres*, The Textile Institute, Manchester, 1993.
76. Lomov, S.V., G. Huysmans, Y. Luo, A. Prodromou, I. Verpoest and A.V. Gusakov. 'Textile geometry preprocessor for meso-mechanical and permeability modelling of textile composites', in *9th European Conference on Composite Materials (ECCM-9)*. 2000. Brighton IOM Communications.
77. Verpoest, I., G. Huysmans, Y. Luo, R.S. Parnas, A. Prodromou and S.V. Lomov, 'An integrated modelling strategy for processing and properties of textile composites', in *Proceedings of the 46th International SAMPE symposium and exhibition May 6th–10th*. 2001 Long Beach, California. 2472–2483.
78. Lomov, S.V., A. Willems, I. Verpoest, Y. Zhu, M. Barburski and T. Stoilova, Picture frame of woven fabrics with a full-field strain registration. *Textile Research Journal*, 2006. **76**(3) 243–252.

5
Optimization of composite forming

W-R YU, Seoul National University, Korea

5.1 Introduction

This chapter is concerned with the optimization of composite forming, especially sheet forming, a manufacturing process that transforms a flat sheet of composite or preform into a three-dimensional shape. Over the last decade, the process of automated sheet forming for composites has emerged from the metal forming process, making use of existing science and technology, and has overcome a prolonged manufacturing process for composites. In the design of the forming process it is vital to predict the deformation behavior of composite sheets, since this helps to prevent defects that can develop as a result of forming tools interacting with the composite materials. Numerous studies have been carried out to develop mechanics models (e.g., Yu, 2002, 2005) that simulate the deformation behavior of composite sheets, and to combine these simulations to analyze the forming behavior of composite sheets. Such models aim to optimize the forming process for manufacturing composite parts with the minimum of defects.

This chapter provides the reader with an overview of the current state of composite forming optimization studies. Optimization methodology in itself is a well documented topic (Luenberger, 1989) and has been applied to many fields ranging from product design and performance to manufacturing processes. In general, the optimization of composite forming includes a cost analysis for materials, energy, tooling and machinery, and a performance analysis, based on variables such as high strength and stiffness considering the minimum weight and thickness of the product. However, this chapter focuses on the optimization of process parameters which can form composite sheets into complex three-dimensional parts with no, or minimal defects. In addition, since many processes are frequently too time consuming or too expensive for optimization based on physical experimentation, researchers have increasingly relied on mathematical models to simulate these processes. Advances in computational power have enabled more extensive use of such models. Therefore, the current chapter discusses optimization of composite forming using models that rely on computational experiments.

118   Composites forming technologies

Since the literature on optimization studies of composite forming is still limited compared to other manufacturing fields such as sheet metal forming, this chapter discusses optimization strategies adopted in sheet metal forming, which may be a good starting point for analyzing optimization strategies for composite forming. In Section 5.2, general aspects of optimization methods are discussed briefly to help the reader understand the mathematical aspect of optimization methodology and identify the limitations of directly applying well-known methods of optimization, such as the downhill simplex and conjugate gradient method, to computational simulation experiments. Section 5.3 discusses objective functions for the optimization of composite forming and the methodological basis for this.

5.2   General aspects of optimization

5.2.1   Optimization methods

This subsection describes the mathematical aspects of optimization methodologies in order to set the basis for the subsequent discussion in this chapter. A thorough explanation of these methods can be found in many text books (Luenberger, 1989; Press, 1992).

The first step in any optimization process is to determine an objective function f that depends on one or more independent variables. The objective function can represent any quantities pertinent to the relevant process, such as manufacturing expenditure, cost and duration, physical properties such as weight and thickness, mechanical performance variables such as strength and modulus, and qualities of the formed product such as wrinkling and defects. The optimization task involves obtaining suitable values for the independent variables where the value of f is maximized or minimized. The tasks of maximization and minimization involve the same process, since the minimization of function f is equal to maximization of $-f$. The objective function can be expressed in an analytic form, although in practice it is often comprised of discrete data which necessitates transformation of the data into a function of independent variables using a regression model. The objective function is very often assumed to be an unknown black box scenario with independent input variables.

After the objective function is defined, the optimization procedure expedites the search for independent variables sets where the objective function has a minimum value. Many algorithms for effective searches have been developed. These algorithms are categorized into two methodologies; first those that rely on function value only, such as the downhill simplex (DS) method; and secondly those that rely on such additional information as the gradient of the objective function, as is the case with the conjugate gradient (CG) method. There follows a brief discussion of the two methods (DS and CG) to understand the iterative

Optimization of composite forming   119

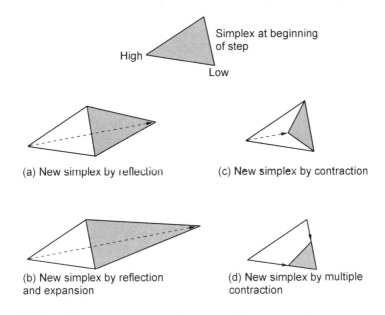

5.1 Possible outcomes for a step in the downhill simplex method.

nature of optimization as well as to help to choose a proper method for process optimization using computational experiments.

Since the DS method requires function evaluation only without derivative information, it may be not very efficient in a number of the function evaluations. However, the DS method is frequently used in optimization procedures because it directly searches for the minimum value of the objective function. A 'simplex' is a geometrical figure consisting, in N dimensions, of N+1 points and all their interconnecting line segments (polygonal faces). In two dimensions, a simplex is a triangle. The search process is described by the successive formation of a new triangle. The search algorithm starts with the function evaluation of any three design points. It is natural to move the highest point in the triangle to another one when searching for the minimum value of a function. The algorithm for the new triangle can be built in numerous ways, although four strategies in particular are usually adopted, and these are shown in Fig. 5.1 (Press, 1992). Visualization of the search process for an analytical function is helpful clearly to understand an optimization scheme. The Rosenbrock function is chosen for the purposes of this discussion.

The Rosenbrock function is characterized by one long curved valley as given in Equation (5.1). See Fig. 5.2 for the mesh and contour plot for the two-dimensional case (i.e., $n = 2$).

$$f(x) = \sum_{n=1}^{n-1} \left(100(x_{i+1} - x_i^2)^2 + (x_i)^2 \right) \qquad 5.1$$

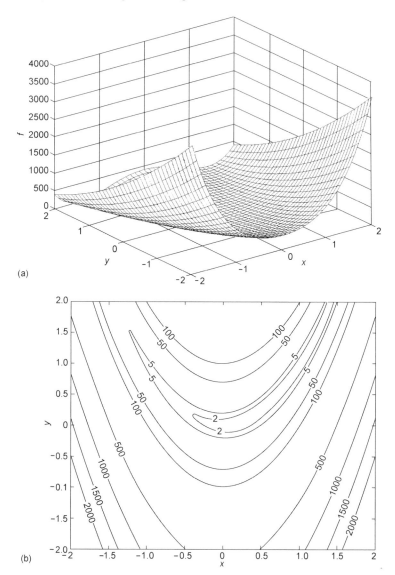

5.2 Rosenbrock functions in two dimensions.

Implementation of the DS method provides the basic knowledge for optimization. The details for computer implementation can be found in many text books (e.g., Press, 1992) and optimization (or minimization) results for the Rosenbrock function are shown in Fig. 5.3. Convergence speed is an important feature for evaluating the performance of an optimization method since slow convergence may indicate that the optimization method requires more function evaluations than other methods, thereby increasing the costs of the optimization procedure.

Optimization of composite forming 121

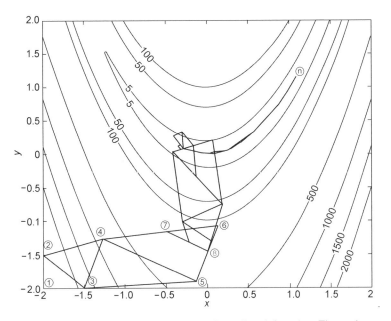

5.3 Implementation of DS for minimizing Rosenbrock function. The optimum (minimum) point is (1,1). Number 4 in the circle is constructed by reflecting number 1, and by forming a new simplex consisting of 2, 3, and 4 points. Numbers 5 and 6 are also generated by reflecting and expansion, while numbers 7 and 8 are searched by multiple contraction, and so on.

As shown in Fig. 5.4, the convergence speed of the DS method is very slow, implying that the method may not be cost-effective.

For process optimization procedures that use computational experiments, it is important to reduce the number of simulations because a single simulation may need a computation time of the order of hours, or sometimes even days, on a modern workstation. Furthermore, parallel simulations, which enable different sets of process parameters to be evaluated on separate computers simultaneously, can be incorporated into optimization schemes to reduce the actual time for performing the optimization procedure. In this regard, the first N+1 points for the first simplex can be obtained using parallel simulation, but the successive points for the new simplex are dependent on the previous results, and cannot benefit from parallel simulation as a result. There are two ways of reducing the computation time of process optimization; one way is to actually reduce the number of simulations, as is the case with the conjugated gradient (CG) method; another way is to utilize parallel simulations that are performed on a clustered computer system.

The conjugated gradient (CG) method requires not just the value of the objective function $f(\mathbf{P})$ at a given N-dimensional point $\mathbf{P}$, but also the gradient $\nabla f(\mathbf{P})$. The CG method consists of setting up a direction on which the

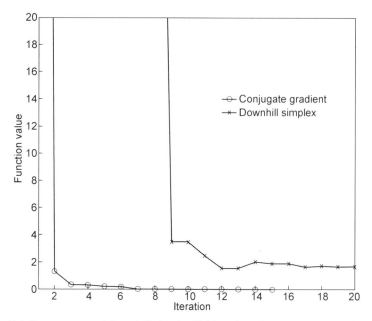

5.4 Convergence of downhill simplex(DS) and conjugate gradient (CG) for the Rosenbrock function in the two dimensional case.

minimization process is performed using a line search. For such directions, a coordinate axis, such as, for example, (1,0) and (0,1) in a two-dimensional case, can be chosen. However, successive minimization along coordinate directions is known to be extremely inefficient since it can take many steps to get to the minimum value through repeated crossing and re-crossing of coordinate directions. To avoid the crossing and re-crossing of a direction, the GC method mathematically defines a 'non-interfering' direction (the so called 'conjugate direction'), along which the minimization process is not spoiled by subsequent minimization along another direction.

Suppose that a movement takes place along some direction **u** to a minimum, and it is now proposed to move along a new direction **v**. The condition that a motion along direction **v** does not spoil the minimization along direction **u** may be that the gradient at the previous minimum point in direction **u** stays perpendicular to **u** because the gradient determines the new direction to be searched. This condition can be mathematically expressed, with x_1 and x_2 being design points (parameter sets) on the direction **v**:

$$\nabla f(\mathbf{x}_1) \cdot \mathbf{u} = 0, \nabla f(\mathbf{x}_2) \cdot \mathbf{u} = 0 \qquad 5.2$$

$$(\nabla f(\mathbf{x}_2) - \nabla f(\mathbf{x}_1)) \cdot \mathbf{u} = \delta(\nabla f) \cdot \mathbf{u} = 0 \qquad 5.3$$

Here, the variation of the gradient may be approximated by the following Equation (5.4):

$$\delta(\nabla f) = \frac{\partial^2 f}{\partial x_i \partial x_j} \delta x_j \mathbf{e}_i = \mathbf{A} \cdot \mathbf{v} \qquad 5.4$$

Finally, the 'non-interfering' condition can be written as the following Equation (5.5):

$$\mathbf{u} \cdot \mathbf{A} \cdot \mathbf{v} = 0 \qquad 5.5$$

When Equation (5.5) holds for two vectors **u** and **v**, it can be said to be conjugate. Therefore, conjugate gradient methods consist of first determining successive conjugate directions and then performing a one-dimensional minimization process along the conjugate direction. For one-dimensional minimization (line search), the 'Golden section method' and 'Brent's method' without or with information of the first derivative of function are frequently used. Furthermore, since the conjugate direction can be evaluated using the first derivative of function (i.e., gradient $\nabla f(\mathbf{x})$) without calculating the second derivative of function as in Equation (5.4) (Luenberger, 1989), this optimization method is called the conjugate 'gradient' method. The searching process of CG method along a new conjugate direction is illustrated in Fig. 5.5 for the two-dimensional Rosenbrock function. As demonstrated in Fig. 5.4, the CG method

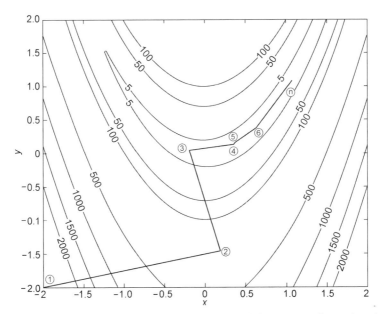

5.5 Implementation of CG method for minimizing a two-dimensional Rosenbrock function. The optimum (minimum) point for the function is (1,1). The first step is to set a conjugate direction on which minimization (line minimization) is performed to generate the number 2. The second conjugate direction at the number 2 is also determined and line minimization along the direction is performed. This procedure continues until successive points get to the minimum point.

is a more effective optimization method than the DS method since it requires a smaller number of function evaluations.

When the CG method is applied to optimization problems using computational experiments, parallel simulations need to be utilized to reduce the optimization time. When determining new conjugate directions, the CG method needs the gradient of function which can be calculated using parallel simulations. In addition, line minimization can be performed based on function evaluations that are provided using parallel simulations. Though a parallel simulation scheme may be incorporated, CG methods may need many numbers of iterations for successive searches, thereby increasing the optimization time. A surrogate model can be used as an optimizer in order to benefit from parallel simulations using computer modeling techniques. This is discussed below.

5.2.2 Optimization using a surrogate model

Process optimization can be performed using a surrogate model (Jakumeit, 2005; Chengzhi, 2005). The surrogate model is referred to an approximate model fitting sample data. In this optimization strategy, the first step is to determine the parameter sets, $\mathbf{X}$ (design points), for which the simulation is performed. In the second step, the objective value $f(\mathbf{X})$ at the design points is used to build a surrogate model as an approximation to the actual data. The third step is to search for the minimum (or optimum) function of the surrogate model. Since the surrogate model approximating sample data is an analytic function, the third step of minimization can be performed quickly using the DS or CG method discussed in Section 5.2.1. If the new optimum is close to the last optimum (convergence) and if the estimate of the surrogate model is close to the results obtained by following a complete simulation, the algorithm then stops, the assumption being that the minimum point has been reached. Otherwise, the surrogate model is improved in a further iteration by adding additional simulations at the new points or by reducing the modeled region. This approach has an advantage over the DS or CG method in that it can maximize the benefit of the parallel simulations.

The responsive surface method (RSM) is a common method of building a surrogate model (an approximate model) in which a polynomial function of varying order (usually a quadratic function) is fitted to sample data using a technique of least square regression (Myers, 2002). It can provide an explicit functional representation of the sampled data which is obtained experimentally or computationally. However, RSM has several limitations. It is unable to predict multiple extrema. In addition, RSM was originally developed to model data obtained from physical experiments which had a random error distribution. Since computational experimentation is a deterministic process (the input produces the same result), in statistical terminology random errors are not present in computational experiments; RSM may not be a proper model for approximating deterministic (computational) data (Sacks, 1989). Researchers have developed the Kriging method to overcome this limitation (Sacks, 1989).

The Kriging method is very similar to RSM but provides an interpolation modeling technique to approximate the results of the deterministic computational experiments. It consists of the sum of many local functions (frequently the Gauss function) and one global function, which may be a polynomial as in RSM. The local function is used to bridge the difference between the global function and the simulation results using a maximum likelihood estimate, which allows for capturing multiple local extrema. A more detailed description of the Kriging method can be found in the work of Sacks (1989). The implementation of the Kriging method can be carried out using the Matlab Kriging Toolbox which is open to the public (Lophaven, 2002).

Figure 5.6 shows the response surface for a two-dimensional Rosenbrock function, which was constructed using the Kriging method. The optimization is now performed easily on the response surface that is analytically expressed, so that any minimization algorithm (DS or CG method) can be utilized. Here, a couple of issues arise when considering the Kriging response surface as an optimizer. The first issue concerns the number of design points (parameter sets), an issue is directly related to the number of simulations. Simply stated, the answer may be 'the more the better' since interpolation in the surrogate model is improved as the volume of data increases. However, it is not possible to increase the volume of data infinitely. Thus, a proper number of design points should be set up for the first time (e.g., eight points) which can be determined by the

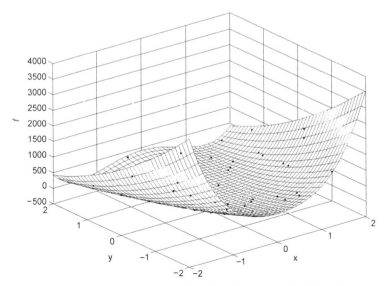

5.6 Response surface constructed using Kriging method for the two-dimensional Rosenbrock function.

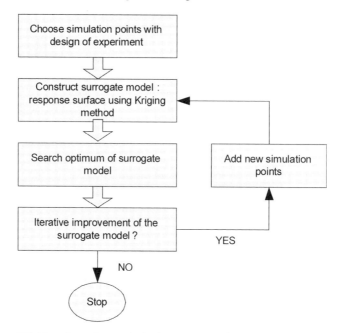

5.7 Flowchart of an optimization strategy using surrogate models.

parallel simulation capability. The optimization is then performed on the surface that was constructed based on the first design points. If the result is not good enough, the surrogate model can be improved by adding new design points with which a new simulation can then be done. Such addition of new design points can continue until it is no longer necessary to improve the response surface. Such an algorithm is depicted in Fig. 5.7. Conceptually, this algorithm is not difficult to understand for implementation. However, many algorithms exist in practice and this indicates that a solid strategy for moving the design boundary and for adding new design points during optimization should be established for its wide application to composite forming.

5.3 Optimization of composite forming

This section describes objective functions and actual optimization methods for composite forming. The first step in optimization study is to identify the objective function. In general, several different objectives may compete with each other, such as, for example, the manufacturing time and cost, and the quality or performance of a formed product. Section 5.3.1 discusses objective functions and provides a brief explanation of the composite forming process. The section then focuses on the process optimization of composite forming, including preform shape and the blank holding force (BHF).

5.3.1 Objective functions in composite forming

In composite forming, various processing parameters are pertinent to the forming process. For thermoplastic and thermoset prepreg forming, the relevant parameters can be classified as heat related, external force related, and material shape related parameters. In the preform forming, heat related factors are not relevant. As thermoforming methods for thermoplastic prepreg, the diaphragm (Delaloye, 1995) and stamp forming (Hou, 1994) processes are well known. Both forming processes consist of three stages; first the stage of heating thermoplastic prepreg to a temperature for matrix flow, secondly the stage of shaping and consolidating heated and flat prepreg into complex and curved parts, and finally the cooling stage. As such, the thermoforming process needs significant processing time for all the three stages, so that the first objective of the thermoforming process may be to reduce the cycle time. Since the majority of process time is related to the heating of prepreg, an objective for process optimization may be to pursue a suitable heating method that can elevate the temperature of prepreg as rapidly as possible. This optimization problem can be solved by developing a heating unit that uses convection, conduction, or infrared or a combination of these methods. Therefore, this optimization may be purely related to the development of a mechanical heating unit.

The last stage of the thermoforming process, which will be discussed briefly before the second stage of thermoforming is discussed in some detail, is the cooling stage where two objectives compete for the optimization of the cooling process. One objective is to allow the correct molecular arrangement for the crystallization of the matrix and another is to shorten the cooling time in order to reduce the cycle time. The optimization problem at this stage can be stated as 'minimizing the cooling time while allowing the resin matrix to orient and crystallizes. The molecular movement and crystallization can be assessed as a function of the cooling time and its profile in order to solve this optimization problem.

In the second stage of thermoforming, the mechanical force by pressure (or a vacuum) in the diaphragm or of the punch in the stamp forming is applied to the heated and flat prepreg, shaping it into curved and complex parts. The fiber reinforcement, especially continuous fiber, experiences compressive stress and fibers may kink or buckle unless a mechanism is adopted that releases this compressive stress. For the release of compressive force, a resin matrix allows fibers to slip within a viscous media, and this requires that the material temperature is maintained high enough to allow matrix flow. As such, the shortening of processing time for the shaping, as well as for minimizing the heat dissipation, may be necessary. In this respect, the objective in process optimization may be 'to find a proper pressure or stamp velocity for maintaining matrix flow', and this optimization can be solved by measuring or calculating heat transfer within prepreg as a function of the stamp or pressure velocity.

Another useful method to prevent or relax compressive force is to apply tension to the prepreg during shaping. In the diaphragm forming process, the biaxial extension of the flexible diaphragm can impart tension to the prepreg, the extent of which is determined by the material properties of the diaphragm. In the stamp forming process, a blank holding force (BHF) is introduced through a blank holder, and this offers a more flexible mechanism for tension control. Therefore, the processing variable to be optimized in the shaping stages is mechanical force and its profile, the extent of which can be determined through a selection of diaphragm material or BHF.

The quality of a formed product can be characterized in terms of its thickness, uniformity, fiber orientation, and by defects such as fiber breakage and in-plane and out-of-plane buckling. The optimization problem in the shaping stage can be rephrased in terms of 'determining the processing variables (mechanical shaping force and BHF) to minimize thickness diversity, fiber orientation diversity, fiber breakage and wrinkling'. As more than one objective function is involved, the optimization of the shaping process is by nature a multi-objective optimization problem. Since the exact or analytic relationship between such objectives and process variables is difficult, or even at times impossible, to derive, the relationship is characterized as a 'black box' relationship, indicating that actual values of the objective functions are only obtained by virtual (computational) or actual experiments. Each objective function in composite forming is described below.

Thickness diversity may be expressed by, for instance, Equation (5.6) as follows:

$$H = \sum_{i=1}(h_i - \bar{h})^2 \qquad 5.6$$

where i indicates material points, and h_i and $\bar{h}$ are its thickness and the average of the thickness, respectively. The degree of fiber orientation distribution in the formed parts is expressed in Equation (5.7) as follows:

$$\Theta = \sum_{i=1}(\theta_i - \bar{\theta})^2 \qquad 5.7$$

where i represents material points, and θ_i and $\bar{\theta}$ are the fiber angle and the average fiber angle in the formed part.

The objective function for fiber breakage is described by fiber strain, providing a constraint for maximum allowable strain expressed in Equation (5.8) as follows:

$$\epsilon_f \leq \epsilon_{max} \quad \text{or} \quad \Xi = (\epsilon_f - \epsilon_{max}) \qquad 5.8$$

where ϵ_f and ϵ_{max} are fiber strain in prepreg and the breaking strain of fiber respectively.

An objective function for representing wrinkling or buckling is more complex because it needs a criterion that predicts its occurrence in formed parts. Little literature has been dedicated to developing a wrinkling criterion. For woven structures, wrinkling was reported to occur as the fiber angle approaches a specific angle (jamming angle) (Prodromou, 1997). Certainly, this may be true in that the main deformation mode in the woven reinforced prepreg or preform is intra-shearing, and the yarns may buckle out as the shear deformation is not possible due to warp and weft fiber touching each other. This is because buckling facilitates the release of build-up stresses. In this case, the objective function for wrinkling minimization may be as set out in Equation (5.9):

$$\Psi = \sum_i (\theta_i - \theta_{jam})^2, \theta_i > \theta_{jam}, \text{ otherwise } \Psi = 0 \qquad 5.9$$

where i represents material points, and θ_i and θ_{jam} are the fiber angle and the jamming angle. However, the above criterion may be not applicable to the unidirectional lamina and other knitted reinforcements. In addition, wrinkling can be observed even where the fiber angle between the warp and weft does not reach the jamming angle, implying that the jamming angle is not the only mechanism for wrinkling.

The optimization procedure can be performed by searching the minimum of those objectives. The multi-objectives set out in Equations (5.6) to (5.9) can be translated into a single objective function using the following weight factors set out in Equation (5.10):

$$f = \rho_i H + \rho_2 \Theta + \rho_3 \Xi + \rho_4 \Psi \qquad 5.10$$

where ρ_i are weight factors for each objective.

The design set of minimizing Equation (5.10) can be obtained using one of the optimization methods described in Section 5.2. Since it is not unusual that the systematic determination of the weight factors is not possible, the single objective function in Equation (5.10) may be not proper; instead, a multi-objective optimization method may be more appropriate.

It may be useful to review objective functions in other fields of forming. The quality of a formed sheet of metal is assessed by severe thinning, which may cause tearing, and by wrinkling. Since sheet metal forming involves the stretching of thin sheets, the principal strains are used to predict where such defects in a forming operation occur. Using experiments or theoretical calculations, the region in which such defects may occur can be identified in terms of principal strains, the limits of which can be expressed in a forming a limit diagram as shown in Fig. 5.8. The optimization process in this case is to 'pursue a processing path and a preform shape in order that one of the material points does not experience strains beyond the forming limit'. Mathematically, the optimization process is equivalent to the minimization of a potential representing the forming limit.

Composite forming has also adopted the technique of a similar forming limit diagram. Researchers (Tucker, 1998) have studied a forming limit diagram for

130   Composites forming technologies

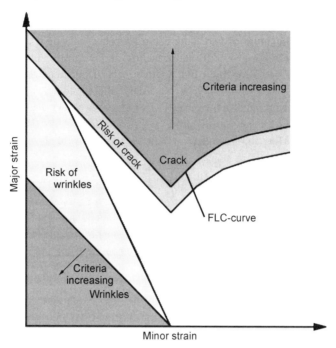

5.8 Schematic formability limit diagram for sheet metal with critical regions for crack, risk of crack, risk of wrinkles and wrinkles (Jakumeit, 2005).

the random fiber mat to develop a shape optimization method. The key premise in this forming limit diagram was that the defects of the forming random mat preform can be quantified with a limit diagram obtained by multi-state loading experiments as shown in Fig. 5.9. The forming limit diagram divides the strain space represented by principal stretches, λ_1 and λ_2, into regions where defects (wrinkle or tears) are likely to occur, and into regions of acceptable deformation without such defects. Several experiments are performed to evaluate the points U, P, and B which mark the boundaries of the acceptable regions of principal strain. Points U, P and B can be determined from a uniaxial elongation test, a pure shear test, and an equi-biaxial stretch test respectively. The boundary PB is found from unequal biaxial stretching experiments performed using various ratios of λ_1 to λ_2. More details can be found in a research paper by Tucker (2003). In this case, the optimization of the forming process can be defined as 'finding process parameters that keep the local strain at every point in the preform away from the boundaries UO and UPB'.

This section has outlined various objective functions involved in composite forming, in particular for thermoplastic composite forming. For preform forming without a resin matrix, the considerations pertinent to the first and third stages of the process are not necessary because preform forming without a resin matrix does not involve heat. Material flow or deformation is dependent on preform

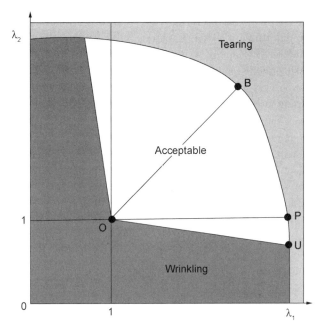

5.9 Qualitative forming limit diagram, showing regions of acceptable deformation, wrinkling, and tearing for random fiber mat (Tucker, 1998).

shape as well as other processing conditions. Both parameters should be optimized for high quality composite forming, and concurrent optimization may be possible using schemes set out in Equation (5.10), adding a shape parameter for preform shape, such as, for example, the coefficients of a parametric curve representing preform shape. However, a direct method for preform (prepreg) shape optimization can be effectively formulated. A separate scheme for optimizing both processing parameters is discussed in Sections 5.3.2 and 5.3.3.

5.3.2 Shape optimization using a direct method

The initial preform (blank) shape is known to determine the deformation behavior of preform in forming. In the metal forming field, researchers have performed many studies to design an initial blank shape with optimum deformation for a specific product shape (e.g., Chung, 1992a, 1992b, 1997). The basic aim of these studies is to define a path which produces a desired homogeneous deformation with minimum plastic work. It is also assumed in the studies that the formability of local material elements is optimum when they deform following a minimum work path. An ideal process is then defined as the one having local deformations that are optimally distributed in the final shape, through which an ideal initial configuration is determined to form a specified final shape.

In composite forming, the preform shape is also an important element in determining the deformation shape. Experimental observation representing the importance of the blank shape design was reported by O'Bradaigh (1997) who formed several specimens modified from a square blank (APC-2 carbon/PEEK) by cutting some corners and showed improved wrinkling behavior. By using a diaphragm forming process a significant and positive correlation was found between the shape of a blank and the wrinkling behaviour. Simulation results have also demonstrated the effect of preform shape on the forming behavior (Yu, 2005). As such, preform shape is an object to be optimized in the composite forming process. A method of optimally designing a blank is now discussed.

In composite forming, limited research has been conducted on the blank design for forming a specified product. One research study (Tucker, 2003) performed an ideal forming analysis for a random fiber reinforcement by utilizing the forming limit diagram. The researchers in this study determined a potential representing the formability of the reinforcement obtained by measuring the stretching behavior of the preform along several deformation paths and then by setting an upper forming limit as shown in Equation (5.11):

$$W = (\lambda_1 - 1)^2 + (\lambda_2 - 1)^2 + \frac{\alpha}{2}\mathbf{c} \cdot \mathbf{c} \qquad 5.11$$

where λ_1 and λ_2 are principal stretches, and α and $\mathbf{c}$ are penalty constant and a vector describing the forming limit in Fig. 5.9.

The first two terms on the right-hand side in Equation (5.11) keep the deformation state as close to the undeformed state as possible, assuming that defects are directly linked to the deformation. This assumption is only valid for composite preform because in sheet metal forming material needs to be stretched to impart work-hardened behavior. The third term in Equation (5.11) is a quadratic penalty function to enforce the forming limits of materials, which were obtained experimentally. The constraint vector $\mathbf{c}$ contains those constraints that are violated, while α is a non-negative penalty parameter controlling the strength with which the forming limits are enforced. The researchers determined a forming limit diagram for Vetrotex CertainTeed's U750 mat experimentally as shown in Fig. 5.10. The potential was then minimized to seek an initial blank shape based on the premise that an initial configuration is ideal if the deformation of all elements within the initial blank resides within the forming limit. The solution procedure for this method is explained briefly here following the works of Chung (1992a, 1992b, 1997) and Tucker (2003).

Consider that a final shape to be formed is characterized by a surface description, for example $x_3 = f(x_1, x_2)$. Optimization for the blank design can be summarized as a task of seeking an initial blank shape in terms of X_1 and X_2 if the initial configuration is assumed as a flat sheet (i.e., $X_3 = 0$). Alternatively, the final shape is conveniently described by a finite element mesh fixed in the global coordinates (x_1, x_2, x_3) using surface coordinates. For the finite element

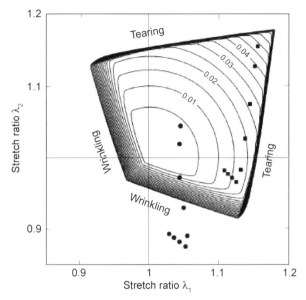

5.10 Contour plot of the formability function W for 750U random-fiber mat where $\alpha = 100$ (Tucker, 2003).

meshes given, the blank shape optimization process involves inversely calculating the initial coordinates of nodes of all elements in the final mesh. Note that the unknown variables are X_1 and X_2 (in the initial configuration), while general FE analysis aims to obtain x_1, x_2 and x_3 based on X_1 and X_2. A mapping function is introduced to calculate the initial configuration of the elements. The mapping function includes non-physical energy representing deviation from the forming limit, and can be expressed using the forming limit diagram formula set out in Equation (5.12) as follows:

$$\Pi = \int (W(\lambda_1, \lambda_2)) dV^0 \qquad 5.12$$

Equation (5.12) can be expressed with the final configuration using Jacobian between the initial and final configuration, as follows:

$$\Pi = \int (\tilde{W}(\lambda_1, \lambda_2)) dV \qquad 5.13$$

where $\tilde{W}(\lambda_1, \lambda_2) = \dfrac{1}{\lambda_1 \lambda_2} W(\lambda_1, \lambda_2)$.

The stretch is determined when the initial (X_1 and X_2) and final (x_1, x_2 and x_3) configurations are given. Since in this case the final shape is given, the stretch is a function of the initial coordinate. Since the mapping function represents a high value for a portion of the blank which falls on the outside of the forming limit, the optimization of the blank shape is mathematically equivalent to a configuration that makes the mapping function minimum. The minimum of Equation (5.14) can be determined by requiring that the derivative of Π with respect to the initial sheet configuration be zero, as follows:

$$\frac{d\Pi}{d\mathbf{X}} = 0 \qquad 5.14$$

where $\mathbf{X} = (X_1, X_2)$. Numerical simulation is performed by discretizing Equation (5.13) using finite element approximation and by translating Equation (5.14) as an *n*-dimensional equation (where *n* equals the total number of nodal points in the final and initial mesh). The Newton-Raphson (NR) method is employed to find the initial coordinates of element nodes corresponding to the final one. Since the NR solution will converge only when a good initial guess for the solution variables is provided, it is critical to choose a proper initial guess. The easiest way for determining the initial shape is to project all the nodes on the final surface onto the X-Y plane. However, this may make it difficult to converge to a solution, particularly for deep curved structures. For the two fiber directional preform, such as a woven structure, a flat pattern can be obtained from the iterative energy minimization code (Long, 2002), which can be tried for the initial guess.

Figure 5.11 shows the final shape (box with flange preform) and the initial blank shape as optimized using the direct method. As shown in Fig. 5.11, large values of the formability function are observed in both the upper corner and in the adjacent flange area, indicating possible defects in these two regions. The optimal preform obtained in this work may not conform to the final shape in real forming because this simulation does not consider processing conditions such as the blank holding force and friction between tools and preform. Iterative simulation could be further performed to include these processing conditions, adjusting the initial blank shape to be more conformable to the final shape (Park, 1999). Recently, a new formulation considering the blank holding force and friction has been developed in the metal forming field and this can also be used in composite forming (Chung, 2000).

In this section, the shape optimization method, which does not rely on manual input to drive the successive iteration for optimization, is discussed. To employ this kind of optimization strategy for optimum blank design, a mapping function, which can be based on formability or deformation theory (Chung, 1993) for sheet metal, should be developed. For the continuous fiber preform, such a mapping function may be formulated based on fiber angle limit, wrinkling criterion, or fiber breakage.

5.3.3 Optimization of process parameters using iterative optimization method

In the previous section, the optimization of preform shape, one of the important process parameters, was discussed. Since other processing parameters are

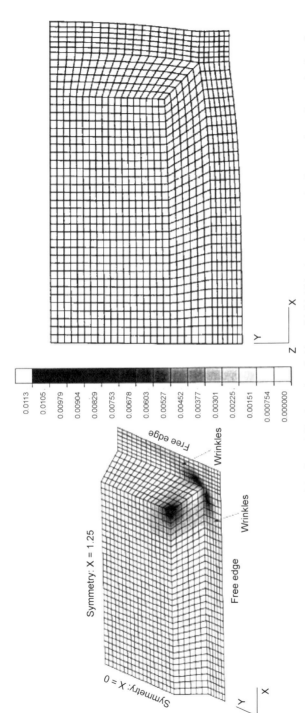

5.11 Target shape with contour plot of the formability function (left) and initial blank preform required to produce the optimal preform (right) for the target shape (Tucker, 2003).

136   Composites forming technologies

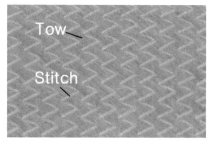
(a) Technical face

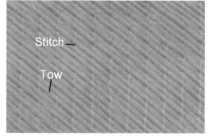
(b) Technical back

5.12 Surface structure of an NCF with tricot stitch.

involved in composite forming besides preform shape, the quality of a formed product can be improved by properly choosing processing parameters. Among the many processing parameters, BHF is recognized as an important factor for composite preform because less tension causes wrinkling in the preform due to compressive stress, while too much tension can bring about fiber breakage. In contrast to the preform shape optimization process, BHF can be optimized using the iterative optimization method explained in Section 5.2. Through experimental observations, the effect of BHF on the forming behavior is first discussed.

The forming of non-crimp fabric (NCF) with tricot stitch is chosen as an example for investigating the effect of BHF on the formed shape (Lee, 2006). As shown in Fig. 5.12, the stitch pattern was introduced to hold the warp and weft yarn together in such fabric. Due to the stitch, the fiber angle between the warp and the weft in as-received NCF is a little deviated from 90°, the actual fiber angle measured being 97°.

Stitch pattern and fiber orientation are known to be crucial factors in determining the forming behavior of NCF. To address the effect of these two factors on forming behavior, two square samples of 350 mm × 350 mm in dimension were prepared cutting glass NCF using a die cutter; one sample (TC_N) with the warp and weft aligned parallel to the edge of a square; the other sample (TC_D) with the warp and the weft aligned in the diagonal of a square. On each sample four different axes (ST90, Warp, ST, Weft) were marked on which fiber angle and node displacement were measured at every 15mm interval in order to quantify the deformation behavior of NCF. Note that the ST90 and ST axes are referred to the directions perpendicular and parallel to the stitches respectively. A hemispherical forming system was designed that is attachable to a laboratory UTM (Universal Testing Machine-MTS). Figure 5.13 shows the schematic diagram of the hemispherical forming station employed in this discussion. The radius of the hemisphere punch was 75 mm, and the area of the blank holder was $0.017\,m^2$.

The punch stamps NCF into a hemispherical shape with a depth of 75 mm during one minute. Then, the formed NCF was hardened for 30 minutes using

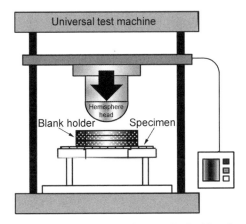

5.13 Schematic diagram of hemispherical forming process.

unsaturated polyester resin. The NCF was removed from the mold in order to measure and characterize the formed shape. The local fiber angle change of the formed NCF was measured by a goniometer, while node displacement was recorded through spatial coordinates obtained using a three-dimensional digitizer. The formed shape of NCF was quantified by measuring the shear angle and arc length along four different axes of the formed NCF. For this forming experiment, 19.4 kg of BHF was applied to both TC_D and TC_N samples. Fiber angles on the warp and weft axes were rarely changed while they were changed considerably on the ST90 and ST directions because compliance to shear deformation in these directions was relatively large. Note that the fiber angle change of ST90 axis was about 30° around the equator where maximum shear deformation was developed while 20° of fiber angle change was observed in the ST direction (see Fig. 5.14). Such an asymmetric shear angle pattern might be caused by stitches because, as observed in the picture-frame shear test (Long, 2002), the stitch structure determines shear resistance according to shear direction.

It is interesting to note that the shear angle of ST and ST90 in TC_N was larger than one of the same direction in the TC_D sample, while the fiber angle in the fiber direction of both samples was little changed. This result clearly shows the effect of blank shape on the forming behavior. If both samples had been prepared in a circular blank shape, the fiber angle changes would have been the same because the fiber orientation and stitches in them were the same. Since a square blank shape was used in this experiment, the excess part of the square had a different fiber orientation; fibers in the TC_N sample were aligned in a ±45° direction based on the square diagonal while fibers in the TC_D sample were aligned diagonally, making the shearing more difficult in the TC_D sample.

Figure 5.15 shows the formed shapes of NCF samples (TC_N and TC_D) according to BHF. Two noticeable features of the BHF effect can be observed in

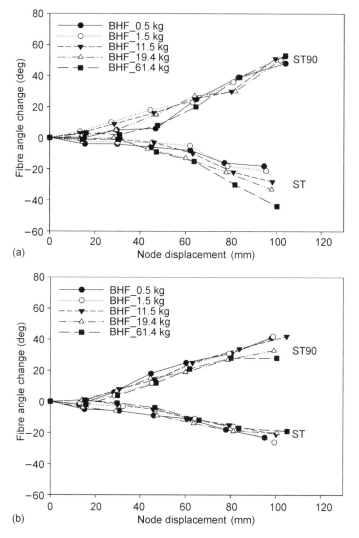

5.14 Fiber angle change in the ST90 and ST directions according to BHF (a) TC_N and (b) TC_D.

both sample cases. For small BHF ((a) and (c) in Fig. 5.15), both samples (TC_N and TC_D) were formed into a hemispherical shape with severe asymmetric deformation and wrinkling around the equator. This was due to insufficient tension applied to the NCFs during forming, in which compression stress was developed in a circumferential direction. As BHF increased, the formed shape changed dramatically into the symmetrical formed shape. From the experiments, it may be concluded first that BHF is a very important factor in determining the forming behavior of NCF and secondly that optimization of BHF may enable a

Optimization of composite forming 139

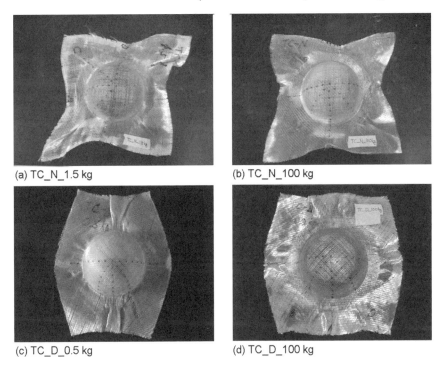

(a) TC_N_1.5 kg (b) TC_N_100 kg
(c) TC_D_0.5 kg (d) TC_D_100 kg

5.15 Formed shapes of NCF (TC_N and TC_D samples).

significant reduction in wrinkling. Wrinkle and in-plane buckling were observed near the equator in the warp and weft fiber direction axes. To clearly investigate wrinkling with high frequency and small magnitude, a slit laser shed light on the surface of the formed sample. As shown in Fig. 5.16, the laser light was more scattered around the equator of a hemispherical formed NCF where the wrinkles were propagated severely. Note that for a region without wrinkles, or with fewer wrinkles, the laser was more visible in a straight line. As BHF increased in the

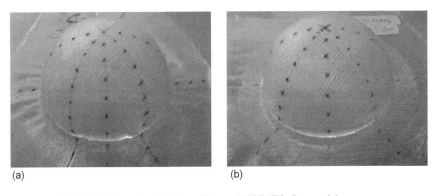

(a) (b)

5.16 Wrinkling visualization of formed NCF (TC_D sample).

140   Composites forming technologies

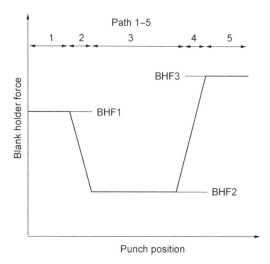

5.17 Figure blank holder force as a function of the drawing depth described by seven parameters (Jakumeit, 2005).

forming operation, laser scattering was reduced which indicated a reduction in wrinkling. With this experiment, it can be confirmed that wrinkling in the formed NCF can be reduced through the optimization of BHF.

As discussed so far, the blank holder force is a crucial factor in determining the forming behavior of composite preform. Optimization of the blank holding force can be performed based on experiments, but systematic optimization of

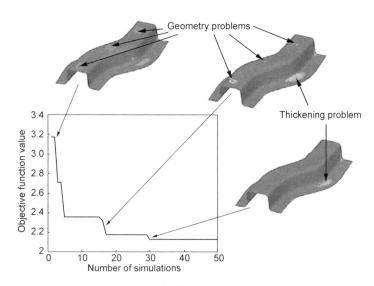

5.18 Convergence behavior of the Kriging strategy for the SRAIL test of a metal sheet (Jakumeit, 2005).

Optimization of composite forming 141

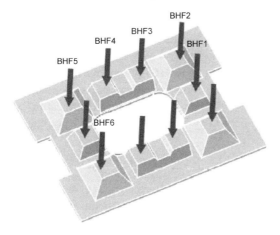

5.19 The schematic diagram of the segmented blank holder (Chengzhi, 2005).

BHF may need to include a variable BHF (Fig. 5.17) and segmented blank holder (Fig. 5.19) whereby numerical simulations or computational experiments are inevitable for including various ranges of blank holding force profiles. In the sheet metal forming field in which the formability function (objective function)

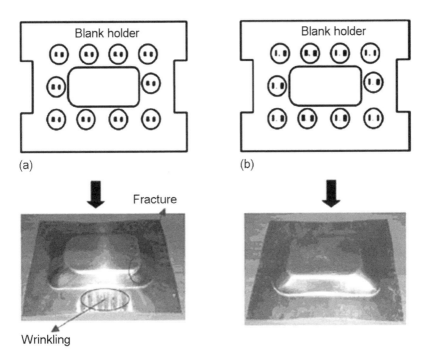

5.20 Stamped sheet metal under initial BHFs (left) and optimal BHFs (right) (Chengzhi, 2005).

was well defined, the optimization of blank holding force has been endeavored to minimize an objective function, which was constructed based on the formability function, using the iterative algorithm described in the previous section. Figures 5.18 and 5.20 demonstrate how well formed parts can be improved by controlling BHF. Since the formability of composite preform has been less established, future study for composite forming optimization should be dedicated to developing an objective function that takes into account fiber angle.

5.4 Conclusions

In this chapter, optimization for composite forming was discussed with objective functions which are essential elements in forming optimization. Among processing parameters in the composite forming process, preform (blank shape) and BHF were identified as essential quantities which should be optimized for better quality of formed preform. Actual methods for optimization were discussed, which can be classified in terms of direct and indirect (iterative) methods. The direct method is to formulate an objective function through a finite approximation of an initial and final blank, and to calculate inversely the initial configuration of preform corresponding to the final one, thereby excluding successive manual inputs to drive an optimization iteration. Though the direct method can save design time because it is not based on trial-and-error practices, it cannot consider actual processing parameters, such as the variable BHF. Indirect (or iterative) methods can be applied to the optimization of processing parameters, though such methods need manual input to drive successive iterations in the optimization process, necessitating an optimization method that can utilize parallel simulations. For the successful implementation of both methods, objective functions should be established that assess the quality of formed parts in terms of processing parameters, and future research should be focused in this direction for the optimization of composite forming.

5.5 References

Chengzhi S, Guanlong C, Zhongqin L (2005), 'Determining the optimum variable blankholder forces using adaptive response surface methodology (ARSM)', *Int J Adv Manuf Technol*, 26, 23–29.

Chung K, Richmond O (1992a), 'Ideal forming-I. Homogeneous deformation with minimum plastic work', *Int J Mech Sci*, 34(7), 575–591.

Chung K, Richmond O (1992b), 'Ideal forming-II. Sheet forming with optimum deformation', *Int J Mech Sci*, 34(8), 617–633.

Chung K, Richmond O (1993), 'A deformation theory of plasticity based on minimum work paths', *Int J Plast*, 9, 907–920.

Chung K, Barlat F, Brem J C, Lege D J, Richmond O (1997), 'Blank shape design for a planar anisotropic sheet based on ideal forming design theory and FEM analysis', *Int J Mech Sci*, 39(1), 105–120.

Chung K, Yoon J W, Richmond O (2000), 'Ideal sheet forming with frictional constraints', *Int J Plast*, 16, 595–610.

Delaloye S, Niedermeier M (1995), 'Optimization of the diaphragm forming process for continuous fiber-reinforced advanced thermoplastic composites', *Compos Manuf*, 6, 135–144.

Hou M, Friedrich K, Scherer R (1994), 'Optimization of stamp forming of thermoplastic composite bends', *Compos Struct*, 27, 154–167.

Jakumeit J, Herdy M, Nitsche M (2005), 'Parameter optimization of the sheet metal forming process using an iterative parallel Kriging algorithm', *Struct Multidisc Optim*, 29, 488–507.

Lee J S, Hong S J, Yu W-R, Kang T J (2006), 'The effect of blank holding force on the stamp forming behavior of non-crimp fabric with a chain stitch', *Compos Sci Technol*, in press.

Long A C, Souter B J, Robitaille F, Rudd C D (2002), 'Effects of fibre architecture on reinforcement fabric deformation', *Plast Rubber Compos*, 31(2), 87–97.

Lophaven S N, Nielsen H B, Sondergaard J (2002), http://www2.imm.dtu.dk/~hbn/dace/.

Luenberger D G (1989), *Linear and Nonlinear Programming*, Massachusetts, Addison-Wesley.

Myers R H, Montgomery D C (2002), *Response Surface Methodology*, New York, John Wiley & Sons.

O'Bradaigh C M, McGuinness G B, McEntee S P (1997). 'Implicit finite element modelling of composites sheet forming processes' in Bhattacharyya D, *Composite Sheet Forming*, Amsterdam, Elsevier, 247–322.

Park S H, Yoon J W, Yang D Y, Kim Y H (1999), 'Optimum blank design in sheet metal forming by the deformation path iteration method', *Int J Mech Sci*, 41, 1217–1232.

Press, W H, Teukolsky, S A, Vetterling W T, Flannery B P (1992), *Numerical Recipes in C*, Cambridge, Cambridge University Press.

Prodromou A G, Chen J (1997), 'On the relationship between shear angle and wrinkling of textile composite preforms', *Compos Part A-Appl S*, 28A, 491–503.

Sacks J, Welch W J, Mitchell T J, Wynn H P (1989), 'Design and analysis of computer experiments', *Statistical Science*, 4(4), 409–435.

Tucker C L, Dessenberger R B (1998), 'Forming limit measurements for random-fiber mat', *Polym Composite*, 19(4), 370–376

Tucker C L, Dessenberger R B (2003), 'Ideal forming analysis for random fiber preforms', *J MANUF SCI E-T ASME*, 125, 146–153.

Yu W R, Pourboghrat F, Chung K, Zampaloni M, Kang T J (2002), 'Non-orthogonal constitutive equation for woven fabric reinforced composites', *Compos Part A-Appl S*, 33, 1095–1105.

Yu W R, Harrison P, Long A (2005), 'Finite element forming simulation for non-crimp fabrics using a non-orthogonal constitutive equation', *Compos Part A-Appl S*, 36, 1079–1093.

6
Simulation of compression moulding to form composites

E SCHMACHTENBERG, Universität Erlangen-Nürnberg, Germany
and K SKRODOLIES, Institut für Kunststoffverarbeitung, Germany

6.1   Introduction

Fibre reinforced compression moulding parts are well established in industry. One can distinguish between two main types of these compounds.

Thermoset compression moulding prepregs (so-called sheet moulding compounds, SMC) were established in the early 1960s. The compression moulding process of these materials starts with the rapid closing of a mould followed by squeezing the prepreg into its final shape. The charge temperature is about room temperature, the mould temperature ranges from 120 to 150°C (Ehrenstein 1992). The temperature of the compound rises during the flow due to the heat transfer from the cavity surfaces. Due to this, the viscosity declines before the cross linking reaction starts (Neitzel and Breuer 1996). Local gradients of pressure in the mould cause different local velocities and fibre orientations during the flow. These orientations affect the anisotropy of the part. Besides the thermal process parameters, it is the flow that has a major influence on the part properties.

Thermoplastic compression moulding compounds (GMT) were invented twenty years after the establishment of SMC. The main difference concerning the process is that these compounds are heated up to melting temperature in a heating station or extruder and put into a 'cold' mould (mould temp. 60–80°C). Due to the withdrawal of heat in the mould the material starts to freeze right after deposition and during the flow in the mould. Thus, the material properties are influenced by the thermal process parameters and the local pressure/velocity gradients, analogously to the thermoset materials.

Hence, the manufacturer of compression moulding parts has a great responsibility to find the optimal process parameters. Therefore, a numeric simulation of the compression moulding process is a very important tool to save time and costs. Only by using this tool is it possible to assess alternative concepts in an early construction phase.

In other areas of plastics processing, e.g. injection moulding, the use of process simulation software is very common. Unlike this, such programs are only little used in the field of compression moulding despite the complexity of the processes and materials. In cooperation between the IKV and M-Base Engineering + Software GmbH, Aachen, the process-simulation software EXPRESS has been developed and constantly been improved (Specker 1990, Heber 1995, Semmler 1998). This program contains several modules: First there was the Newtonian flow simulation, which gives information on the flow in the cavity (Specker 1990). Afterwards other modules were integrated like the fibre orientation computation with an interface to FE-programs. With this, one gets the possibility of a FE-analysis regarding the anisotropic rigidity and the possibility of failure analysis. With this knowledge a calculation of shrinking and warpage of the parts could be done.

A disadvantage of the Newtonian flow is that the material properties (e.g., the temperature dependence of the viscosity) are not described. Therefore, a non Newtonian flow model was integrated (Heber 1995) which is based on a Hele-Shaw flow-law as it is successfully used in various injection moulding simulation programs. With the use of a temperature dependent non Newtonian flow model the local pressure ratios in the cavity can be calculated at each time step. This leads to an exact calculation of the local velocities and the velocity gradients, which are an input for the fibre orientation computation. Furthermore, the local pressure is used to calculate the pressing force and the movement of the centre of force.

These results give the manufacturer an exact knowledge of the whole process. Additional to the existing know-how of the manufacturer further conclusions can be drawn depending on the results of the simulation (Schmachtenberg and Ritter 2002). This opens new prediction possibilities for parts with high requirements on surface quality. Additionally, it can be estimated if the existing presses can be used to manufacture the new parts or if a new investment is necessary.

6.2 Theoretical description of the simulation

6.2.1 Flow and heat transfer simulation

The following section gives an overview of the methods used to simulate the fluid and energy flow during processing. At first the calculation of an isothermal flow will be shown, that will be extended to a non-isothermal flow calculation. This mode can be used for the flow simulation for both thermoplastic and thermoset materials if the freezing of material is taken into account.

With the isothermal flow calculation you have a quantitatively less accurate prediction of the flow behaviour compared to the non-isothermal simulation. However, with the advantage being that computational time is significantly decreased. In the case that the properties of the simulated material are highly temperature dependent and you are concerned with obtaining a more accurate prediction of shrinkage, warpage and the final material properties for the part, a non-isothermal flow calculation is recommended.

Isothermal flow simulation

The simplified forms of the conservation of mass (eq. 6.1) and momentum (eq. 6.2) given below describe the flow of isothermal, Newtonian, incompressible fluids, and serve as the rheological foundations of the flow simulation (Renz 1989).

$$\nabla \bar{v} = 0 \qquad 6.1$$

$$\rho \left(\frac{\partial \bar{v}}{\partial t} + (\bar{v}\nabla)\bar{v} \right) = \rho \bar{g} - \nabla \bar{\sigma} - \nabla p \qquad 6.2$$

The governing equations above can be further simplified for the compression moulding process based on the application of the Hele-Shaw model (Hele-Shaw and Stirling 1899), for which the following assumptions must be fulfilled (Osswald 1987):

1. The inertial forces are insignificant with respect to the viscous forces and can be neglected.
2. The flow is isotropic, incompressible, and the velocities at the mould walls are always equal to zero (no-slip assumption).
3. The effects of gravity and surface tension can be neglected.
4. The elastic forces are insignificant with respect to the viscous forces and can be neglected.
5. Drag flow is insignificant with respect to pressure induced flow and can be neglected.
6. The fluid can be assumed to have Newtonian behaviour and is dominated by shear flow.
7. The assumptions required to apply laminate theory are valid for the considered geometries (the part thickness is significantly smaller than the other part dimensions).

Applying the assumptions above to eqs. 6.1 and 6.2 and arranging the variables in a specific form leads to the following differential equation describing the time-dependent pressure distribution during processing (Osswald 1987, Heber 1995):

$$\frac{\partial}{\partial x}\left(S \frac{\partial p}{\partial x} \right) + \frac{\partial}{\partial y}\left(S \frac{\partial p}{\partial y} \right) - \dot{h} = 0 \qquad 6.3$$

The variable S in eq. 6.3 is defined as the flow conductivity and is a function of the material viscosity as well as the transient flow channel height. For

Simulation of compression moulding to form composites   147

isothermal, Newtonian fluids the flow conductivity reduces to the following expression:

$$S = \frac{h^3}{12\eta} \qquad 6.4$$

Equation 6.3 can be solved numerically for the pressure distribution at each time step from which the instantaneous velocities can be derived. The following boundary conditions are required to obtain a solution to eq. 6.3:

1. The pressure at the flow front is equal to zero in all unfilled regions.
2. The pressure gradient perpendicular to the top and bottom halves of the mould wall is zero (the material is not allowed to squeeze out of the mould cavity where the two mould halves come together).

Equation 6.3 can be solved by using the finite element method. A widespread approach to describe the flow of the material in the calculation is the control volume approach (Osswald 1987). This approach accounts for the movement of the flow front by defining a control volume mesh that is based on a static mesh. The flow front progression is followed into and out of the control volumes by defining a filling factor. Figure 6.1 shows a simplified representation of the relationship between the finite element mesh and the control volume mesh along with the behaviour of the filling factor as the flow front progresses through the control volumes.

Contrary to the assumptions above, process conditions in the compression moulding process are not isothermal. In the case of thermoset materials, the temperature of the charge is about 25°C and the mould temperature about 120°C to 150°C. Thermoplastic materials were put with melt temperature (approx.

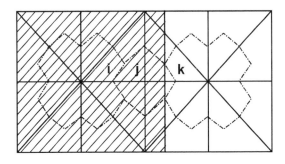

Melt

Element edge

Control volume edge

Fill Factor: f_i, f_j, f_k where:

$f_i = 1$ $0 < f_j < 1$ $f_k = 0$

6.1 Control volume principle.

148 Composites forming technologies

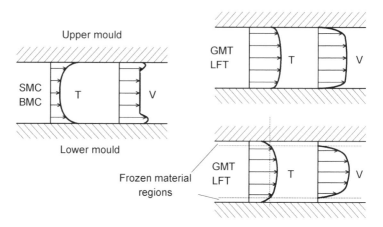

6.2 Compression moulding flow phenomena for SMC-BMC and GMT-LFT.

200°C to 250°C at materials with a PP matrix) in a cold mould (temperature 60°C to 80°C). Due to the strong dependency of the viscosity on the material temperature there is a big difference between the results of an isothermal simulation and the non-isothermal process. The temperature and velocity distribution for these two polymer families are shown in Fig. 6.2. Therefore a non-isothermal flow simulation considering the temperature change of the material during the flow and the dependency of the viscosity on the material temperature and the shearing rate as well is more accurate. Only with the non-isothermal simulation predictions of the local pressures are the pressing force and the local fibre orientation distribution helpful.

Non-isothermal flow simulation

As mentioned previously, a non-isothermal flow simulation models the actual process more accurately compared to an isothermal flow simulation. One model to describe the influence of the temperature and shear rate on the viscosity of the material is the Carreau-WLF model (eq. 6.5 and 6.6) (Heber 1995, Carreau 1968, Williams *et al.* 1955).

$$\eta = \frac{P_1 a_T}{(1 + a_T P_2 |\dot{\gamma}|)^{P_3}} \qquad 6.5$$

$$\log(a_T) = \frac{8.86(T_B - T_S)}{101.6°C + (T_B - T_S)} - \frac{8.86(T - T_S)}{101.6°C + (T - T_S)} \qquad 6.6$$

The shear rate, $\dot{\gamma}$, corresponds to the velocity gradient through the flow channel height, while the temperature shift coefficient, a_T, accounts for the variation of the viscosity at various temperatures. At this point it becomes apparent that modelling the heat flow during the compression moulding process is necessary.

6.2.2 Heat transfer simulation

Energy transport during the filling stage

Because the filling stages of the compression moulding process are strongly temperature dependent, the calculation of the temperature distribution is an essential step in the simulation of the process. Eq. 6.7 gives the simplified form of the energy equation used in the simulation of the heat transfer:

$$\rho c_p \frac{\partial T}{\partial t} = \gamma \frac{\partial^2 T}{\partial z^2} \qquad \text{Conduction}$$
$$- \rho c_p \left(v_x \frac{\partial T}{\partial x} + v_y \frac{\partial T}{\partial y} \right) \qquad \text{Convection} \qquad 6.7$$
$$- \tau_{xz} \frac{\partial v_x}{\partial z} - \tau_{yz} \frac{\partial v_y}{\partial z} \qquad \text{Diffusion}$$

The equation above was derived using the following assumptions:

1. Conduction heat transfer only occurs in the through-thickness direction.
2. Convection heat transfer only occurs in the direction of material flow.
3. The temperature profile is symmetric.
4. The heat transfer between the material and the mould walls is ideal, resulting in a constant mould temperature throughout the filling stage.
5. In the case where a thermoset is being simulated, it is assumed that the material begins to cure only after the entire mould has been filled.

The validity of the assumptions used in the application of the Hele-Shaw model as well as the simplification of the energy equation are validated in (Heber 1995). The simplified energy equation is then solved in conjunction with the governing flow equation, where the non-linear viscosity acts to connect the two governing equations to one another. This, however, increases the complexity of the simulation, meaning that the calculation time significantly increases.

Energy transport during and after the filling stage

After the mould has been completely filled the convection and diffusion terms in the above energy equation drop out, leaving the single conduction term on the right-hand side of the equation. If a thermoset is being simulated, then the source term (responsible for accounting for the heat generation during curing) must be reintroduced. However, overall the energy equation becomes simpler, meaning that a less rigorous method can be applied to solve this equation, saving the unnecessary expenditure of computational time.

150   Composites forming technologies

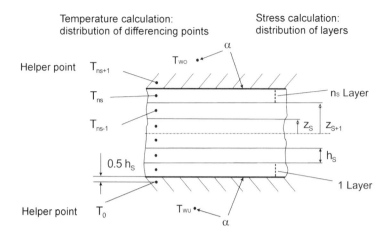

6.3 Through-thickness discretisation.

Based on this argument, the energy transport in the second part of the simulation is solved using one-dimensional implicit finite difference method. In order to use this method, additional finite difference points need to be defined in the through-thickness direction of the part. These points are located in the middle of each of the ten defined stress layers (this will be discussed later in the section on thermomechanical theory), along with two extra points, called the 'helper points', located outside of the geometry. These 'helper points' will be used in the shrinkage and warpage module to calculate the additional warpage due to the influence of corners within a part. Figure 6.3 schematically shows the placement of these points through the thickness of the part geometry.

The solution of the differential equation requires that it is first converted into a finite difference form using the following substitutions:

$$\frac{\partial T_S}{\partial t} = \frac{T_{S,k+1} - T_{S,k}}{\Delta t} \qquad 6.8$$

and:

$$\frac{\partial^2 T_S}{\partial z^2} = \frac{\frac{T_{S+1,k+1} - T_{S,k+1}}{h_S} - \frac{T_{S,k+1} - T_{S-1,k+1}}{h_S}}{h_S} \qquad 6.9$$

$$\frac{\partial^2 T_S}{\partial z^2} = \frac{T_{S+1,k+1} - 2T_{S,k+1} + T_{S-1,k+1}}{h_S^2} \qquad 6.10$$

In the case of thermosets the development of heat due to curing must also be included in the solution of the energy transport. The degree of cure $c_{S,k}$ for a stress layer S at a time k is defined as the current heat energy released divided by the total heat energy available. The rate of heat generation can then derived by differentiating the rate of cure with respect to time, or symbolically as:

Simulation of compression moulding to form composites 151

$$c_{S,k} = \frac{Q_{S,k}}{Q_g} \rightarrow Q_{S,k} = Q_g c_{S,k} \rightarrow \dot{Q}_{S,k} = Q_g \frac{\partial c_{S,k}}{\partial t} \qquad 6.11$$

It has also been observed that the reaction kinetics of the degree of cure are temperature dependent. The temperature dependency of the rate of cure is defined using the model developed by Kamal and Sourour (1973):

$$\frac{\partial c_{S,k}}{\partial t} = f(c_{S,k}, T) = (d_1 + d_2 c_{S,k}^m)(1 - c_{S,k})^n \qquad 6.12$$

where $d_1 = a_1 e^{-b_1/RT_k}$ $d_2 = a_2 e^{-b_2/RT_k}$

where a_1, a_2, b_1, b_2, m and n can be obtained by applying regression analysis to experimental data from DSC (differential scanning calorimetry) analysis for a given material. After experimentally determining the values for these parameters $\dot{Q}_S$, the rate of heat generation, can be redefined as follows:

$$\dot{Q}_{S,k} = Q_g(d_1 + d_2 c_{S,k}^m)(1 - c_{S,k})^n \qquad 6.13$$

To simplify the form of the energy equation, the following grouping parameter is defined:

$$\Omega = \frac{\lambda \Delta t}{c_p (h_S)^2} = a \frac{\Delta t}{(h_S)^2} \qquad 6.14$$

Substituting this grouping parameter into the energy equations puts this equation in the following tri-diagonal form:

$$-\Omega T_{S+1,k+1} + (2\Omega + 1)T_{S,k+1} - \Omega T_{S-1,k+1} = T_{S,k} + \frac{\Delta t}{\rho c_p} \dot{Q}_{S,k} \qquad 6.15$$

Example: four stress layers

The following equations exemplify the derivation of the governing equations for the energy transport, which can be solved at each time step for each element in the mesh to give the temperature at the finite difference point for a given stress layer. In this example for the sake of brevity only four stress layers are used. However, the method is the same for the case of more than four stress layers.

Equation 6.15 for the four stress layers becomes:

$$S = 1: \; -\Omega T_{0,k+1} + (2\Omega + 1)T_{1,k+1} - \Omega T_{2,k+1} = T_{1,k} + \frac{\Delta t}{\rho c_p} \dot{Q}_{1,k} \qquad 6.16$$

$$S = 2: \; -\Omega T_{1,k+1} + (2\Omega + 1)T_{2,k+1} - \Omega T_{3,k+1} = T_{2,k} + \frac{\Delta t}{\rho c_p} \dot{Q}_{2,k} \qquad 6.17$$

$$S = 3: \; -\Omega T_{2,k+1} + (2\Omega + 1)T_{3,k+1} - \Omega T_{4,k+1} = T_{3,k} + \frac{\Delta t}{\rho c_p} \dot{Q}_{3,k} \qquad 6.18$$

$$S = 4: \; -\Omega T_{3,k+1} + (2\Omega + 1)T_{4,k+1} - \Omega T_{5,k+1} = T_{4,k} + \frac{\Delta t}{\rho c_p} \dot{Q}_{4,k} \qquad 6.19$$

In the above equations $T_{0,k+1}$ and $T_{5,k+1}$ represent the distinct reference temperatures (the 'helper points'). One can then set the equations in the following matrix form, from which the desired temperatures $T_{1,k+1}$ to $T_{4,k+1}$ can be solved for in conjunction with the solution of the equation for the 'helper points' (defined further on in this section):

$$\begin{bmatrix} 2\Omega+1 & -\Omega & 0 & 0 \\ -\Omega & 2\Omega+1 & -\Omega & 0 \\ 0 & -\Omega & 2\Omega+1 & -\Omega \\ 0 & 0 & -\Omega & 2\Omega+1 \end{bmatrix} \begin{Bmatrix} T_{1,k+1} \\ T_{2,k+1} \\ T_{3,k+1} \\ T_{4,k+1} \end{Bmatrix} = \begin{Bmatrix} T_1^* + \Omega T_{0,k+1} \\ T_2^* \\ T_3^* \\ T_4^* + \Omega T_{5,k+1} \end{Bmatrix} \quad 6.20$$

where

$$T_1^* = T_{1,k} + \frac{\Delta t}{\rho c_p}\dot{Q}_{1,k} \qquad T_2^* = T_{2,k} + \frac{\Delta t}{\rho c_p}\dot{Q}_{2,k}$$

$$T_3^* = T_{3,k} + \frac{\Delta t}{\rho c_p}\dot{Q}_{3,k} \qquad T_4^* = T_{4,k} + \frac{\Delta t}{\rho c_p}\dot{Q}_{4,k}$$

The temperatures T_1^* to T_4^* represent the temperatures at the new time step in each of the respective four stress layers including the increase in temperature due to the reaction energy released during cure.

Energy transport after part ejection

EXPRESS also considers the heat transfer that occurs when the part is removed from the mould cavity and set aside to cool to ambient temperature. Additionally, upon ejection of the part warpage phenomena characteristic to the compression moulding process are accounted for by modifying the boundary conditions for the energy equations. The necessary modification to the above described one-dimensional heat transfer analysis is to modify the coefficient of thermal diffusion to reflect the transport of energy from the part to the cooler surroundings.

$$\alpha_O = \alpha_U = \alpha_{Air} \qquad \text{Coefficient of thermal diffusion (in still air)}$$
$$T_{WO} = T_{WU} = T_U \qquad \text{Temperature of the surroundings}$$

An energy balance on the upper edge of an element gives the temperature $T_{5,k+1}$ at the upper 'helper point' node (for the example where $n_S = 4$) as:

$$\dot{q} = \left(\frac{T_{n_S+1,k+1} + T_{n_S}^*}{2} - T_{wo}\right)\alpha_O \equiv \left(T_{n_S}^* - \frac{T_{n_S+1,k+1} + T_{n_S}^*}{2}\right)\left(\frac{\lambda}{\frac{h_S}{2}}\right) \quad 6.21$$

where the following relation simply represents an interpolation between the node in the uppermost stress layer and the upper 'helper node', giving the temperature at the upper edge of an element:

$$\frac{T_{ns+1,k+1} + T^*_{ns}}{2} \qquad 6.22$$

Finally, one can solve the following expression for the temperature at the upper 'helper point' node:

$$T_{ns+1,k+1} = T^*_{ns}\left[\frac{\left(\frac{2\lambda}{\alpha_o h_S}-1\right)}{\left(\frac{2\lambda}{\alpha_o h_S}+1\right)}+\frac{2T_{wo}}{\left(\frac{2\lambda}{\alpha_o h_S}+1\right)}\right] = T^*_{ns} + \frac{2(T_{Wo}-T^*_{ns})}{\left(\frac{2\lambda}{\alpha_o h_S}+1\right)} \qquad 6.23$$

Following the same procedure for the bottom edge of the element one derives the following expression for the lower 'helper point' node:

$$T_{0,k+1} = T^*_1\left[\frac{\left(\frac{2\lambda}{\alpha_u h_S}-1\right)}{\left(\frac{2\lambda}{\alpha_u h_S}+1\right)}+\frac{2T_{wu}}{\left(\frac{2\lambda}{\alpha_u h_S}+1\right)}\right] = T^*_1 + \frac{2(T_{Wu}-T^*_1)}{\left(\frac{2\lambda}{\alpha_u h_S}+1\right)} \qquad 6.24$$

6.2.3 Calculation of fibre orientation

The simulation of the fibre orientation during the compression moulding simulation is essential for an accurate prediction of the thermomechanical behaviour and the final mechanical properties of complex, fibre-reinforced parts. The fibre orientation is influenced by the flow of the material (Fig. 6.4). For this reason a process simulation has to calculate the orientations and redistribution of the fibre reinforcement from the very beginning of the process until the end of the filling stage. Additionally the ability to define any charge pre-orientation that might have developed during the processing of the mat or bulk charge is necessary. The following section provides an overview of the models and theory, which make up the foundation of the fibre orientation simulation.

The fibre orientation distribution is described with an angle class system. Figure 6.5 shows how the distribution function is discretised into various angle classes. These angle classes are defined with respect to a local coordinate system for each element. The various class discretisation acts as the foundation upon which the fibre orientation function can be calculated. At the bottom half of Fig. 6.5, two examples of the distribution function can be seen. The distribution on the left-hand side shows an isotropic fibre distribution, while that on the right-hand side displays an anisotropic fibre distribution where the main fibre orientation direction occurs at 90°. Another method to describe the fibre orientation is the use of orientation tensors.

The consideration of fibre transport

Due to the flow process, the fibres not only rotate but are also transported beyond the element boundaries. This demands that both a different fibre

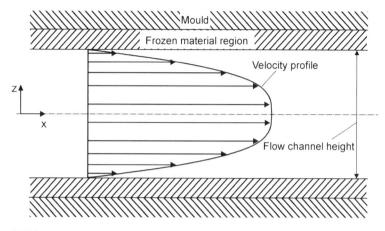

6.4 Velocity profile in the through-thickness direction.

orientation as well as different fibre volume content must be recorded in a given element at each time step. It is therefore necessary to access the fibre orientation and fibre volume content of the neighbouring elements to satisfy the continuity conditions. Therefore the average flow rate for each element over all of the layers is calculated. The distribution function for each of the neighbouring elements is then weighted for each layer and added to the previous distribution function for the corresponding layer. Figure 6.6 clarifies the transfer of material and the corresponding orientation between two neighbouring elements. The value of the fibre distribution function must be converted from the local fibre angle coordinate system used in element A to that used in element B.

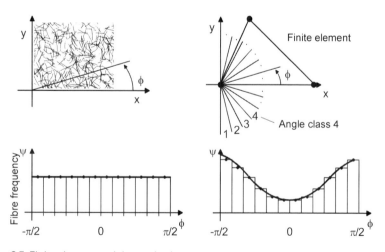

6.5 Finite element and the angle class system.

Simulation of compression moulding to form composites   155

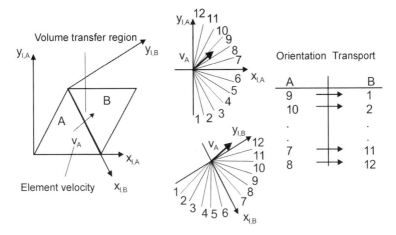

6.6 Fibre orientation transport between elements.

Model for calculation of fibre rotation

The first notable attempt to model the fibre orientation in composite parts was developed by Jeffery (1922). The model uses two equations to describe the three-dimensional rotation of a single ellipsoid in a high viscous Newtonian matrix. The rotation of the ellipsoid results from the difference in velocity at its ends. An important prerequisite for the validity of this model is that the matrix has a lower fibre concentration than the critical fibre concentration, while Jeffery's model does not account for fibre-fibre interaction. The critical fibre concentration describes the point at which an individual fibre is no longer oriented solely as a result of the varying velocity field, but whose orientation is also influenced by the presence of neighbouring fibres. The critical fibre concentration is defined to be 1% of the total part volume. Equation 6.25 describes the rotation of a fibre for the 2-dimensional case in which all of the fibres lie in a single plane, where r_e represents the axis length ratio and $v_{x,x} \ldots$ represent the velocity distributions respective to the various flow directions (Advani 1987).

$$\frac{\partial \phi}{\partial t} = \frac{r_e^2}{r_e^2 + 1} \left(\sin \phi \cos \phi (v_{y,y} - v_{x,x}) + \cos^2 \phi v_{y,x} - \sin^2 \phi v_{x,y} \right)$$
$$- \frac{1}{R_e^2 + 1} \left(\sin \phi \cos \phi (v_{y,y} - v_{x,x}) - \sin^2 \phi v_{y,x} + \cos^2 \phi v_{x,y} \right) \quad 6.25$$

For fibre reinforced materials with a fibre volume fraction of more than 1% (usually the case) the preferred model becomes the Folgar-Tucker model (Folgar 1983, Folgar and Tucker 1984). This model is based on Jeffery's model, but includes the consideration of interaction between the individual fibres, thereby ensuring that the fibres cannot become oriented 100% in any direction.

Although the Folgar-Tucker fibre orientation model is much more flexible

156   Composites forming technologies

than the aforementioned Jeffery model, and therefore, more applicable for complex parts, it has its disadvantages. Namely, to produce a solution with stable results, the equation that describes the fibre rotation needs the help of an additional numerical procedure. Furthermore, the use of this model requires the ability to access required input parameters from the flow simulation in order to solve the governing fibre orientation equation (i.e. the description of the shear and extensional velocities in the flow field).

The following assumptions were made in the derivation of this model (Folgar 1983, Folgar and Tucker 1984):

1. The fibres can be considered as rigid bodies with a uniform length and diameter.
2. The fibres are much larger than the molecules of the surrounding material such that the effect of Brownian motion can be negated.
3. The fibre-matrix compound is incompressible.
4. The viscosity of the material matrix is so high that inertial effects can be negated.
5. The fibres are randomly distributed throughout the matrix material.
6. There are no externally applied forces or moments on the fibres.
7. Interaction between two fibres takes place when the centres of gravity of these fibres move past one another within a distance that is smaller than the length of the fibres.
8. All of the interactions between the fibres are irreversible and statistically evenly distributed. Each interaction causes a modification of the fibre angle, which is independent of the current orientation angle of the fibre.

Because it is not practical to consider each single fibre interaction separately for complex geometries, Folgar and Tucker considered the fibre orientation using a statistical approximation applicable to the entire domain. This statistical approximation, the fibre distribution function ψ, corresponds to a Gaussian probability distribution of the fibre orientation. This function must satisfy the following continuity equation, which encompasses all of the fibres rotating into and out of an arbitrary control volume within the flow field:

$$\frac{d\psi}{dt} = \frac{\partial \psi}{\partial t} + v_x \frac{\partial \psi}{\partial x} + v_y \frac{\partial \psi}{\partial y} = -\frac{\partial(\psi \dot\phi)}{\partial \phi} \qquad 6.26$$

The above equation is simplified as the material convection terms drop out with reference to a stationary coordinate system.

Folgar and Tucker extended Jeffery's equation for the rotational speed of a single fibre with the addition of a dampening term (eq. 6.27).

$$\frac{\partial \phi}{\partial t} = -C_I \dot\gamma \frac{1}{\psi} \frac{\partial \psi}{\partial \phi} - \sin \phi \cos \phi \, v_{x,x}$$
$$- \sin^2 \phi \, v_{x,y} + \cos^2 \phi \, v_{y,x} + \sin \phi \cos \phi \, v_{y,y} \qquad 6.27$$

In the above equation, the variable $\dot{\gamma}$ represents the second invariant of the rate of deformation tensor (eq. 6.28), and C_I represents the fibre interaction coefficient that acts as a material parameter to describe the material orientation behaviour.

$$\dot{\gamma} = \sqrt{2v_{x,x}^2 + (v_{x,y} + v_{y,x})^2 + 2v_{y,y}^2} \qquad 6.28$$

The frequency of the fibre-fibre interaction is described within the model proportional to $\dot{\gamma}$.

Equations 6.26 and 6.27 lead to the following differential equation (eq. 6.29) describing the time dependency of the fibre orientation function as a function of the fibre interaction coefficient ψ, as well as the shear and extensional velocities present within the flow field.

$$\frac{\partial \psi}{\partial t} = C_I \dot{\gamma} \frac{\partial^2 \psi}{\partial \phi^2} - \frac{\partial}{\partial \phi} \left[\begin{array}{l} \psi(-\sin\phi\cos\phi\, v_{x,x} - \sin^2\phi\, v_{x,y} + \\ +\cos^2\phi\, v_{y,x} + \sin\phi\cos\phi\, v_{y,y} \end{array} \right] \qquad 6.29$$

The following section briefly discusses the validity of the assumptions made to derive the above equation with respect to the compression moulding of fibre-reinforced parts. The consideration of the fibres as rigid bodies within compression moulding materials is a reasonable assumption. One could argue that the presence of very long fibre filaments or bundles may deviate from the rigid body assumption. In this case it has been shown that one can also consider a long fibre as smaller, broken off, inflexible fibres, linked together one after the other (Diest 1996). The assumption that Brownian motion can be negated is valid for the considered material since the relative fibre size is much larger than the size of the matrix molecules.

The third and fourth assumptions have been validated many times in various studies in the field of rheological theory (Heber 1995, Michaeli 1990). The random distribution of the fibres in a material matrix, along with the assumption of a z-symmetric velocity profile, implies that the volume of fibres remains constant above and below the thickness half-plane of the part and that no mixing occurs between these two regions during the flow process. This assumption is no longer valid for complex geometries involving significant material flow in all three coordinate directions. The sixth assumption that no external forces are applied is appropriate under the consideration that external forces can only develop along the mating edges of the mould. When the charge positioning is done carefully, such that the majority of the flow front reaches these edges at approximately the same time, the influence of these external forces on the flow can be overlooked.

The seventh and eighth assumptions having to do with the fibre interaction are the core assumptions made for this model. Folgar (1983) and Folgar and Tucker (1984) show that these assumptions have been confirmed in many different experimental works with fibre-reinforced, thermoset compression moulding

158   Composites forming technologies

materials. A translation of these results for thermoplastic materials is found in Semmler (1998). The consideration of the thermoplastic specific properties (e.g., discrete layer fibre orientation) will be mentioned later in this chapter.

Solution of the differential equation

Because the differential equation for the calculation of fibre rotation (eq. 6.29) is not solvable using analytical methods, numerical methods must be employed that are both computationally stable and efficient. Therefore, the governing equations for the fibre orientation must be discretised and rewritten in implicit form.

The implicit method requires that the value of the function for the current time step is ascertained from the value of the function of the neighbouring nodes at the current time step (Fig. 6.7, left side). This also means that the solution of the corresponding finite difference equation cannot be attained without the use of an explicit finite differencing method at the boundary values (angle classes 1 and 25) (Fig. 6.7, right side).

In order to make use of the implicit method, eq. 6.29 must first be rewritten in the following form:

$$\frac{\partial \psi}{\partial t} = C_I \dot{\gamma} \frac{\partial^2 \psi}{\partial \phi^2}$$
$$- \frac{\partial \psi}{\partial \phi} (\sin \phi \cos \phi (v_{y,y} - v_{x,x}) - \sin^2 \phi \, v_{x,y} + \cos^2 \phi \, v_{y,x}) \qquad 6.30$$
$$- \psi \frac{\partial (\sin \phi \cos \phi (v_{y,y} - v_{x,x}) - \sin^2 \phi \, v_{x,y} + \cos^2 \phi \, v_{y,x})}{\partial \phi}$$

To put this equation into an implicit form, the substitutions outlined in eqs. 6.31–6.34 are made in eq. 6.30.

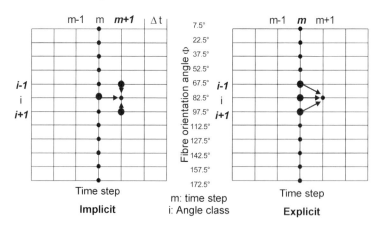

6.7 Implicit and explicit finite element methods.

Simulation of compression moulding to form composites

$$\psi = \psi_i^{m+1} \tag{6.31}$$

$$\frac{\partial \psi}{\partial t} = \frac{\psi_i^{m+1} - \psi_i^m}{\Delta t} \tag{6.32}$$

$$\frac{\partial \psi}{\partial \phi} = \frac{\psi_{i+1}^{m+1} - \psi_{i-1}^{m+1}}{2\Delta \phi} \tag{6.33}$$

$$\frac{\partial^2 \psi}{\partial \phi^2} = \frac{\psi_{i+1}^{m+1} - 2\psi_i^{m+1} + \psi_{i-1}^{m+1}}{\Delta \phi^2} \tag{6.34}$$

The resulting finite difference equation is solved for ψ_i^{m+1}, converted into tri-diagonal form, the boundary conditions are explicitly defined, and the system of equations is solved by using a Gaussian-based matrix solving algorithm (Schwarz 1986).

Thermoplastic discrete layer calculation of the fibre orientation

The calculation of the fibre rotation and transport for thermoplastic materials must be approached using discrete layers. Dividing the geometry into layers allows consideration of the difference in the velocity profile with respect to the flow channel height as well as the effect of the transient flow channel height due to the cooling and eventual freezing of the material. Figure 6.4 schematically displays a typical velocity profile during the compression moulding of a thermoplastic material and shows that the fibres become fixed in the regions close to the mould walls as the material freezes. Figure 6.8 shows the principle applied to ascertain the orientation using a discretisation of layers. One can see from this figure that two different layer discretisations are used. The variable layers are allowed to change their respective thickness and position during the process, whereas the stationary layers maintain the same thickness and position throughout the process.

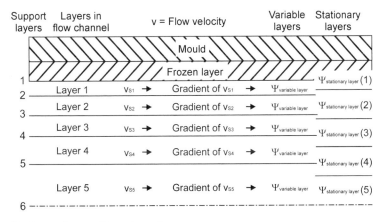

6.8 Layer discrete fibre orientation.

In comparison to the middle of the flow channel, the velocity profile gradient is clearly higher in the regions closer to the mould walls (see Fig. 6.4). For this reason, a finer discretisation of the variable layers is applied closer to the mould walls. The thickness of the variable layers and their respective positions are dependent upon the height of the flow channel (which varies during the filling stage; a major difference between modelling compression moulding in comparison to injection moulding), as well as the thickness of the layer that is considered to be completely frozen.

The fluid velocity for the variable layers is calculated from the shear velocity present for a given layer as follows:

$$v_\mu = \dot{\gamma} h_{vS} + v_{\mu+1} \qquad 6.35$$

Interpolation is used to obtain the average velocity located at the middle of each variable layer for every element. The velocities at the respective nodes for an element are then calculated using the weighted system based on the element area, from which the gradients at the element centre can be ascertained through differentiation.

These gradients are used as the input parameters for eq. 6.29, the computation of the fibre orientation. For each time step, the corresponding distribution function, whose value is dependent upon the current position and thickness of the frozen layers, is averaged and saved in each of the five static layers. These stationary layers, defined at the beginning of the computation, have a tenth the thickness of the part, and their positions remains constant during the entire process. At the end of the flow process, the fibre orientation in both the melt and frozen material is saved in the stationary layers, which can then be accessed in displaying the final orientation of the part.

The consideration of fibre pre-orientation

As a result of the process used in the production of fibre-reinforced materials for compression moulding, these materials contain a fibre pre-orientation, even before the compressing moulding process begins. For example, pre-orientation originates in the production of the fibre mat for SMC/mat-GMT materials (where the fibres tend to become oriented in the machine direction for mat-weave lay-ups), as well as in the plastification process of bulk materials as a result of the screw geometry. In the past, these material characteristics, as well as the definition of charge positions, were usually ignored in the design process. However, practical investigation into the influence of pre-orientation of charges and their positioning show these variables can strongly influence the mechanical and thermo-mechanical properties of the part. The corresponding algorithm for considering fibre pre-orientation is outlined in Fig. 6.9.

Like the fibre orientation, the pre-orientation is defined using a distribution factor. The main direction of the orientation as well as the position of the

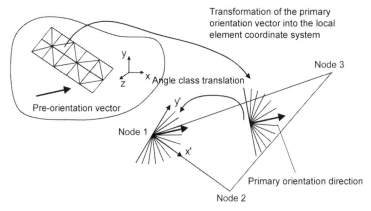

6.9 Translation of fibre orientation.

charges must also be defined. The positioning of the charges is accomplished by selecting certain elements of the geometry mesh and defining them as a group. The main orientation vector is converted from global into element local coordinates at the beginning of the calculation. Finally, the distribution function, as shown in Fig. 6.9 containing 25 discrete angle classes, is converted into the local element angle class system.

The pre-orientation distribution factor is defined as a material parameter, which requires that the distribution is measured for each charge. This measurement is most easily obtained using a program, which can calculate the fibre orientation from digitised images of actual charges (e.g., FiberScan).

6.3 Examples of use of the simulation

In the following some examples should show the capabilities of the simulation. At first a non-isothermal simulation of the flow fronts is presented that helps the user to detect weld lines and non filled areas due to a too early freezing of the material. Based on this simulation the fibre orientation simulation could be done that enables a stiffness and failure analysis of long fibre reinforced parts. Last but not least the simulation can help to build up an easy quality control system that is based on the calculation of the centre of force. All the process simulations were done with the simulation program EXPRESS, developed by the Institute of Plastics Processing (IKV), Aachen, Germany in cooperation with M-Base Engineering and Software GmbH, Aachen. The mechanical simulations were done by the finite element program Abaqus.

6.3.1 Flow simulation

The flow simulation results of three geometries are shown: the first one is a door panel made out of SMC which is a rather shell-like part with very high

162   Composites forming technologies

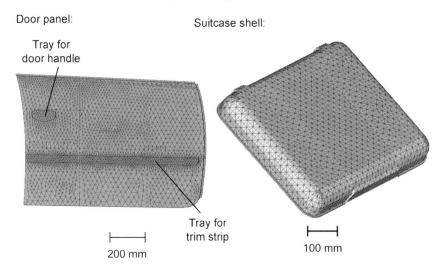

6.10 Example geometries (I): door panel and suitcase shell.

requirements on the surface quality. The second one is a suitcase shell made out of GMT. The third one is a dashboard made out of long fibre reinforced thermoplastics (LFT) in the long fibre extrusion compression moulding (ECM) process, which should show, that rather complex parts with undercuts could be simulated, too. These parts are shown in Figs 6.10 and 6.11.

Figure 6.12 shows the calculated flow fronts for the door panel in comparison to short shots that were made in the compression moulding process. The dark grey area represents the calculated flow front and the coloured lines represent the measured flow fronts. For these measurements the closing process was stopped at a defined stroke. With these short shots one gets

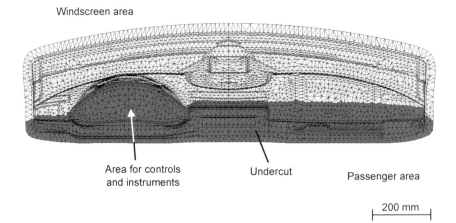

6.11 Example geometries (II): dashboard.

Simulation of compression moulding to form composites   163

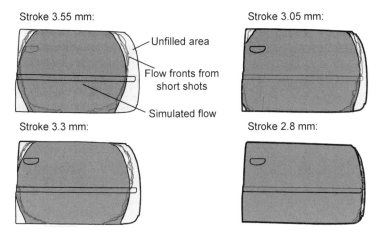

6.12 Comparison of the flow front with short shots.

unfilled parts representing the flow front at the stopped stroke. Figure 6.12 shows that there is a certain deviation of the flow fronts measured in three different tests where the closing process was stopped at the same stroke. This may come from the variation in the process, e.g. charge size or charge position. However, the calculated flow of the material is in good accordance with the measurements.

Beneath the flow fronts the calculated press force is a very important process parameter, because with its knowledge one can estimate whether the existing presses can be used to manufacture new parts or if a new investment is necessary. Figure 6.13 shows the measured press force for the suitcase shell made out of GMT in comparison to the calculated press force for a non-isothermal

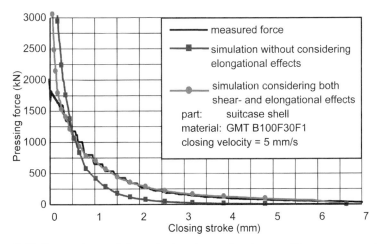

6.13 Comparison of the calculated pressing force with measurements (material: GMT).

164   Composites forming technologies

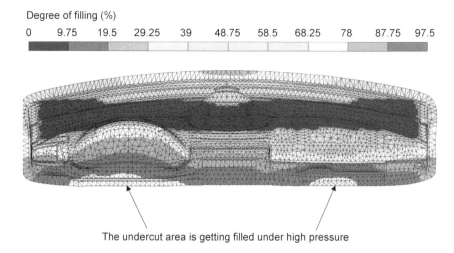

6.14 Flow fronts at the end of the filling process.

compression moulding process. The calculation was done first with the assumption of a pure shear flow without wall slippage and additionally an advanced simulation under consideration of both strain effects and wall slippage was done. The result shows that for both simulations the calculation agrees sufficiently on the measured data. Furthermore, the advanced simulation considering the strain flow of the material is more precise. The results show clearly that the process simulation is able to predict the press force for the non-isothermal compression moulding process of long fibre reinforced thermoplastics.

Usually there are not just parts having a shell-like geometry. In Fig. 6.14 the calculated flow front of a thermoplastic dashboard is shown. This is a very complex part with undercuts as shown in Fig. 6.11. Nevertheless a simulation is possible.

6.3.2   Fibre orientation simulation

The fibre orientation simulation is very useful to detect areas with a very high fibre orientation in the part. These areas do usually have very poor mechanical properties perpendicular to the fibre orientation and they tend to high warpage due to the different thermo-mechanical properties in and perpendicular to the fibre orientation. Additionally, with the knowledge of the local fibre orientation distribution a prediction of the local anisotropic material behaviour is possible.

For this example the door panel was used (Fig. 6.10). In order to assess the possibilities of the simulation three different charge positions were defined to get different local fibre orientation distributions. The calculation was verified with experiments where the fibre orientation was measured by analysing x-ray

Simulation of compression moulding to form composites 165

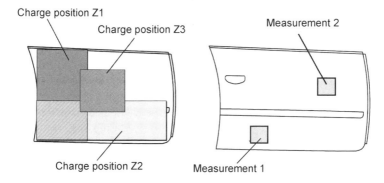

6.15 Positions of the charge and measurement of the fibre orientation distribution.

pictures with the image processing software FibreScan. The charge positions and the location where the orientation was measured are shown in Fig. 6.15. Figure 6.16 shows the calculated fibre orientation distributions in comparison with measurements for both positions. The diagram shows a clear accordance of the simulated orientations with the measurements. Thus a prediction of the fibre orientation is possible for shell-type parts made out of SMC. This also leads to very good simulation results of the calculation of the mechanical part properties as shown below.

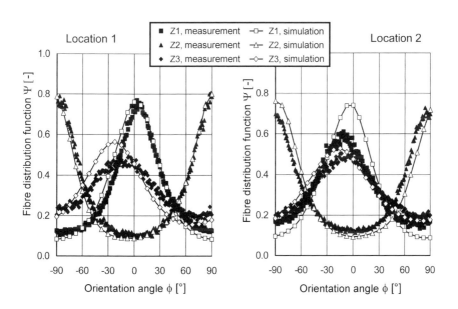

6.16 Comparison of the calculated fibre orientation with measurements.

6.3.3 Stiffness and failure analysis

This section briefly describes the basic composite theory to archive a calculation of the local material properties at each point of the component using initial numbers for the matrix and fibre properties. The combination of micro- and macromechanical theory simplifies the analysis process without performing any mechanical testing. These models require the calculated fibre orientation. This calculation process runs in two stages as outlined in Fig. 6.17. In the first step a micromechanical model is applied to determine the anisotropic stiffness parameters for one layer. This model assumes unidirectional fibres parallel and perpendicular to the principal orientation of the material.

The laminate theory is then applied to calculate the element stiffness properties accumulating the stiffness from all layers in the element regarding the differing directions of the principal orientation in each layer. The thickness of a single layer corresponds to the value of a particular angle class from the fibre distribution function and takes on the same directional alignment. The model effectively combines the individual layers using the values determined from the micromechanical model for each individual laminate layer and their respective alignments, giving the resultant elasticity tensor for an element.

Micromechanical model

With the application of a micromechanical model one can calculate the fibre-matrix material properties by separately considering the material as being constructed from unidirectionally oriented fibres and a matrix material. Halpin and Tsai developed the micromechanical model for unidirectional fibre reinforced materials. However, the components produced in compression

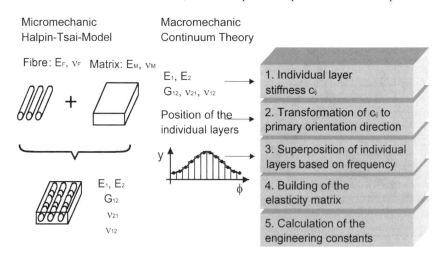

6.17 Calculating the engineering constants.

moulding are not oriented unidirectionally. Therefore, 25 different layers are assumed each having a slightly different principal orientation.

Using eqs. 6.36–6.41, the elastic moduli, shear moduli and Poisson's ratios are calculated for a single orientation angle 'layer' using the following relationships:

$$E_1 = E_M \frac{1 + \xi\kappa\Phi}{1 - \kappa\Phi} \quad \text{where} \quad \xi = 2\frac{l_F}{d_F} \qquad 6.36$$

$$E_2 = E_M \frac{1 + \xi\kappa\Phi}{1 - \kappa\Phi} \quad \text{where} \quad \xi = 2 \qquad 6.37$$

$$\kappa = \frac{\frac{E_F}{E_M} - 1}{\frac{E_F}{E_M} + \xi} \qquad 6.38$$

$$v_{21} = v_F \Phi + v_M(1 - \Phi) \qquad 6.39$$

$$v_{12} = v_{21} \frac{E_2}{E_1} \qquad 6.40$$

$$G_{12} = G_M \frac{1 + \xi 0.5\Phi}{1 - 0.5\Phi} \quad \text{where} \quad \xi = 1 \qquad 6.41$$

Macromechanical model

These 25 layers are then combined using the principle of superposition to calculate the anisotropic material stiffness for each element (or each stress layer of each element). This method is known as the Continuum Theory, developed by Puck and Halpin (Puck 1967). The specific thickness for a given imaginary layer is defined using the value of the probability parameter, which is determined from the fibre distribution function. In this way, the influence of the fibre orientation of a continuous material is transformed into a calculation for a laminate material composed of 25 layers with unidirectional material properties. The combination of the micro- and macromechanical models leads to the numerical calculation of the composite material properties. In the following section, models for the above material properties are introduced to show how the temperature and pressure dependency of these properties can also be included in the calculation of the final composite properties.

For a failure analysis a global failure criterion is used. It evaluates within the analysis whether the current stress state causes a failure or not. Up to now, distinct failure modes for fibre fracture and inter fibre fracture have not been available for long fibre reinforced plastics. The used global failure criterion (eq. 6.42) describes the failure factor f_E with a polynominal formulation such as the Norris criterion (Echaabi *et al.* 1996). If this factor is equal or larger than 1 failure of the material is calculated.

168   Composites forming technologies

$$f_E = \left(\frac{\sigma_1}{R_1}\right)^2 + \left(\frac{\sigma_2}{R_2}\right)^2 + \left(\frac{\sigma_1\sigma_2}{R_1R_2}\right)^2 + \left(\frac{\tau_{21}}{R_{21}}\right)^2 \qquad 6.42$$

The parameters R of this equation are the tensile and the compression strength in principal fibre orientation and perpendicular to it as well as the shear strength. The different strength values depend on the fibre orientation distribution and thus they are locally different and have to be measured for different fibre volume content and orientation degrees (Piry 2004).

Example of mechanical simulation

To show the possibilities of the coupled flow and mechanical simulation an analysis of the door panel mentioned above was done. For the mechanical simulations the charge positions Z1 and Z2 (Fig. 6.15) were assumed. With the given numbers of the local fibre orientation distribution and fibre content the local anisotropic material data was calculated. To prove the calculations a mechanical, linear distributed load of 200 Newtons was assumed. The calculated deformations of the part were compared to measurements. The test set up is shown in Fig. 6.18. The comparison shows, that the stiffness analysis of SMC parts closed to experiment (Fig. 6.19). In order to check the influence of the fibre orientation on the local stiffness the deformation along the section A-A drawn in the figure was analysed in more detail. Figure 6.20 shows the comparison of three calculations with different assumptions of the material behaviour with the measurements. The isotropic calculation shows a big deviation of the calculated deformation to the measurement in opposition to calculation with the assumption of anisotropic material behaviour. However, the difference between the linear calculation and the more complex calculation with the assumption of a non-linear material behaviour is very small in the stiffness analysis. This changes in the failure analysis. Here the failure was defined with the start of surface cracks. Figure 6.21 shows the calculated failure factor of the linear and the non-linear calculation along the section A-A. The charge position for this analysis was position Z1. Here a failure factor greater than 1 assumes the appearance of surface cracks in the area. Additionally the area where these surface cracks were detected on the section is shown in the figure. It is shown that the more complex simulation of the mechanical properties leads to a better

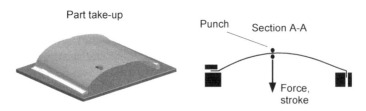

6.18 Set up of the mechanical tests.

Simulation of compression moulding to form composites 169

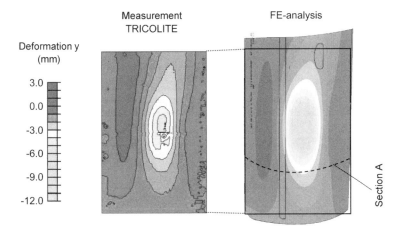

6.19 Comparison TRICOLITE-measurement and FE-analysis (Z2, $F = 200$ N).

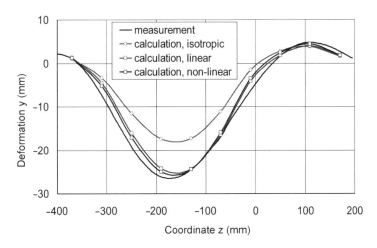

6.20 Comparison measurement and FE-analysis (blank Z1, section A).

prediction of the failure behaviour and gives the part designer the opportunity to tap the whole material potential. On the other hand, the non-linear simulation is much more complex and needs higher efforts to measure the material data.

6.3.4 Quality control

Process simulation programs are opening up new possibilities in quality assurance. By using process simulation one can determine and visualise process data (e.g., local pressures, temperatures, velocities, etc.) on the whole part. With this knowledge the sensor position can be optimised in order to measure the process data at the critical spots.

170   Composites forming technologies

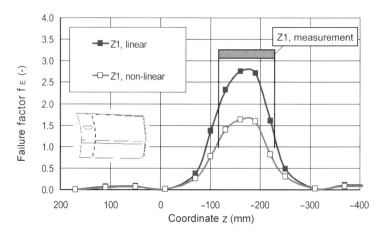

6.21 Failure factor f_E for crack failure.

The most important parameters that can be measured on the press without the sensors in the mould are: press force, press work, forces of parallelism, cylinder press stroke and closing velocity. One easy value that characterises the moulding process is the time dependent behaviour of the centre of force.

From the four forces of the parallelism cylinders $F_{Parallel}$ one can calculate a resultant force F_{Res} (Fig. 6.22). The vectorial change of this resultant force regarding the centre of gravity of the mould describes the flow of the material. Thus, this is an easy method to monitor several process parameters like charge position, charge size, closing velocity, etc. A compressed form of this description is the norm of the vector. This can be easily calculated and monitored by the data logging of the press.

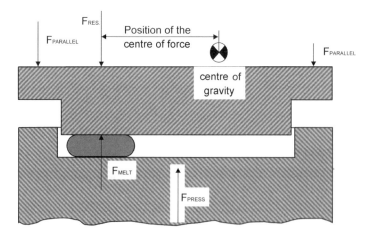

6.22 Determining the center of force.

Simulation of compression moulding to form composites

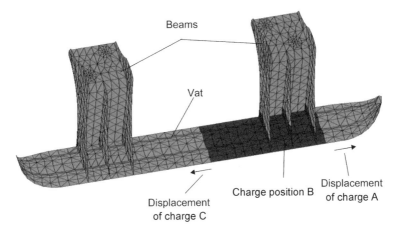

6.23 Charge positions for the sled runner.

Additionally it can be computed by simulation in order to get target data and to define specific setpoint values for the centre of force. Practical studies were carried out for a sled runner manufactured on a parallelism-controlled 2000 kN press. In these studies the charge position on the vat differs by 10 mm (Fig. 6.23). Figure 6.24 shows the variation of the centre of force for different charges, with different positions in the cavity. The deviation of the curve with different blank positions comes out very clearly. The results from a process simulation with charge position B is in good agreement with the corresponding measurements. The other parts made with charge positions A and C do not meet the required tolerances for charge position. This example shows that with the aid of process simulation software the manufacturer is able to derive set point values for a quality control system, which is the basis for a part assessment.

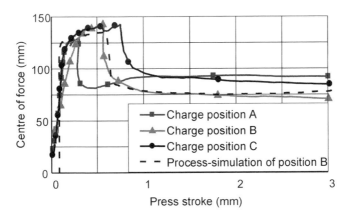

6.24 Influence of the charge position on the centre of force.

172   Composites forming technologies

6.4   Measurement of the material data

Unfortunately, there is a wide variety of possible mistakes that could lead to incorrect results of the simulation. Besides the assumptions of wrong process parameters or bugs in the simulation software the assumption of suboptimal material data is a reason for incorrect results. Therefore the user has to keep in mind that without good material data one could not achieve the full potential of the simulation. One of the key material parameters is viscosity. A screening analysis on the deformation energy, which is the integral of the press force, shows that the viscosity parameters have a major influence on the calculated pressing force (Fig. 6.25). The viscosity of long fibre reinforced materials could be measured with special rheometer types called squeeze flow rheometers. One type of squeeze flow rheometer is the Press rheometer developed by Oelgarth (1997). The Press rheometer is a small mould with an open shear edge where the material can be withdrawn during the closing procedure. The set up of the rheometer is shown in Fig. 6.26. During different isothermal compressions, each with different temperatures and closing speeds, the press forces were measured. An analytical model is able to calculate the progression of the press force for this rather easy flow. Thus the viscosity parameters, which are a part of the analytical model, were determined by a non-linear regression analysis of these force progressions (Fig. 6.27). A detailed description of the proceeding is shown in (Oelgarth 1997, Ritter 2003). Alternative concepts for viscosity measurements are shown in (Bush *et al.* 2000, Orgeas *et al.* 2001).

The result of the measurement continues to improve, if one evaluates the measurement under consideration of the wall slip effect at the mould surfaces.

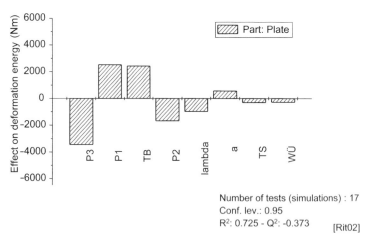

6.25 Influence of the material parameters on the deformation energy (thermoset material).

Simulation of compression moulding to form composites 173

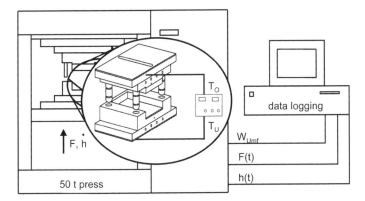

6.26 Set up of the press rheometer Rheopress.

Figure 6.27 shows the measured forces for a SMC and the fitted curves with and without consideration of the wall slip effects (Ritter 2003, Schmachtenberg and Skrodolies 2005). Figure 6.28 shows clearly that the fitted curves under consideration of wall slip effects deviate much less from the measured data than the fitted curve with the assumption of wall stick behaviour even by the analysis of the isothermal compression of thermoplastics. With this developed procedure it is possible to determine the viscosity parameters as well as wall slip coefficients.

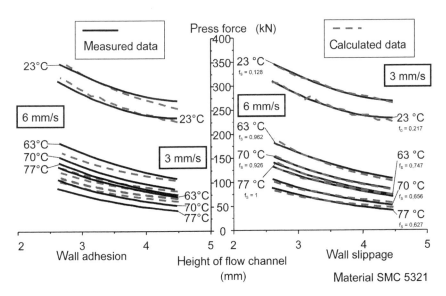

6.27 Press rheometer analysis: comparison of the analysis with wall adhesion and wall slippage effects (SMC).

174 Composites forming technologies

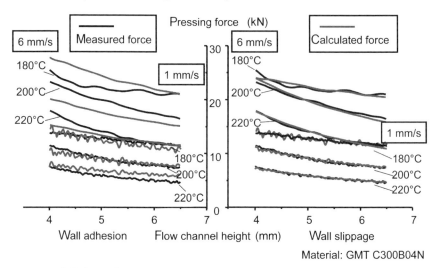

6.28 Press rheometer analysis: comparison of the analysis with wall adhesion and wall slippage effects (GMT).

6.5 References

Advani S G (1987), *Prediction of fibre orientation during processing of short fiber composites*, PhD Thesis at the University of Illinois at Urbana-Champaign.

Bush S F, Torres F G, Methven J M (2000) Rheological characterisation of discrete long glass fibre (LGF) reinforced thermoplastics, *Composites: Part A*, 31, 1421–1431.

Carreau P J (1968), *Rheological Equations from Molecular Network Theories*, PhD Thesis at the University of Wisconsin, Madison.

Diest K V (1996), *Prozeßsimulation und Faserorientierungserkennung von GMT-Bauteilen*, Aachen, Shaker Verlag.

Echaabi J, Trochu F, Gauvin R (1996), Review of failure criteria of fibrous composite materials, *Polymer Composites* 17, 786–798.

Ehrenstein G (1992), *Faserverbund-Kunststoffe, Werkstoffe-Verarbeitung-Eigenschaften*, München, Wien, Carl Hanser Verlag.

Folgar F (1983), *Fiber orientation distribution in concentrated suspensions: a predictive model*, PhD Thesis at the University of Illinois at Urbana-Champaign.

Folgar F, Tucker C L (1984), Orientation behavior of fibers in concentrated suspensions, *Journal of Reinforced Plastics and Composites*, 3, 98–119.

Heber M (1995), *Modell zur rheologischen Auslegung faserverstärkter, thermoplastischer Preßbauteile*, Aachen, Verlag Mainz.

Hele-Shaw H S, Stirling J (1899), The motion of a perfect liquid, *Nature*, 7 Sept. 1899, 107.

Jeffery G B (1922), The motion of ellipsoidal particles immersed in a viscous fluid *Proceedings of the Royal Society of London*, Series A, 102, 161–179.

Kamal M R, Sourour S (1973), Kinetics and thermal characterisation of thermoset cure, *Polymer Engineering and Science*, 13 (1), 59–64.

Michaeli W (1990), *Extrusionswerkzeuge für Kunststoffe*, München, Hanser Verlag.

Neitzel M, Breuer U (1996), *Die Verarbeitungstechnik der Faser-Kunststoff-Verbunde,* München, Wien, Carl Hanser Verlag.

Oelgarth A (1997), *Analyse und Charakterisierung des Fließverhaltens langfaserverstärkter Preßmassen,* Aachen, Verlag Mainz.

Orgéas L, Le Corre S, Dumont P, Favier D, Tourabi A (2001), Influence of the fibre volume fraction and mechanical loading on the flow behaviour of sheet moulding compounds (SMC), *4th ESAFORM conference on material forming,* Liège, Belgium.

Osswald T (1987), *Numerical methods for compression mold filling simulation,* PhD Thesis at the University of Illinios at Urbana-Champaign.

Piry M (2004), *Mechanische Auslegung von SMC-Bauteilen und Charakterisierung der relevanten Werkstoffeigenschaften,* Aachen, Verlag Mainz.

Puck A (1967), Zur Beanspruchung und Verformung von GFK-Mehrschichtenverbund-Bauelementen Teil 1. Grundlagen der Spannungs- und Verformungsanalyse, *Kunststoffe,* 57, 4, 284–290.

Renz U (1989), *Grundlagen der Wärmeübertragung,* RWTH Aachen.

Ritter M (2003), *Materialcharakterisierung von Langfaserverstärkten Pressmassen und Beschreibung des Pressprozesses durch Simulation und Messung des Kraftschwerpunktverlaufs,* Aachen, Verlag Mainz.

Schmachtenberg E, Ritter M (2002), Quality assurance system based on the process simulation for the compression molding process, *SAMPE Conference, Long Beach California (USA).*

Schmachtenberg E, Skrodolies K (2005), Process-simulation of compression moulding compounds, *Annual Technical Conference of the Society of Plastic Engineers (Antec),* Boston, MA, USA.

Schwarz H R (1986), *Numerische Mathematik,* Stuttgart, Teubner Verlag.

Semmler E (1998), *Simulation des mechanischen und thermomechanischen Verhaltens faserverstärkter thermoplastischer Pressmassen,* Aachen, Verlag Mainz.

Specker O (1990), *Pressen von SMC: Computersimulation zur rechnerunterstützten Auslegung des Prozesses und zur Ermittlung der Bauteileigenschaften,* Aachen, Fotodruck Mainz.

Williams M L, Landel R F, Fery J D (1955), The temperature dependence of relaxation mechanisms in amorphous liquids, *Journal of the American Chemical Society,* 77 (7), 3701–3706.

6.6 Symbols

a [m/s] Thermal diffusivity
a_T [–] Temperature shift factor
$c_{s,k}$ [–] Degree of cure
d_F [m] Fibre (or fibre bundle) diameter
f_E [m] Failure factor
g [m/s] Gravitation
$\dot{h}$ [m/s] Closing velocity
l_F [m] Fibre length
p [Pa] Pressure
r_E [–] Axis length ratio

t	[s]	Time
v	[m/s]	Velocity
x	[m]	Coordinate direction
y	[m]	Coordinate direction
z	[m]	Coordinate direction
C_I	[–]	Fibre interaction coefficient
E	[N/m^2]	Youngs modulus
G	[N/m^2]	Shear modulus
H	[m]	Flow channel height
P_1	[Pas]	Carreau parameter 1
P_2	[s^{-1}]	Carreau parameter 2
P_3	[–]	Carreau parameter 3
Q	[J/m^3]	Heat
$\dot{Q}$	[J/(m^{3}s)]	Rate of heat generation
R	[N/m^2]	Strength
S	[s/m^2]	Flow conductivity
T	[°C]	Temperature
T_B	[°C]	Reference temperature (WLF equation)
T_S	[°C]	Standard temperature (WLF equation)
α	[W/(m^{2}K)]	Heat-transfer coefficient
λ	[W/(mK)]	Thermal conductivity
ρ	[kg/m^3]	Density
η	[Pas]	Viscosity
ν	[–]	Poissons ratio
ξ	[–]	Material and geometry parameter in the Halpin-Tsai Model
σ	[N/m^2]	Stress
τ	[Nm^2]	Shearing stress
ϕ	[°]	Angle
ψ	[–]	Fibre frequency
Φ	[–]	Fibre volume content
Ω	[–]	Grouping parameter
∇	[–]	Nabla operator

7
Understanding composite distortion during processing

M R WISNOM and K D POTTER, University of Bristol, UK

7.1 Introduction

Composites distort during manufacture, causing difficulties assembling structures, and leading to increased costs. An iterative approach to tool design is often used, with the geometry modified on a trial and error basis in order to produce the correct cured shape. In this chapter the fundamental mechanisms that cause the residual stresses producing these distortions are first considered. The ways in which these mechanisms cause distortion in flat plates and spring-in in curved composites are then considered, together with the ability of models to predict the changes in geometry. Finally distortion in more complex structures is discussed briefly. Much of the research that has been done on this topic has been for autoclave curing of thermosetting resins, but the same phenomena are directly relevant to other manufacturing processes.

7.2 Fundamental mechanisms causing residual stresses and distortion

7.2.1 Differential thermal contraction

The expansion coefficient of polymer matrix materials is usually much higher than that of the fibres. Also the expansion coefficients of many fibres are orthotropic. For example, carbon fibres have very low or slightly negative expansion coefficients in the fibre direction, but higher values in the transverse direction. This leads to residual stresses at the micro-scale during cooldown even in unidirectional material. Figure 7.1 shows stresses calculated by micromechanical finite element analysis based on a repeating array of fibres for carbon fibre/epoxy subject to a 100°C temperature drop.[1] Compressive stresses are generated along the fibres, with tension in the matrix in the fibre direction. Hoop tensile stresses arise in the matrix as it shrinks around the fibres. There are also radial stresses which are compressive where the fibres are close together, and tensile where they are furthest apart. This produces Von Mises stresses of nearly 30 MPa in the matrix in this case as a result of the combination of

178   Composites forming technologies

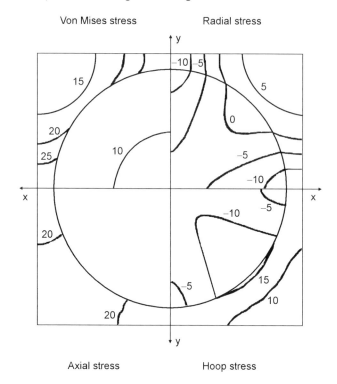

7.1 Residual stresses in carbon-epoxy due to 100°C temperature drop.[1]

stresses, which is a substantial fraction of the strength of the material. These residual stresses can affect the stress-strain behaviour and failure, but do not normally cause distortions because they arise on a very local scale and any distorting effects tend to average out over a larger volume of material.

The second effect is the difference in ply-level expansion coefficients in the fibre and transverse directions that causes in-plane residual stresses in laminates, which can be analysed by classical laminated plate theory. These can lead to distortion in flat plates when lay-ups are not balanced and symmetric.[2] For example, it is well known that unsymmetric cross-ply laminates exhibit curvature, and this is often used to study residual stresses, see, e.g., ref. 3. The presence of angle plies can lead to twist. This mechanism normally only produces significant stresses after vitrification, and is therefore most important on the cooldown after high temperature processing. Close study of unidirectional prepreg shows a significant level of variability in fibre content on a local scale that can generate small differences in distortion between nominally identical laminates.

For components with double curvature, the necessary rearrangements of the reinforcement fibre trajectories to accomodate the geometry in each ply make it very unlikely that the laminate will be balanced and symmetrical at all points

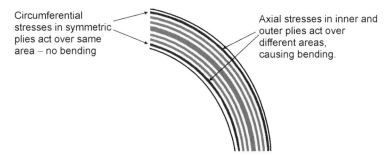

7.2 Pattern of residual stresses normal to plane in a curved balanced symmetric laminate.

across the surface. This lack of symmetry will lead to distortions in the part geometry. To make predictions of these geometrical distortions, thermoelastic and fibre path modelling need to be coupled.

It is less well appreciated that ply-level residual stresses can also produce distortion in curved plates even when the lay-up is balanced and symmetric, due to the effective shift of the neutral axis due to curvature. Figure 7.2 shows the pattern of in-plane residual stresses through the thickness of a balanced symmetric lay-up determined by finite element analysis. The values in this curved section are very similar to those in an equivalent flat lay-up. Consider the two plies shown in dark colour, which have the fibres in the circumferential direction. The circumferential stresses act over the same area, i.e. the ply thickness times the dimension normal to the plane of the figure. Therefore stresses of equal magnitude in the two plies in this direction would not produce any bending. However in the axial direction normal to the section the stresses in the inner ply act over a smaller area than those in the outer ply due to the shorter circumferential length. The forces are no longer in balance with respect to the mid-plane, causing bending and a shift in the neutral axis. This also affects the bending in the other direction, known as spring-in, via the Poisson ratios, although the magnitude is usually small. The same phenomenon causes twist in curved sections with angle plies.[4]

A third effect is the higher through-thickness expansion coefficients (matrix dominated) compared with in-plane values (fibre dominated). This causes a change in curvature of curved laminates with temperature for any lay-up, and is the main origin of the spring-in phenomenon[5] illustrated in Fig. 7.3. This is a geometrical effect that can produce distortions in the absence of large stresses whilst the material is in the rubbery state as well as after vitrification.[6]

7.2.2 Chemical shrinkage

Polymers shrink during the cure producing an additional volume change to that caused by thermal effects. This can be a very substantial effect, with a volume

180   Composites forming technologies

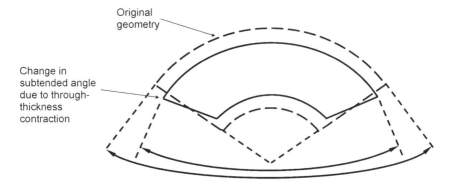

7.3 Illustration of spring-in mechanism.

change of 7% being typical for an epoxy resin.[7] Chemical shrinkage can cause similar effects to thermal contraction, with stresses at the micromechanical level, in-plane stresses and distortions in laminates, and changes in curvature in curved plates. However, for many resins much of the shrinkage occurs before gelation where the material is effectively a liquid. This may be compensated by resin flow, and may cause changes in thickness or volume fraction, adding to consolidation-related geometry changes, but does not directly produce stresses causing distortion. Crystallisation in materials such as PEEK can also cause volume changes leading to residual stresses.[8]

7.2.3 Tool-part interaction

Stresses can arise during the cure due to differential strains or frictional forces between the part and the tooling on which it is manufactured. Aluminium or steel tools have much higher expansion coefficients than composites, and tend to stretch the parts as they heat up. This can happen as a result of small shear stresses at the tool interface sustained by autoclave pressure causing tension in the part. If the stresses are not uniform through the thickness, they can lead to distortion.[9] During stamp-forming of thermoplastic matrix composites, much larger pressures can be generated than those seen in shaping thermoset matrix prepregs, leading to potentially very high frictional shear and associated residual stresses.

A second tool-part interaction mechanism is due to locking, where the geometry of the part forces it to move with the tool as it expands. The simplest example is a filament wound tube, where substantial tensile stresses can be induced due to expansion of the mandrel. For example, Ganley *et al.* estimated stresses at the surface of 250 MPa for unidirectional IM7/977-2 carbon fibre-epoxy prepreg hoop wound on a steel mandrel,[10] which led to very large spring-in when the tubes were sliced longitudinally.

Understanding composite distortion during processing

7.2.4 Other mechanisms

Moisture causes swelling of the matrix which produces similar effects to thermal or chemical volume changes, for example changing the curvature of unsymmetric laminates.[11]

Volume fraction variations through the thickness can arise, especially where there is resin bleed during the cure. These can produce distortion even in flat unidirectional composites.[12]

Fibre movement during the cure can cause changes in geometry and properties leading to stresses and distortion. Consolidation and resin flow at radii can lead to corner thickening with concave tooling and corner thinning with convex tooling,[13] and the effect is especially pronounced with tooling that contains both concave and convex regions close to each other. This can cause volume fraction variations and resin-rich regions which produce distortion. Wrinkles that can arise at corners, particularly with convex tooling, and separation between the part and tool may also affect residual stress development. Differential curing through the thickness due to temperature lags and exotherms can be important in thick parts, and may cause additional residual stresses and gradients through the thickness.[14]

7.3 Distortion in flat parts

7.3.1 Distortion due to asymmetric layups

If the layup is not symmetric, then the differential contractions during cooldown can produce deformation of the laminate. The classic example of this phenomenon is a (0/90) unsymmetric laminate, which will deform into a circular arc, as shown in Fig. 7.4. A (45/–45) laminate will twist on cooldown, whilst more complicated asymmetric laminates will exhibit combinations of bending and twisting.

Experiments with carbon-epoxy have shown that the development of curvature with temperature is approximately linear,[15] and the distortion can be predicted quite accurately based on the room temperature thermoelastic properties and the difference between the cure temperature and room temperature. If the glass transition temperature is below the cure temperature, then curvature

7.4 Curvature in (0/90) unsymmetric laminate due to residual thermal stresses.

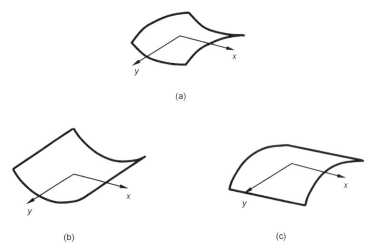

7.5 Initial saddle shape of (0/90) unsymmetric laminate (a) becomes unstable and switches to one of two stable cylindrical shapes (b) or (c).

only develops from that point until room temperature. Distortion will tend to reduce with time due to uptake of moisture, which causes a compensating swelling of the resin.

The deformation of even a simple flat (0/90) unsymmetric laminate is in fact quite complex. When the length to thickness ratio is small, a saddle shape is formed, with equal and opposite curvatures in the two directions. However, as this ratio increases, the saddle shape becomes unstable, and changes to one of two energetically favourable cylindrical shapes, as shown in Fig. 7.5. The geometry can be snapped between these two stable states. Solutions based on the Rayleigh-Ritz technique are available to predict the resulting shape of simple plates.[16] Finite element techniques can also be used,[17] and are effective for more complex shapes.

Significant residual stresses are required to sustain these distortions, which is why they only arise on cooldown below the glass transition temperature. For the same reason chemical shrinkage does not usually contribute significantly to distortion due to asymmetric layups because it mainly occurs before vitrification. This has been demonstrated by experiments reheating cured (0/90) unsymmetric laminates of AS4/8552 carbon-epoxy. The laminates became flat at a temperature only slightly above the original cure temperature, indicating that most of the distortion was thermoelastic.[15]

7.3.2 Distortion due to resin bleed and volume fraction gradients

Flat plates are often observed to distort away from the tool with convex up curvature, especially when they are relatively thin. One mechanism that can

Understanding composite distortion during processing

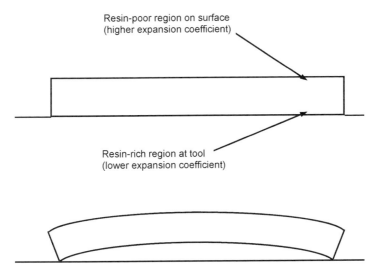

7.6 Distortion due to volume fraction gradient through the thickness.

cause this is variations of volume fraction through the thickness due to resin flow in materials with relatively low viscosity, or when resin is deliberately bled from the surface during consolidation. This can produce resin-poor regions at the surface, whilst near the tool interface, resin-rich layers can form. The thermal expansion coefficients increase with decreasing volume fraction and so tend to be higher on the tool side, leading to convex bowing during the cooldown from cure, as shown schematically in Fig. 7.6.

For example Radford reported bowing of 4 mm on 445 mm long T300/948A1 carbon-epoxy laminates 12 plies thick, and 70 mm bowing for 5-ply laminates 810 mm long.[12] Micrographs and fibre volume fraction measurements through the thickness showed a gradual gradient, but with marked differences in the surface plies. The bulk of the material had a fibre volume fraction between 48% and 52%, but on the surface ply it reached 58%, compared with only 41% for the ply in contact with the tool.

Expected curvatures were calculated using classical laminated plate theory based on measured volume fractions and thermoelastic properties derived from micromechanics, and gave similar magnitudes to the measured curvatures. Since the origin of this distortion is thermoelastic, it is reversible, and will decrease on reheating the part, or as moisture is absorbed.

7.3.3 Distortion due to tool-part interaction

An alternative mechanism that can cause bowing of flat laminates is tool-part interaction. Small shear stresses at the part surface due to tool expansion can build up over a long distance to cause significant tension in the part. If the part is

184 Composites forming technologies

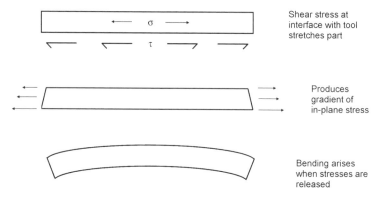

7.7 Distortion due to shear interaction at tool interface.

uniformly stretched, the stresses may be released after manufacture without causing any distortion. A variation of stress through the thickness is required to produce a change in shape. If a stress gradient arises in the part early in the cure, it will be locked in when vitrification occurs, causing bending when the stresses are released, as illustrated schematically in Fig. 7.7. In contrast to bowing caused by volume fraction gradients, this mechanism is non-thermoelastic, and so is not affected by subsequent temperature changes.

Stresses arising due to tool-part interaction have been measured on flat strips of unidirectional AS4/8552 prepreg 650 mm long, and one-ply thick (0.25 mm) cured on an aluminium base plate with a layer of FEP release film.[18] Spots of the prepreg were precured and strain gauges bonded, allowing strains to be measured throughout the standard autoclave cure cycle with dwells at 120°C and 180°C. Figure 7.8 shows the results.

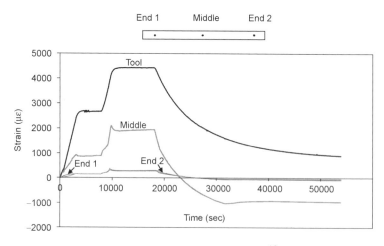

7.8 Strains on flat unidirectional strips during the cure.[18]

Strains started to develop straight away, with about 900 microstrain in the first dwell, before gelation. The thermal expansion coefficient in the fibre direction is very small, so this indicates tensile stresses are developing, with a magnitude of about 120 MPa. This is apparently a result of friction between fibres and the tool as the resin is effectively a liquid and would not be expected to transfer any significant stress before gelation. The readings from the gauges near the ends of the strip showed that stress builds up approximately linearly, with a shear stress of around 0.1 MPa at the tool surface. The strain continued to increase, reaching about 2000 microstrain after vitrification on the second dwell, corresponding to about 250 MPa, and an equivalent shear stress of approximately 0.2 MPa. This is a very high stress at a point in the cure when the material is normally assumed to be stress free. Strips with four unidirectional plies of the same material with the strain gauges attached to cured spots on the top ply showed about a quarter of the strain of the single-ply case, confirming that stresses are due to a constant shear stress at the tool-part interface. Similarly high stresses induced by tool-part interaction have been inferred from experiments with an instrumented tool plate.[19]

The stresses in these flat plates did not lead to significant bowing, presumably because they were fairly uniformly distributed through the thickness. However, if stress gradients arise through the thickness, they can produce considerable distortion. For example Twigg *et al.* measured convex up bowing of up to 40 mm on 1200 mm long flat unidirectional T800H/3900-2 carbon-epoxy plates four plies thick, which was attributed to tensile stresses near the tool surface decaying through the thickness.[9]

One reason for a stress gradient is frictional effects before gelation. Twigg *et al.* postulated that the inter-ply friction coefficient in their material is less than the friction coefficient between the tool and part.[9] This means that stresses in the surface fibres will not all be transferred through the thickness, giving rise to a stress gradient and curvature. In contrast ply pull-out tests on AS4/8552 subjected to the same process conditions as the autoclaved laminates showed a higher resistance to shear before gelation between layers of prepreg than between prepreg and aluminium.[20] This scenario would not give rise to a stress gradient, and is consistent with the low level of bowing seen on the flat plates of that material.[18]

The layup can have a significant effect on the stresses built up before gelation. Although unidirectional flat plates of AS4/8552 did not bow, similar tests with a $(0/90)_s$ layup on specimens 1000 × 50 mm produced a consistent convex up curvature, with a maximum bowing of about 8 mm.[21] This compares with maximum bowing within the range −2 to +2 mm for unidirectional specimens of the same dimensions manufactured in the same way. The difference is believed to be because only the surface 0° ply carries the full tension induced by shear at the tool before gelation. These stresses cannot all be transferred through the 90° ply, because the fibres more easily slip over each other in this direction, and the effective frictional shear stress is reduced. There is therefore a stress gradient through the thickness, resulting in curvature.

This mechanism has been modelled by Bapanapalli and Smith in a finite element simulation assuming that a surface layer of material of constant thickness is subject to a certain tensile strain.[22] The thickness and strain parameters were deduced by fitting the curvature results of unidirectional strips of different thicknesses, and these values were then used to predict the curvatures of cross-ply specimens of different layups and thicknesses, giving good correlation. Very different values of both parameters were required to fit results for Invar and aluminium tools. The material used was T800HB/3900-2, which gave significant curvatures for both unidirectional and cross-ply layups. However, the same approach would not be expected to predict the curvatures of AS4/8552 due to the different behaviour of the pre-preg, as discussed above. This illustrates the difficulties of modelling cure distortions, due to the number of interacting factors that control the response.

7.4 Spring-in of curved parts

7.4.1 Thermoelastic spring-in

The higher through-thickness expansion coefficient of composites such as carbon-epoxy compared with the in-plane expansion coefficients causes curved composites to spring-in on cooldown from cure. The effect is reversible, with the spring-in reducing if the part is reheated, and is referred to as thermoelastic spring-in. The magnitude can be predicted from the equation:

$$\frac{\Delta\theta}{\theta} = (\alpha_I - \alpha_T)\Delta T \qquad 7.1$$

where $\Delta\theta$ is the change in angle θ, α_I and α_T are the in-plane and through-thickness expansion coefficients and ΔT is the change in temperature.

This mechanism can be demonstrated from measurements on unidirectional and cross-plied L shaped parts heated in an oven.[23] AS4/8552 carbon-epoxy with a lay-up of 0_4 and a thickness of 1 mm was laid up and cured on the inside of an Invar tool with radius 20 mm, where the 0° direction follows the curvature. The width was 100 mm and arm length 100 mm. After cure, one arm was clamped to the base of the oven, and the change in the 90° angle was measured with the help of a laser reflected from a mirror on the other end of the specimen as it was heated.

Figure 7.9 shows the results for the angle as a function of temperature. The response is approximately linear, and the slope corresponds to a change in angle of 0.47° for a temperature change from 20° to 180°.

The in-plane expansion coefficient α_I from manufacturer's data was $0.02 \times 10^{-6} K^{-1}$. The transverse coefficient α_T was measured independently to be

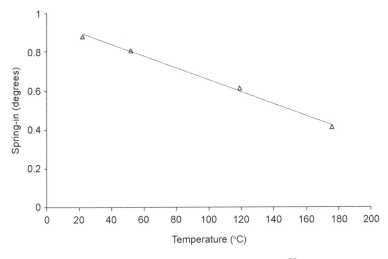

7.9 Thermoelastic response of unidirectional L specimens.[23]

$32.6 \times 10^{-6} K^{-1}$ using strain gauges. Assuming transverse isotropy, and using this value for the through-thickness expansion coefficient in equation (7.1) gives a change in angle of 0.469°, the same as the measured result. Using laminated plate theory, the in-plane and through-thickness expansion coefficients for a (0/90)$_s$ cross-ply laminate were calculated to be $2.73 \times 10^{-6} K^{-1}$ and $45.7 \times 10^{-6} K^{-1}$ respectively. Putting these into equation (7.1) gives a predicted spring-in of 0.619°, which again was found to be very close to the measured value of 0.63°, confirming the validity of the thermoelastic calculations.

7.4.2 Chemical shrinkage after vitrification

Chemical shrinkage of the resin can be treated in the same way as thermal contraction, by adding an additional term to equation (7.1):

$$\frac{\Delta\theta}{\theta} = (\alpha_I - \alpha_T)\Delta T + (\phi_I - \phi_T) \qquad 7.2$$

where ϕ_I and ϕ_T are in-plane and through-thickness chemical shrinkages. This is a simplified version of the equation given by Radford and Rennick.[24] As discussed below, equation (7.2) is in fact only valid after vitrification, but can be used to calculate spring-in during this phase by taking ϕ_I and ϕ_T as the chemical shrinkage after vitrification. However, for many resins chemical shrinkage after vitrification is small. This was studied for AS4/8552 by measuring the curvatures of flat unsymmetric cross-ply laminates after interrupted cure cycles.[15] Specimens which were rapidly quenched just after the material vitrified showed similar final curvature to those quenched from the same temperature after the full cure cycle was completed. This shows that chemical shrinkage after

188   Composites forming technologies

vitrification is negligible for this material, and would not be expected to contribute significantly to spring-in.

7.4.3 Spring-in due to shrinkage before vitrification

Most current models such as equation (7.2) do not distinguish between shrinkage before and after vitrification. However, in the phase between gelation and vitrification the composite is in a rubbery state, and can shear through the thickness. If it could shear freely, this would provide a mechanism to take up the change in circumferential length due to through-thickness contraction without inducing stresses in the fibre direction, eliminating any additional spring-in.

This is illustrated in Fig. 7.10. The schematics on the left show what happens when the material is stiff in shear, as it is when fully cured. With a through-thickness contraction while the part is constrained to the tool, there is no change in shape, and in-plane stresses are generated. When the part is released from the

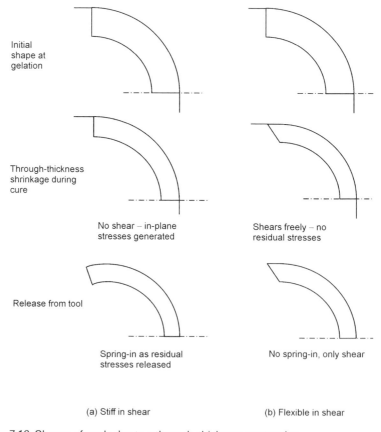

7.10 Change of angle due to a through-thickness contraction.

tool these stresses are released, and the part springs in. The schematics on the right illustrate the situation when the material is compliant in shear, as it is early in the cure. Now a through-thickness contraction leads to shearing of the part, with no in-plane stresses generated. There is therefore no spring-in, only through-thickness shearing.

In practice the material is not completely free to shear and this can be treated by a shear lag analysis.[25] If the rubbery shear modulus is high, or the composite is long compared with its thickness, the same result is obtained as with equation (7.2). However, as the shear modulus reduces, or as the length to thickness ratio decreases, the predicted amount of spring-in reduces.

Results are shown in Fig. 7.11 for AS4/8552 cross-ply laminates with different ratios of radius to thickness using a measured through-thickness shrinkage of 0.98%[26] and estimated through-thickness shear modulus of 43 MPa. The in-plane shrinkage was assumed to be negligible. When the thickness is large compared with the radius, the specimen is able to shear, and the predicted spring-in approaches the thermoelastic value of 0.62° given by equation (7.1). As the thickness becomes small compared with the radius, the predicted spring-in approaches the value of 1.50° given by equation (7.2).

This effect has been demonstrated experimentally by measuring spring-in of segments of 270° of cross-plied $(0/90)_{ns}$ AS4/8552 tubes from 1 to 4 mm thick. They were cured using a pressure bag simulating autoclave conditions on the inside of a 50 mm internal diameter carbon-fibre tube to eliminate any effects due to tool interaction.

The outer diameters of the composite parts were measured at room temperature and equivalent spring-in angles were calculated for a 90° angle with respect to the tube inner diameter measured at 180°C. The results confirm

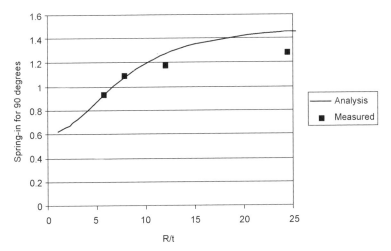

7.11 Shear lag effect on spring-in for different thickness curved cross-ply sections.

the predicted trend of reducing spring-in with increasing thickness, with values varying from 1.28° at 1 mm thickness to 0.93° at 4 mm. The shear lag analysis correlates quite well with the experimental results.

It is the total change in through-thickness strain between gelation and vitrification that controls spring-in during this phase, including thermal effects. If gelation occurs at a lower temperature than vitrification, there will be a through-thickness expansion on heating up to vitrification which will reduce the effect of chemical shrinkage. Both these components were included in the experimental value of shrinkage quoted here. Note that shrinkage before gelation will not contribute to this particular mechanism since the part geometry is only fixed when the resin gels.

When a curved section continues into adjacent flat sections such as with an L-shaped geometry, then the flat sections will have a similar constraining effect to a longer arc length, increasing the predicted spring-in towards the value given by equation (7.2). The resulting strain gradients in the flat sections will also cause further distortion due to bowing of the arms, complicating the deformed shape and making simple comparisons of spring-in angles difficult.

7.4.4 Spring-in due to fibre wrinkling and tool-part interaction

As discussed in Section 7.3.3, the higher thermal expansion of many tool materials compared with the composite causes the tool to stretch the part. If a stress gradient arises through the thickness, it will produce distortion that could either increase or decrease spring-in, depending which side the tooling is. However, on a curved part, there is a further effect as the stretching from the tool can interact with wrinkles formed in the curved sections. When prepreg is laid up around a radius, the path difference between inner and outer surfaces leads to fibre wrinkling on the inner surface such that any tool-part interaction stresses will be disproportionately carried by the fibres on the outside of the radius. This stress gradient opposes the effects of the other spring-in mechanisms, and can even cause spring-out in very thin parts.

This is illustrated by the results from the unidirectional L specimens with different thicknesses mentioned in Section 7.4.1. As well as the thermoelastic response, the total spring-in angles were measured with the help of a coordinate measurement machine.[21] Straight lines were fitted to the data from points at the ends of the radius to the ends of the arms to avoid the influence of arm bowing. Results are shown in Fig. 7.12 for specimens from 4 mm to 0.5 mm thick. The two thicker specimens gave similar results, but the 1 mm one showed a reduction in spring-in, whilst the 0.5 mm specimen actually sprang out. A similar trend was observed with cross-ply samples, although the magnitude of the effect was reduced.

The tool material was Invar, which still expands sufficiently to produce significant tool-part interaction. An even larger effect was found with an

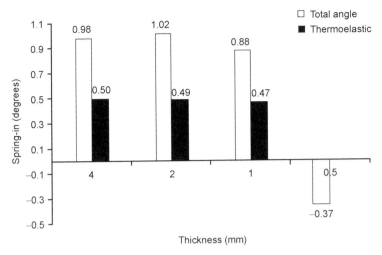

7.12 Effect of thickness on spring-in of unidirectional L sections.[23]

aluminium tool, with spring-out of as much as 16° measured in single-ply L sections. Note that this trend of reducing spring-in for thin parts is the opposite to that due to shear lag discussed in Section 7.4.3.

These effects have been analysed using a simple analytical model assuming the material at the surface expands with the tool, but that there is a linear stress gradient through the thickness.[27] The extent of the gradient was deduced from the kinematics of the change in circumferential length arising from the varying radius through the thickness when the plies are laid up in the corner. The analysis was able to explain the observed trends and give the approximate magnitude of the effect.

7.4.5 Effect of consolidation on spring-in

Consolidation due to resin flow and cure shrinkage prior to gelation causes reductions in thickness. The inextensible fibres tend to reduce this effect in curved sections cured on concave tooling, leading to higher corner thickness in the absence of ply-ply slip. This change of geometry can cause a small increase in thermoelastic spring-in, as can be seen from the results in Fig. 7.12. L sections 4 mm thick with corner thickening gave a thermoelastic spring-in of 0.50° compared with 0.47° for similar parts only 1 mm thick with relatively constant thickness. Although this is a fairly small effect, it was consistent. Cross-ply specimens with layup $(0/90)_{ns}$ exhibited a similar increase in spring-in from 0.63° at 1 mm thickness to 0.67° at 4 mm.

This difference can be predicted by finite element analysis taking account of the different geometry and volume fractions. Figure 7.13 shows a model used for this,[23] where the inner radius was increased to reproduce the corner thickness

192 Composites forming technologies

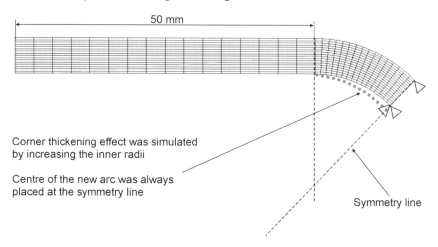

7.13 Finite element model of the effect of corner thickening on spring-in.[23]

measured experimentally. Material properties were adjusted for the change in volume fraction using micromechanics equations and assuming the thickness change was accompanied by resin flow into the corner. The results showed an increase in predicted spring-in from 0.47° to 0.49° for unidirectional material and from 0.62° to 0.65° for cross-ply, close to the measured values. The resin flow into the thickened corner can cause thinning of the adjacent laminate which may also have an influence on the distortion.

There are also other more complex effects due to consolidation which are still not properly understood. For example, stresses may be set-up due to the autoclave pressure on inside radii being reacted by tension in the fibres, which can contribute to spring-in. Separation between the part and tool may occur in the corner despite the autoclave pressure, and can increase spring-in.

7.5 Distortion in more complex parts

The same mechanisms discussed above can cause more complex distortions in larger scale structures. For example, measurements on long C-shaped components showed that in addition to the expected spring-in, there was twist along the length of the channel, which was attributed to tool-part interaction causing tension in the surface 45° ply.[21] There was also bowing of the web, which was again attributed to tool-part interaction.

Where the geometry of the part causes it to lock to the tool, high stresses can be generated, which can cause large distortions where stress gradients are present. For example, substantial distortion has been demonstrated in tests where a single ply of prepreg was wound around an aluminium tool to force interaction.[28] Tensile stresses cause bending near the corners because of wrinkling, which means most of the load is carried in the straighter fibres on the outside. In

Understanding composite distortion during processing 193

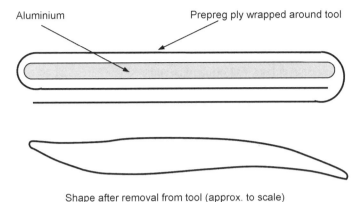

7.14 Schematic of distortion due to forced interaction with tool.[28]

addition, where the ply overlaps, there is a large stress gradient caused by the change in neutral axis. These effects produced the complex distortion shown in Fig. 7.14, with up to about 30 mm deflection on a part 500 mm long. Dropped plies in tapered layups similarly cause bending if there is tension from tool-part interaction. Analagous effects due to the part locking to the tool can be caused by the shape of the component, for example with features such as joggles or steps.[28] In these regions the influence of the fibre wrinkling effect depends on the extent to which the reversal of corner curvature leads to the reduction of wrinkling. These effects are rather complex and appear to depend, for example, on the quality of layup and fit of reinforcement to the tool.

Constraint between deformations of different parts of built-up structures can cause more complex distortions. For example, in a curved C spar with web stiffeners, the curved sections will tend to spring-in, but will be prevented from doing so by the stiffeners.[29] The shape arising can be analysed by finite element modelling including the mechanisms of thermal stress, cure shrinkage and tooling constraint. Other sophisticated finite element programs have been developed specifically for analysing manufacturing distortion, such as COMPRO,[30] which allows the whole process to be simulated including heat transfer, cure kinetics, resin flow, tool-interaction and residual stress development. This can model in detail the deformations induced in 2-D sections, and has also been used successfully in conjunction with 3-D models to simulate distortion in complex aircraft parts such as the T-45 box rib and 777 trailing edge fairing sandwich structures.[31]

Where large deformations of the material occur during forming of highly shaped parts, significant changes in fibre angles occur, which can greatly change the thermoelastic material constants and hence the residual stresses and distortions. Temperature gradients may also occur, which again can influence part distortion. These mechanisms have been successfully modelled by finite element simulation of parts formed on square and cylindrically shaped punch tools, and an automotive body panel.[32] Similar effects have been seen on thermoplastic parts stamped out using rubber pad forming.[33]

Lastly, whilst the research emphasis has rightly been on understanding the causes of process-induced deformations and residual stresses there is also the potential for deliberately inducing residual stress and distortion states to achieve specific ends such as a multiplicity of stable geometries.[34]

7.6 Conclusions

The mechanisms causing distortion during the cure have been discussed, the most important being differential thermal contraction during cooldown, cure shrinkage and tool-part interaction. Distortion of flat plates due to asymmetric layups can be satisfactorily predicted from laminated plate analysis, with the contribution from chemical shrinkage usually being small. Where resin bleed occurs there can be significant changes of volume fraction through the thickness, and the effect of these on distortion can also be analysed with laminated plate theory. Thermal expansion of the tool can cause tensile stresses in the surface of the part leading to bowing away from the tool. This is harder to predict because it depends on the gradient of stress through the thickness, which is a function of the frictional behaviour of the material at the tool surface and between plies of different orientations.

Curved parts exhibit a thermoelastic spring-in on cooldown that can be predicted using laminated plate theory. In addition there is a non-thermoelastic component of spring-in due to shrinkage occurring primarily between gelation and vitrification. This depends on the way the material shears through the thickness, and so is a function of the constraint due to part geometry and ratio of in-plane dimensions to thickness as well as material properties. It can be predicted for simple curved geometry, but its effect on real parts is much more complex. Tool-part interaction can cause further distortion that can either increase or decrease spring-in, depending which side the tooling is on. This can interact with the effect of fibre wrinkling on the inside of the radius, causing a stress gradient through the thickness which tends to reduce the spring-in angle, and can even cause spring-out in very thin parts. Corner thickening associated with consolidation tends to slightly increase spring-in.

The same mechanisms are responsible for distortions in larger scale structures, with the constraint between different parts that tend to distort differently being crucial. Tool-part interaction mechanisms, in particular, can become complex due to features of the geometry causing the part to lock to the tool. Some successes have been achieved in modelling distortions of more complicated parts, but there is still a need for further understanding of the mechanisms and advances in the analysis tools before the behaviour of real components can be predicted with confidence.

7.7 References

1. Nedele, M. and Wisnom, M. R., Micromechanical modelling of a unidirectional carbon fibre reinforced epoxy composite subjected to mechanical and thermal loading, *Proc. American Society for Composites Seventh Technical Conference, Pennsylvania State University*, pp 328-338, October 1992.
2. Hyer, M. W., Some observations on the cured shape of thin unsymmetric laminates, *Journal of Composite Materials* 15:175–194, 1981.
3. Kim, K. S. and Hahn, H. T., Residual stress development during processing of graphite/epoxy composites, *Composites Science and Technology* 36:121, 1989.
4. Hyer, M. W., Rousseau, C. Q. and Tomkins, S. S., Thermally induced twist in graphite-epoxy tubes, *Journal of Engineering Materials and Technology* 110:83–88, 1988.
5. Nelson, R. H. and Cairns, D. S., Prediction of dimensional changes in composite laminates during cure, *34th International SAMPE Symposium* pp 2397–2410, May 1989.
6. Wisnom, M. R., Gigliotti, M., Ersoy, N., Campbell, M. and Potter, K. D., Mechanisms generating residual stresses and distortion during manufacture of polymermatrix composite structures, *Composites Part A*, 37:522–529, 2006.
7. Li, C., Potter, K. D., Wisnom, M. R. and Stringer, L. G., In-situ measurement of chemical shrinkage of MY750 epoxy resin by a novel gravimetric method, *Composites Science and Technology* 64:55–64, 2004.
8. Unger, W. J. and Hansen, J. S., The effect of cooling rate and annealing on residual-stress development in graphite fiber reinforced peek laminates, *Journal of Composite Materials* 27:108–137, 1993.
9. Twigg, G., Poursartip, A. and Fernlund, G., Tool-part interaction in composites processing Part I: experimental investigation and analytical model, *Composites* 35A:121–133, 2004.
10. Ganley, J. M., Maji, A. K. and Huybrechts, S., Explaining spring-in in filament wound carbon fiber/epoxy composites, *Journal of Composite Materials* 34:1216–1239, 2000.
11. Wu, Y. J., Takatoya, T., Chung, K., Seferis, J. C. and Ahn, K., Development of the transient simulated laminate (TSL) methodology for moisture ingression studies using unsymmetric laminates, *Journal of Composite Materials* 34:1998–2015, 2000.
12. Radford, D. W., Cure shrinkage-induced warpage in flat uniaxial composites, *Journal of Composites Technology & Research* 15:290–296, 1993.
13. Hubert, P. and Poursartip, A., Aspects of the compaction of composite angle laminates: An experimental investigation, *Journal of Composite Materials* 35:2–26, 2001.
14. Bogetti, T. A. and Gillespie, J. W., Process-induced stress and deformation in thick-section thermoset composite laminates, *Journal of Composite Materials* 26:626–660, 1992.
15. Gigliotti, M., Wisnom, M. R. and Potter, K. D., Development of curvature during the cure of AS4/8552 [0/90] unsymmetric composite plates, *Composites Science and Technology* 63:187–197, 2003.
16. Dano M. L. and Hyer M. W., Snap-through of unsymmetric fiber reinforced composite laminates, *International Journal of Solids and Structures* 39:175–198, 2002.
17. Gigliotti, M., Wisnom, M. R. and Potter, K. D., Loss of bifurcation and multiple shapes of thin [0/90] unsymmetric composite plates subject to thermal stress, *Composites Science and Technology* 64:109–128, 2004.
18. Potter, K. D. Campbell, M. and Wisnom, M. R, Investigation of tool/part interaction

effects in the manufacture of composite components, *14th International Conference on Composite Materials, San Diego*, July 2003.
19. Twigg, G., Poursartip, A. and Fernlund, G., An experimental method for quantifying tool-part shear interaction during composites processing, *Composites Science and Technology* 63:1985–2002, 2003.
20. Ersoy, N., Potter, K. and Wisnom, M. R., An experimental method to study the frictional processes during composites manufacturing, *Composites Part A* 36:1536–1544, 2005.
21. Garstka, T., Ersoy, N., Potter, K. and Wisnom, M. R., The effect of tool-part interaction on the shape of curved composite laminates, to be published.
22. Bapanapalli, S. K. and Smith, L. V., A linear finite element model to predict processing-induced distortion in FRP laminates, *Composites Part A* 36:1666–1674, 2005.
23. Garstka, T., Potter, K. and Wisnom, M. R., Thermally induced spring-in in composite L-shaped and U-shaped laminates, to be published.
24. Radford, D. W. and Rennick, T. S., Separating sources of manufacturing distortion in laminated composites, *Journal of Reinforced Plastics and Composites* 19:621–641, 2000.
25. Wisnom, M. R., Ersoy, N. and Potter, K. D., Shear lag analysis of the effect of thickness on spring-in of curved composites, *Journal of Composite Materials*, in press, 2006.
26. Garstka, T., Ersoy, N. and Potter, K. D., In-situ measurements of through-the-thickness strains during processing of AS4/8552 composite, *Composites Part A*, submitted.
27. Garstka,T., C. Langer, K. D. Potter and M. R. Wisnom, Combined effect of tool-part interaction and fibre wrinkling on the shape of curved laminates, *Proc. CANCOM, Vancouver*, August 2005.
28. Potter, K. D., Campbell, M., Langer, C. and Wisnom, M. R., The generation of geometrical deformations due to tool/part interaction in the manufacture of composite components, *Composites Part A*, 36A:301–308, 2005.
29. Svanberg, J. M., Altkvist, C. and Nyman, T., Prediction of shape distortions for a curved composite C-spar, *Journal of Reinforced Plastics and Composites* 24:323–339, 2005.
30. Johnston, A., Vaziri, R. and Poursartip, A., A plane strain model for process-induced deformation of laminated composite structures, *Journal of Composite Materials* 35:1435–1469, 2001.
31. Fernlund, G., Osooly, A., Poursartip, A., Vaziri, R., Courdji, R., Nelson, K., George, P., Hendrickson, L. and Griffith, J., Finite element based prediction of process-induced deformation of autoclaved composite structures using 2D process analysis and 3D structural analysis, *Composite Structures* 62:223–234, 2003.
32. Hsiao, S. W. and Kikuchi, N., Numerical analysis and optimal design of composite thermoforming process, *Computer Methods in Applied Mechanics and Engineering* 177:1–34, 1999.
33. Wijskamp, S. and Akkerman, R., Tool-part interaction during rubber forming of composites, *8th ESAFORM Conference On Material Forming. Cluj-Napoca, Romania*, April 27–29, 2005.
34. Potter, K. D. and Weaver, P. M., A concept for the generation of out-of-plane distortion from tailored FRP laminates, *Composites A* 35A, 1353–1361, 2005.

8
Forming technology for composite/metal hybrids

J SINKE, Technical University Delft, The Netherlands

8.1 Introduction

Today, we are living in a multi-material society. This is true for consumer products, but also for engineering artefacts. Many structures are assembled from a selection of different materials. Materials for engineering purposes are, for example, metal alloys, composite materials, ceramics, and natural materials. These different materials are not only used side by side, at the level of structural elements, but also in mixed or so-called hybrid materials. Typical examples of hybrid materials are sandwich materials, metal matrix composites, and fibre metal laminates.

Ideas for new hybrid materials arise when one constituent alone cannot fulfil a specific set of requirements. Creating a hybrid material could be an option to solve this problem: by combining different materials, advantages of the constituents can be added and the new hybrid can comply with the intended objective. That does not imply that hybrid materials are super-materials, but that they have good to excellent values for a specific set of properties.

Sandwich materials, for example, offer a high bending stiffness in combination with a low weight. In addition, sandwich materials also have very good acoustic damping and thermal isolation properties. The sandwich concept involves a beam-like element with load carrying facings and a lightweight core for the shear loads and the support of the facings. When combined with high tensile fibres, the sandwich structures could be used for very specific applications, for which no other material (or combination of materials) is currently available. Typical examples are the 'Voyager' (1986), the first aircraft that flew non-stop around the world and the 'Space Ship One' winner of the X-price competition (2004). Both aircraft are designed and built by Scaled Composites of Burt Rutan (CEO). But also commercially available composite aircraft are made of sandwich materials, like the 'Starship' and the 'Extra 400'.

Metal Matrix Composites are hybrid materials made of ceramic fibres or whiskers in a metal matrix. Examples of ceramic fibres are alumina or silicon carbide fibres, and the matrix is often a metal alloy with a rather low melting

198   Composites forming technologies

8.1 The FML-concept: a 3/2 lay-up of a fibre metal laminate (three layers of metal alloy and two layers of composite – each composite layer consisting of two to four prepreg layers).

point like magnesium or aluminium alloys. The combination of these materials offers engineering materials with good mechanical properties at elevated temperatures, up to 400°C. Typical MMC-applications are engine parts and parts subjected to excessive wear.

The last examples to be mentioned here are the fibre metal laminates (FML). These laminates have been developed for their excellent damage tolerance properties, such as fatigue resistance, residual strength, and impact resistance. These laminates consist of alternating layers of thin metal sheet and composite layers (see Fig. 8.1). Current applications of these laminates are fuselage panels of the Airbus A380, leading edges of the tail planes of the A380, and blast-resistant containers.

In this chapter an overview is presented about the forming technology of composite/metal hybrids, in particular the fibre metal laminates (FML). The main reason for this focus is that sandwich materials, based on thermoset materials, deform elastically only, and MMC are generally cast. A second reason is that FML offer a good example to illustrate the typical phenomena of hybrids.

In the second section a brief outline will be given about the development of FML, followed by a section about the properties of these laminates. In Sections 8.4 and 8.5 the appropriate forming processes and the modelling of these processes are described. The last two sections give some concluding remarks and references.

8.2 Development of composite/metal hybrids

The different hybrids consisting of composite and metal constituents have their own history. This is illustrated by a brief description of the development of the sandwich concept, and a more extensive one about the development of fibre metal laminates.

8.2.1 Sandwich materials[1,2]

The concept of sandwich materials has been known since the middle of the 19th century. The first published example was a bridge, built in 1846 in Wales, for which sandwich-like structural elements were used. These elements were made of steel faces and wooden core materials riveted together. Later, in Germany as well as in England, new sandwich materials were developed. The main reason for this development was obtaining lightweight structures with a high bending stiffness. In 1943 the honeycomb structure was invented, so after the Second World War there was a wide variety of possible material combinations: face sheets made of metal alloys, composites and plastics, and core materials made of Balsa wood, synthetic foams, honeycomb, etc. Among the most commonly used sandwich materials are the ones with composite facings and metal (aluminium) honeycomb, used in space structures, and sandwiches made of metal facings and a synthetic foam or honeycomb, used in aircraft structures. However, the sandwich concept is most widespread in full composites, with composite facings and a foam or honeycomb for the core (no metallic constituents).

8.2.2 Overview of the development of FML[3]

The history of FML starts with the bonding technology Fokker applied on their F-27 aircraft launched in 1955. The main reasons for the introduction of bonded structures were costs (related to lack of investment capital) and structural stability. But the layered structures also offered another advantage: bonded structures had significantly better fatigue resistance. In the following decades the bonding technology was further developed. One of the items of that research was that Fokker tried to improve the fatigue resistance by applying fibres in the bond layers. The fatigue resistance was further improved but not significantly and Fokker stopped further research. Other reasons were that Fokker had no new aircraft where they could apply the new material, and replacing existing materials was too expensive.

The Technical University of Delft continued the research in the late 1970s, and started changing the materials and the thickness of the layers. The reduction of metal layer thickness, in particular, resulted in a big improvement of the fatigue resistance (see Fig. 8.2).

In 1979 the first prototype of a fibre metal laminate was tested at the University. This laminate was named ARALL: *AR*amid *AL*uminium *L*aminate, and was based on aramid fibres. From the early 1980s onwards, the research on the laminates expanded and some industries showed their interest. Companies like ALCOA (US) and AKZO (NL) sponsored the research. The first large application came with the cargo-door of the C17 military transport aircraft. The door performed well but later a metal one replaced it since the ARALL-door was too expensive.

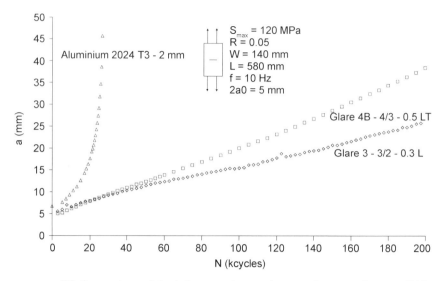

8.2 Comparison of the fatigue crack growth curves between the monolithic aluminium alloy 2024-T3 and GLARE 3. Note: until the 1990s, 2024-T3 had been the baseline alloy for aircraft fuselages; it was the best choice with respect to fatigue resistance.

In 1986 research started on a second FML-variant: GLARE, based on glass fibres. The main reason to switch to glass fibres was that the aramid fibres failed under particular loading conditions. But fibre failure is unacceptable: the fatigue resistance relies on the mechanism of crack bridging. The glass fibres in GLARE do not have this disadvantage, and therefore GLARE became the most important FML-type. During the development of GLARE another big problem was solved. By so-called 'splicing' of the metal layers, very large skin panels could be manufactured without (riveted) joints. In the ARALL- period the laminates were treated as metal sheets: first, flat laminates were manufactured which were subsequently formed and joined to larger structures. For GLARE, however, the laminates were made like composites: the large skins (and reinforcements) are made by lay-up processes.

In the early 1990s Airbus Industry started a design study for a very large aircraft. This aircraft should complete the 'family of aircraft' offered to customers. After a lot of market and feasibility studies, the final design took shape around 1996 and was called 'A3XX'. At the same time GLARE was mature enough to be regarded as a potential candidate for the fuselage of this aircraft. After discussions with Airbus, the research in Holland (University of Delft, the National Aerospace Laboratory and Stork Fokker) increased significantly. The government was willing to support this basic research, which had the objective to make GLARE ready for application in full-size structures. As the final result of this process the GLARE laminates are applied on significant parts of the A380 fuselage (large front and aft sections). Stork Fokker produces most of these skin panels in a specially built factory. Besides the fuselage panels GLARE laminates are also applied in the leading edges of the vertical and horizontal tail planes of the A380.

8.3 Properties of fibre metal laminates

8.3.1 Mechanical properties

The mechanical properties of fibre metal laminates (FML) depend on the constituents of the laminates: the fibres, the metal alloy, and the adhesive or resin. The properties of the fibres and the metal alloy are dominant for most properties like strength and stiffness. The properties of the adhesive are important only in the laminate properties in which a shear component is involved. A typical example is the interlaminar shear strength (ILSS).

The laminates are anisotropic or orthotropic, depending on their composition. Therefore, as a result the in-plane mechanical properties are presented for the longitudinal (L) direction (parallel to the rolling direction of the metal layers) and the long-transverse (LT) direction (in plane but perpendicular to the rolling direction). For an A-version of a particular laminate the (main) fibre direction and the rolling direction coincide; for a B-version the (main) fibre direction and rolling direction are perpendicular to each other.

Tension test

In Table 8.1 the most important mechanical properties for a few FML-types are presented, with the values of aluminium alloys for comparison. In the following text a few comments are made about these properties.

The Young's modulus or elastic modulus (E) depends on the fibre directions: this is obvious for the GLARE-2 laminate, which only has fibres in L-direction. GLARE-3 is a cross-ply laminate with 50% of the fibres in L-direction and 50% of the fibres in LT-direction and the E-modulus in L- and LT-directions for GLARE-3 laminates is equal. The metal alloys, the fibres, and their respective volumes determine the magnitude of the E-modulus. The classic rule of mixtures that is used for full composites can also be applied for FML.

The yield stress (σ_y) is dictated by the metal alloys, which is the only constituent with plastic yielding. The value of the yield stress, however, is also influenced by the fibres, by the stiffness of the fibres and the fibre content of the laminate. The ultimate tensile stress (σ_{ult}) is dictated by the failure of the fibres. Fibres deform elastically until failure. The failure of the fibres is abrupt and accompanied by a significant release of energy, causing fracture and delamination of a significant part of the specimen.

The fibres also dominate the failure strain (ϵ_f) of the laminate. When fibres

Table 8.1 Mechanical properties for some GLARE-laminates and some aluminium alloys

Property	Sym	Dir	Dim	GLARE-1	GLARE-2	GLARE-3	2024-T3	7475-T6
Young's modulus	E	L	GPa	65	66	57.5	72.5	71
		LT		49	50	57.5	72.5	71
Yield stress σ_y		L	MPa	550	400	320	324	483
		LT		340	230	320	290	469
Ultimate tensile stress	σ_{ult}	L	MPa	1300	1230	755	440	538
		LT		360	320	755	435	538
Failure strain	ϵ	L	%	4.5	5	5	13.6	8
		LT		7.5	13.5	5	13.6	8

Sym = symbol
Dir = orientation/direction
Dim = dimension
GLARE-1: UD laminate; Al-7475-T6 alloy
GLARE-2: UD laminate; Al-2024-T3 alloy
GLARE-3: CP-laminate (50/50); Al-2024-T3 alloy

are running in loading direction the failure strain of the laminate is equal to the (small) failure strain of the fibres, which is in the range of 1–5%, depending on the fibre system. Only for UD-laminates with fibres in one direction (L-direction) the failure strain in LT-direction is somewhat larger, and comparable to the failure strain of the metal alloy.

The strain hardening beyond the yield stress is almost linear. Since the fibres continue to deform elastically, their contribution to the stress, once the metal alloy yields, is significant. As a result, the stress-strain relation beyond the yield point is almost linear, which results in a typical bilinear stress-strain curve.

Fatigue

The high fatigue resistance is the most important property of FML. In Fig. 8.2 crack growth curves are presented for GLARE-3 and aluminium 2024-T3. From that figure it is obvious that the crack growth rate in FML is much smaller. For aluminium alloys about 80–90% of the total fatigue life is required for crack initiation, and about 10–20% for crack growth. Once a crack is initiated, the crack grows relatively fast. For FML it is the opposite: about 10–20% (or less) of the fatigue life is related to crack initiation, and about 80–90% to crack growth. The number of cycles to initiate a fatigue crack in both aluminium and FML is comparable.

The slow crack growth in FML is caused by a mechanism called 'fibre bridging' (see Fig. 8.3). The metal layers may have fatigue cracks, but the fibre

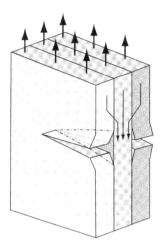

8.3 Schematic presentation of the crack bridging mechanism in FML.

layers stay intact, bridging the crack. The fibres bypass the stresses of the cracked aluminium layers, whereas in monolithic aluminium the stresses have to 'flow' around the crack tip. In the latter case the stress intensity at the crack tip and the crack growth rate increase significantly. An essential part for the crack bridging is the local delamination of the fibre and metal layers. For a slow crack growth there should be a good balance between the crack size, the delamination and the properties of the constituents.[4]

Residual strength

The residual strength of a material is the remaining strength, when damage is present in the material. In general, this property is tested using sheet specimen with a specified damage to it (e.g., a saw-cut of specific length). FML have high residual strength values when compared with aluminium alloys. Again the main reason is the crack bridging by the fibres. When the damage is caused by fatigue, the crack bridging is maximised; when the fibres are cut by the damage source, some crack bridging will still occur during the stable crack extension preceding the final failure.[5]

Impact and blast resistance

The combination of metal plasticity and a high strain hardening, which is facilitated by the fibres, also give the FML high impact and blast resistance. In particular, the blast resistance is high, since a large portion of the plastic deformation capacity of the material is used in applications like the bomb-proof luggage container that is made by ECOS (US).

8.3.2 Some physical properties

Density

The density of the laminates depends on the constituents and composition of the laminates. The density of FML is usually smaller than the density of the related metal alloy. By adding composite layers, which have densities in the range of 1.5–2.0 kg/dm^3, the overall density of, e.g., GLARE is about 10% lower than for the aluminium alloys.

Electric and magnetic properties

The electric and magnetic properties of a laminate are related to its constituents: often, metal alloys are good electrical conductors, but polymers are not. For the fibres: the aramid and glass fibres have poor conductive properties, but carbon fibres are good conductors. So, depending on the combination of metal alloys, fibres, and polymers, the laminates can have a wide range of different characteristics. In addition, the in-plane properties are often much different from the properties in thickness direction, where the composite layers may act as barriers.

Internal stresses

FML are made by alternating metal sheets and prepreg layers. The layers are stacked, sealed in a vacuum bag and cured in an autoclave at elevated temperature and high pressure. The temperature depends on the polymer that is used as matrix material. Most resins for GLARE-type laminates are toughened epoxies, and have curing temperatures of about 100–150°C. Curing different materials at these temperatures causes internal stresses, since the coefficients of thermal expansion differ. Therefore, in a GLARE laminate, after curing the aluminium layers are in tension and the composite layers are in compression. The magnitude of these residual stresses depend on the thermal properties, the composition, the number and thickness of the layers, fibre orientations, etc. Typical values are 10–30 MPa (tension) for the metal layers and 40–80 MPa (compression) for the composite layers.

Due to their composition FML offer excellent resistance to flames, corrosion, etc. The alternating layers protect each other and penetration of flames, moisture, etc., is very difficult because of the thickness. This is explained in the following two examples.

Fire resistance[6]

FML do have good fire resistance, in particular the thicker laminates (4/3-lay-up and larger). When conventional aluminium alloys are exposed to flames, the flame penetrates the sheet in a very short time (less than two minutes). FML based on aluminium sheets require a much larger time before flames penetrate through the laminate (in the order of 10–15 minutes). The reason is that the composite layers carbonise and delaminate, thereby shielding the heat from inside layers. At the same time the aluminium, due to its conductivity, dissipates the heat over a large area.

Corrosion resistance[7]

Another protection mechanism is active when the laminate is exposed to moisture. In metal alloys this will result in corrosion: the metal dissolves locally, and pits and holes arise in the metal. In FML, the composite layers stops further corrosion attack, and the corrosion is limited to the outside metal layer. On the other hand, the metal layers protect the composite layers. These layers can deteriorate under the influence of ultraviolet radiation and moisture. Metal layers cover the composite layers (except for the edges), and therefore, there is almost no deterioration of the composite layers. Only minimum moisture ingress is possible via the edges.

8.4 Production processes for fibre metal laminates

Production processes for hybrids are related to the properties of the constituents. In this section both machining and forming processes are described, with emphasis on the forming processes. Many production processes have a resemblance to either metal or composite production processes.

8.4.1 Cutting processes for FML

Cutting of FML is not too difficult, but the constituents and the composition of the laminates require procedures different from the cutting of metal alloys or composites. For cutting of FML, the following issues should be kept in mind:

- The highly abrasive fibres cause rapid tool wear, unless specific tooling materials are used. This is true for glass and carbon fibres. For aramid fibres wear resistant tooling is also required, but for another reason. The aramid fibre is neither hard nor abrasive, but very tough, and only sharp cutting edges are able to cut these fibres.
- Dull cutting tools, due to tool wear, produce heat. This heat can affect the laminate, and is specifically detrimental to the laminate causing delaminations and/or matrix damage.
- Forces perpendicular to the laminates may cause delaminations. These forces act perpendicular to the interfaces between the different layers, and should be controlled carefully. Part of this control is the limitation of tool wear, since

the decreasing effectiveness of wearing cutting tools is often compensated with an increase in the cutting force.

These problems are related to the wear of the tool bit; therefore, adapting the tool material to the resist the wear is a very important measure. Tests have revealed that tool bits made of solid cemented carbide/hard metal (HM) or poly crystal diamond (PCD) tipped tool bits offer the best solutions. Unfortunately, these tool bits are also brittle so they can only be used in stable machines. For manual operations, a combination of a tough HSS tool body and HM or PCD cutting edges is required.

Some cutting processes, such as (abrasive) water jet cutting or laser jet cutting, are not in contact with the laminate. Abrasive water jet cutting is used for cutting edges or making cutouts, although the latter is more complicated. The edge quality is not very high: the edge has a sandy appearance, which is not acceptable in every application. The laser jet cutting introduces a small heat-affected zone in the laminates: a zone in which the condition of the metal alloy has been changed and where the matrix material is damaged. Despite the fact that this zone is very small, for aerospace applications, the existence of this zone is not acceptable.

Cutting takes place prior to a forming process and after a forming or lay-up process. For the cutting prior to the forming process no close tolerances are required, but for the cutting or trimming after the forming or lay-up process, the tolerances and edge qualities are high. This will influence the selection of the cutting process. Some common cutting processes for FML are discussed below.

Shearing

Shearing is a process used for rough blanking of metal sheets or blanks used in subsequent forming processes. The mechanism of this process is based on indentation and shearing of the material in its thickness direction. Typical processes applying this principle are punching, shearing and slitting. FML can also be sheared but the allowable thickness of the laminate is limited. The main reason is the high local forces, which increase with increasing thickness, and result in delaminations.

Routing or edge milling

The most common process to machine the edges of sheet materials is routing. In this process a simple mill trims the laminate to the right dimensions. Routing does not cause any problems when sharp and wear-resistant tool bits are used. When the tool bit becomes dull, the chips and debris can be pushed between the layers of the laminate. An additional requirement for routing is that the helix angle of the tool should be small, otherwise the top layer of the laminate is peeled off. Alternative processes for routing are the water and laser jet-cutting processes. As stated before the edge quality for these processes is not as high as for a milling process.

Drilling

Most structures are made by joining structural elements like shells, skins, beams, and profiles. A favourable joining method is mechanical fastening, using rivets or bolts. When this joining technique is selected, holes should be drilled in the structural elements. As for the other cutting processes, the drill bits should be sharp and remain sharp. When the drill bit is sharp, the feed force, perpendicular to the laminate, is small. A small feed force is necessary to prevent the delamination of the layers. As for routing, the helix angle of the drill bit is important: that angle should not be too large, in order to prevent delamination (peeling) of the top layer.

8.4.2 Formability aspects of FML

In-plane deformations

Metal alloys can be deformed plastically and the failure strain is usually in the order of 10–50%. Metal alloys also exhibit some elastic deformation, resulting in spring back, residual stresses, or a combination of the two. The achievable combinations of strains are large; they can be decomposed in in-plane tensile strains, in-plane compressive strains, in-plane shear strains, bending deformations, etc.

Composite layers have a very limited formability. The failure strain of the fibres is small: from 1% for carbon fibres, and 2–2.5% for aramid fibres up to 4–5% for glass fibres. The deformation of fibres is pure elastic, so after deforming, the fibres cause some kind of spring back. The failure strain of the matrix or resin in a prepreg is limited, too, in particular for thermoset polymers. In addition, the embedded fibres create stress concentrations in the matrix, further reducing the failure strain. Thermoplastic matrix materials offer a (much) larger failure strain, but the high glass-transition temperature T_g, causes high internal stresses (see Section 8.3.2).

The laminates can be divided in two groups: uni-directional (UD) laminates, all fibres are placed in one direction, and laminates with fibres in two orthogonal directions, cross ply (CP) laminates. UD-laminates are applied in profiles like stringers. These prismatic elements are bent with the fibres running in longitudinal direction. In that case the fibres are not deformed, and the small failure strain of the fibre does not limit the deformations. In case of CP-laminates, the stretching and bending in fibre directions is severely limited by the limited strain of the fibres. The only way to have some deformation capability is stretching in the bias-directions of the laminate. In that case the so-called 'trellis-effect' is activated.

Bending

Deformations in FML are also influenced by the lay-up of the laminates. Thin laminates are bent easily: the distance of the deforming fibres to the neutral axis is small. For a 2/1 lay-up the fibres are at the neutral axis and for a 3/2-lay-up the distance of the fibres to the neutral axis is still quite small. Depending on the fibre orientation (UD-laminates are favourable for bending) the minimum bend radius increases with increasing thickness. When fibres are loaded during bending, the minimum bend radius increases exponentially, and the minimum bend radius is a function of the (small) failure strain of the fibres:

$$r_{min} \approx t/2\epsilon \qquad 8.1$$

where t is the distance of the most outer fibres layers to the centre line and ϵ the failure strain of the fibres.

When the fibres are parallel to the bend line (no fibre deformations) the increase of the minimum bend radius is comparable to metal alloys. However, for thicker laminates (1.5–2.0 mm and larger), the failure mode changes from fracture of the outer surface to delamination of the interfaces just outside the bend zone. Once the delamination is the dominant fracture mode, a rapid increase in the minimum bend radius/thickness ratio is the result.

Elastic recovery

When FML are deformed into structural elements, part of the applied work is elastic. This elastic energy causes spring back, residual stresses, or a combination of these two. For bending of profiles or flanges the spring back dominates; for shallow three dimensional shapes, the largest part of the elastic energy is stored as residual stresses. FML often show larger spring back angles or residual stresses than (reference) metal alloys. The main reason is the elastic energy, which is absorbed by the fibres that deform elastically only.

8.4.3 Forming processes for FML

For the manufacture of FML parts conventional metal forming processes are suitable options. This is true for small to medium-size parts like profiles, beams, clips and cleats. For large-size shells like the skins of aircraft wings and fuselages the lay-up technology is more appropriate (see next section). These shells are too large to be formed on conventional equipment and the formability of the laminates is insufficient. In this section some forming processes are described briefly. In this overview a distinction is made between processes for laminates and processes for the forming of individual layers, before being processed in lay-up and curing processes.

The following processes are in use or their feasibility has been demonstrated: press brake bending, roll bending, roll forming of profiles, press forming and stretch forming.

Forming technology for composite/metal hybrids 209

Press brake bending

The bending of profiles with a press brake is similar to the bending of metal profiles. Since the fibres are parallel to the bend line, the metal constituent dominates the minimum bend radius and spring back angles for thin laminates (up to 1.5–2.0 mm) (see Fig. 8.4 for examples). For the bending of thicker laminates, the bend radius increases rapidly as discussed in Section 8.4.2. One option to manufacture thick profiles is to assemble preformed profiles in a subsequent bonding step (see Fig. 8.4). Applying this method very small bend radius/thickness ratios are achievable. However, for this option the design freedom of the profile geometry decreases.

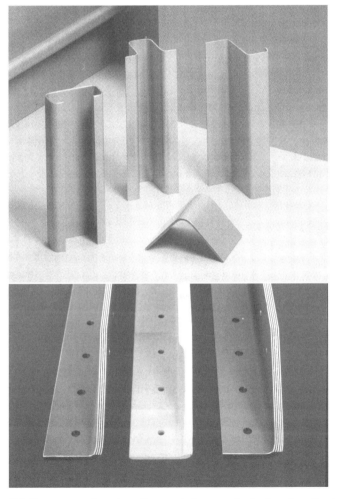

8.4 Some examples of profiles made by press brake bending (top) and two examples of thick stringers made by bending and bonding of several profiles, bottom left and right (the stringer in the center is a machined stringer).

Roll bending

The roll bending of laminates can be used to create a single curvature in a shell or skin. Due to roll bending some internal stresses are introduced and the spring back requires multiple passes through the set of rolls to obtain the final radius. Currently, lay-up techniques are used for the manufacture of single curved shells.

Roll forming of profiles

Roll forming is a forming process for the manufacture of symmetrical profiles from strip. It has been demonstrated that this process is feasible for thin laminates, but the investment costs for the tooling are high.

Press forming

The applicability of press forming processes like rubber forming is limited. The main reason is that in-plane deformations for FML are limited by the failure strains of the fibres (depending of the fibre system: 1–5%). Only shallow shapes can be press formed like stiffening beads used in web plates or panels. The manufacture of curved flanges is not really feasible: for shrink flanges the laminates wrinkle and delaminate, and for stretch flanges the strain is limited by the maximum strain for the fibres and the complicated spring back compensation for the tools.

Stretch forming

Finally, the use of the stretch-forming process is also limited. When the stretching forces are applied in one of the fibre directions, the maximum achievable strain is small and there will be significant spring back. The best option is to stretch form these laminates in bias-direction, activating the trellis-effect in the composite layers. However, this is only valid for applications where the fibre orientation is of minor importance such as for impact-dominated shells (leading edges of wings, cockpit and canopy).

A second option for the manufacture of FML-parts is first to form the metal layers and to assemble them into laminate, applying a subsequent lay-up and curing cycle. The forming of the thin metal layers is a delicate process, due to the small thickness of the layers; the layers may buckle and wrinkle. If possible, the layers should be formed in stacks, because then the layers can support each other, it reduces the number of forming operations, and it improves the matching of the layers. A few manufacturing processes are tried to demonstrate this production route.

Roll forming

The roll forming process for the manufacture of profiles is used for stacks of metal and prepreg layers (uncured laminates). The stack is prepared, then roll formed into the required shape and cured after the forming operation. The matching of the layers is perfect and the spring back and residual stresses are very small. However, the thickness in the radii of the profiles might not be constant due to some resin squeeze out from the composite layers during roll forming.

Stretch forming

For the manufacture of slightly double curved panels it is possible to stretch form several metal layers in one cycle. The benefits of this route are the increase of the formability limits (further increase is possible using heat treatments), and the minimisation of residual stresses and spring back; a drawback is the surface treatments that have to be applied to double curved sheets.

Press forming (of metal layers)

The press forming for individual layers was demonstrated in the JSF-program. In this program a prototype of a weapon bay door was manufactured. These doors consisted of two skin panels and an inner structure. One skin panel is almost flat and the other skin panel has a deep drawn shape. Rubber forming of the metal layers followed by film infusion process of the fibre reinforcement and curing in an autoclave made the latter skin panel. This manufacturing procedure was applied successfully.

Resin infusion

The last issue for this production route is about the application of the composite layers prior to the curing cycle. There are several ways to place the composite layers. One method is the lay-up of prepreg, but also infusion processes like resin infusion and film infusion of dry fabrics are feasible. Which method is the best, depends on the specified fibre architecture and the applied polymers.

8.4.4 Lay-up techniques for FML shells

Large size FML shells cannot be made by forming processes, but are made by lay-up processes, which are related to the manufacturing of composites.

The lay-up processes are used for large single or slightly double curved shells, like skin panels for aircraft structures, such as the Airbus A380. The benefits of lay-up processes are the size of the skin panel, which is only limited

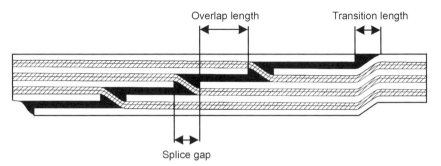

8.5 The configuration of an overlap splice.

by the autoclave (not by the dimensions of the raw materials), and the lay-up of (local) details, which makes the panel an integral part. For example: during the lay-up of the fuselage skin panels of the Airbus A380 additional layers, so-called doublers are added to reinforce the panel. Typical areas for these doublers are the window belt area and the door areas.

The fact that the size of the autoclave determines the panel size is common for composites, but is not obvious for FML. Since the width of metal sheets is limited, splices (see Fig. 8.5) are required to make larger panels. A splice is a kind of bonded joint made during the lay-up process. The discontinuity in the individual layers is placed at different locations. The best option used today is the overlap splice. The overlap gives only a small increase in laminate thickness, but is very cost-effective since no close lay-up tolerances are required.

Single curved laminates

Fuselage skin panels and leading edges of wings are typical examples of single curved shells. These laminates are curved in one direction, which means that during the lay-up process the metal sheets are subjected to bending deformations but not to in-plane deformations. Since the sheets are thin (0.2–0.5 mm) the forces required to place the sheets in the curved moulds are negligible; the same is true for the prepreg layers.

During the lay-up of the panels, splices are made in the panels if the panel dimensions are larger than the width of the metal sheets. The splices can be oriented in length or width direction.

Double curved laminates

Besides single curved panels, fuselages also have double curved shells: shells with smooth curvatures in two orthogonal directions. The lay-up of the composite layers is easy: the prepreg layers are flexible and more or less pliable. The lay-up of the metal layers is much more difficult. The sheets have to match the double curvature of the lay-up tool and this is only feasible by elastic in-plane deformations. To force the sheet in the mould, relatively high autoclave pressures are needed. Once the pressure is applied internal stresses exist in the metal sheet: compressive stresses at the edges and tension stresses in the centre of the sheet. The magnitude of these stresses is related to the width of the sheet; the smaller the width the smaller the stresses. When a sheet becomes too wide, the pressure is not sufficient to suppress the wrinkling at the edges and the sheet width should be decreased. The applicable width of a metal sheet in a double curved panel also depends on the curvatures or bend radii. The elastic stresses contribute to the spring back of the panel after curing.

8.5 Modelling of FML

The modelling of the behaviour of FML and in particular modelling of forming processes does not have a long tradition yet. Therefore, no extensive experience of modelling can be presented in this section. The main focus thus far has been on the modelling of the mechanical behaviour of laminates. Special interest has been in the modelling of fatigue behaviour and the failure of FML under different loading conditions. For some topics finite element models, for others, analytical models are developed. The main objective for the model development is to give the designers, who apply the materials in aircraft structures, the necessary tools for designing with FML. Some models are developed for particular forming processes, like bending.

In this section a brief overview is given of the efforts of the modelling of FML.

8.5.1 Analytical modelling of mechanical properties

Classic laminate theory

For a first approximation of the elastic properties the classic laminate theory (CLT) can be used, just as for composite materials. In general the FML are orthotropic laminates with one or two fibre directions. The properties in 1- and 2-directions can be calculated using the properties of the constituents, the metal alloy and the fibre composite or prepreg. Other simple rules that can be applied for FML are the Rules of Mixtures (RoM).

Once the material properties in the plastic area have to be calculated, the plasticity of the metal layers becomes important. There are several options to model this behaviour. The first option is: the stress strain curve beyond the yield stress is assumed to be a linear relationship, by neglecting the strain hardening of the metal. The strain hardening of the metal is small compared to the stress increase in the composite layers. The limit of the stress-strain curve is the failure strain of the fibres. A second option, by a much more complex approach, involves the strain hardening behaviour of the metal alloy. In that case special algorithms are required to calculate the stress-strain curve.

In these methods the implementation of the residual stress system further increases the complexity of the required formulas due to the curing process.

The MVF approach[8]

A simple tool for designers to calculate some basic properties is the 'metal volume fraction' (MVF) approach. This MVF represents the relative metal contribution in a FML and is defined as:

$$\text{MVF} = (\Sigma\, t_{metal})/t_{lam} \qquad 8.2$$

where t_{metal} is the thickness of one metal layer; t_{lam} the thickness of the laminate)

The MVF approach is based on a linear relationship between the material properties at MVF = 1, which represents a pure metal, and MVF = 0, which represents the property of the composite layer. Tests revealed that the MVF method is applicable between 0.45 < MVF < 0.85. FML like GLARE and ARALL have a MVF in this range. The accuracy of this tool is about five percent. When a knockdown factor of 1.1 is applied to the final result, the MVF-tool results are safe and slightly conservative. During the development of GLARE the airworthiness authorities accepted this approach for several typical material properties like tensile strength, compression strength, bearing strength, and blunt notch strength.

Stress intensity factors for fatigue

Alderliesten[4] proved that an extension to the classical theory of stress intensity factors, as used for the fatigue in metal structures, could also be applied for FML. In his model the stress intensity at the tip is the superposition of the far field stress intensity and a (negative) stress intensity caused by the fibre bridging:

$$K_{tip} = K_{far\,field} + K_{bridging} \qquad 8.3$$

Since the fibre bridging stress is dependent on the delamination, the K_{tip} and $K_{bridging}$ values also depend on the delamination sizes and delamination shapes. Based on a range of tests, the delamination growth has been modelled using a Paris-type equation for the energy release during delamination. Using this equation the delamination growth results are related to the crack opening, the crack growth, and the bridging stresses, which are in balance with each other.

Comparing test data, for through cracks under constant amplitude loading, with the numerical results shows good results for the crack growth in different FML (for an example see Fig. 8.6). A step further would be to apply a similar approach to fatigue of FML with a part-through crack.

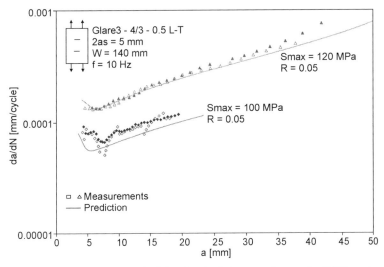

8.6 Experimental and numerical data for the fatigue crack growth of FML under constant amplitude loading and through cracks.[4]

8.5.2 FE modelling of FML structures

FE-codes are also used for numerical stress calculations of FML structures. For the modelling of the FML there are several options.

The first option is to calculate the laminate properties for every section of the structure. As stated before, FML skins are often reinforced locally with doublers, and the lay-up of the skin can change, depending on the applied loads. Thus the laminate can be tailored by adding or removing layers, and/or by changing fibre orientations. For stress analysis of aircraft structures the required composition is determined for each 'bay': a section confined by two stringers and two frames (bay-size is in the order of 200×500 mm). For that section, the laminate properties can be calculated by the classic laminate theory or by the MVF approach. The laminate properties are used during the optimisation of the structure. When the structure has to be adapted, new laminate properties for the sections involved are supplied.

FE-models can also use more sophisticated modelling: the different layers of the laminate are modelled, and each layer has its material model. This type of model requires additional modelling features such as the modelling of the interfaces between the layers. Since the number of elements and the computation time will increase significantly, this modelling is not applied to large or full-size structures, but to structural details only. Current models are very accurate in predicting, e.g., failure of riveted joints (see Fig. 8.7).

216 Composites forming technologies

8.7 A riveted joint, tested (top) and modelled (bottom). The model ABAQUS was used; the layers were modelled individually.

8.5.3 Modelling of forming processes[9]

FML are hybrid laminates made of composite layers and metal layers. Due to the plasticity of the metal layers, it is possible to give the laminates some permanent deformations. In Section 8.4.2 the formability aspects of FML were discussed, which can be summarised as: small permanent deformations in fibre directions, somewhat larger failure strains perpendicular to the fibres, and often a significant elastic recovery after forming, either by means of spring back or internal stresses.

In-plane deformations

The applicable in-plane deformations are very small and deep drawing operations cannot be applied. Only some limited stretching is feasible, in particular when the laminates are not directly stretched in the fibre directions but at an off-axis angle (e.g., in bias-directions). In that case the principle of the forming limit curve (FLC) can be used to predict the failure of the laminate. This curve presents the forming limits, e.g. expressed in strains for in-plane strain combinations. For that case the limits in fibre directions are set by the failure strains of the fibres.

Bending

For the manufacture of FML profiles the cheapest method is to bend prefabricated laminates. However, the minimum bend radii for bending laminates depend primarily on the fibre orientation and the number of layers. A model has been developed to evaluate the manufacturability of a particular

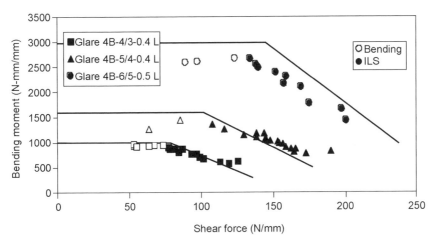

8.8 Forming limits expressed in maximum applicable combination of bending moment and shear forces for different FML.[9]

laminate/radius combination. This model deviates from elastic models in the sense that plastic behaviour is incorporated. Also the shear stresses and strains in thickness directions cannot be neglected, which makes the modelling rather complex. An important parameter in bending of FML is the transverse shear stress, which should be equal at both sides at each interface. This results in different values for the transverse shear strain (since the materials properties are different), and subsequently straight cross sections does not remain straight during bending. During research several models for the local shear stresses have been investigated. The ultimate result of the model is related to a plot where the bending limit is a combination of the bending moment and the maximum transverse shear load (see Fig. 8.8). The laminate will fail in tension, by failure of the fibres or the outside metal layer, or in delamination (shear failure). With increasing shear force, the maximum applicable bending moment decreases, and thereby the minimum bend radius increases. The model has been validated with test results and good agreement has been obtained.

8.5.4 Modelling of lay-up processes[10]

One of the methods to manufacture large skin panels is by lay-up techniques. The panel lay-up can be modelled and optimised when the skin panel is designed. The need for optimisation comes from the use of metal layers in the hybrid laminate. These layers have discrete edges and special design rules are applied for the interaction of the edges with other features like rivet joints, cutouts, etc. In a software design environment special design routines are developed for the optimum panel design when related to costs or weight. The number of possible combinations is reduced by these procedures.

8.6 Conclusions

In this chapter an overview is presented of the characteristics, properties and production of FML. Since the laminates are hybrid materials, the characteristics and properties are related to metal alloys and to composite materials. The same is true for manufacturing processes.

Properties and characteristics

The most important properties of FML are the damage tolerance properties: the FML are better than metal alloys in fatigue and residual strength, and better than composites in impact, formability and stress redistribution at stress concentrations. Although only a limited number of FML have been developed yet, this concept will be extended to hybrid materials containing a wide range of metal volume fractions.

Manufacturing

The manufacturing processes of FML are a mixture between the processes for metals and for composites. Cutting of FML, like composites, offer difficulties like tool wear and delaminations, but FML can be machined by conventional processes. The manufacture of the reinforced skin and structural elements is based on lay-up technologies (composites) and some forming processes (metal alloys). Both processes have restrictions: the complexity and flexibility of the lay-up processes is limited by the rigid, non-pliable metal layers, and the formability of FML by the small failure strains of the fibres and the shear strain of the adhesive (causing delaminations).

Future trends are the development of cheaper manufacturing processes, based on alternatives for the lay-up of the metal layers, and innovation of methods to exploit the formability of the laminates. Other options for cost reductions are changes in pretreatment and curing processes.

Modelling

Due the short history of FML, the modelling of its behaviour and the development of simulation tools is only at the beginning.

For the property calculations a number of reliable tools are already available. The simulations of processes or material behaviour, involving plasticity, are more complex and future developments are aimed to bring the simulations to a mature level like the models for metals and composites.

8.7 References

1. Gordon J.E., *Structures*, England, Penguin Books, 1978.
2. Tooren M.J.L. van, *Sandwich fuselage design*, Delft, Delft University Press, 1998.
3. Vlot A., *Glare, History of the development of a new aircraft material*, The Netherlands, Kluwer Academic Publishers, 2001.
4. Alderliesten R.C., *Fatigue crack propagation and delamination growth in GLARE*, Delft, Delft University Press, 2005.
5. Vries T.J. de, *Blunt and sharp notch behaviour of GLARE laminates*, Delft, Delft University Press, 2001.
6. Hooijmeijer P.A., *Burn-through and lightning strike, in Fibre Metal laminates, an introduction*, ed. Vlot., A. and Gunnink, J.W., The Netherlands, Kluwer Academic Publishers, pp. 399–408, 2001.
7. Borgonje B., Ypma M.S., Hart W.G.J., *Corrosion, in Fibre Metal laminates, an introduction*, ed. Vlot., A. and Gunnink, J.W., The Netherlands, Kluwer Academic Publishers, pp. 427–439, 2001
8. Roebroeks G.H.J.J., *The metal volume fraction approach*, TD-R-00-003, Delft, SLI, 2000.
9. Jong T.J. de, *Forming of Laminates*, Delft, Delft University Press, 2004.
10. Vermeulen B., Tooren M.J.L. van, Peeters L.J.B., Knowledge based design method for fibre metal laminate fuselage panels, *ASME Conference*, Long Beach, USA, 2005.

9
Forming self-reinforced polymer materials

I M WARD and P J HINE, University of Leeds, UK and
D E RILEY, Propex Fabrics, Germany

9.1 Introduction

The aim of this chapter is first to provide an outline of the postforming methodology for self-reinforced composites, using as an illustration the materials produced by the Leeds hot compaction process, and secondly to describe examples of the present commercial applications of this new technology. This will be preceded by an introductory section introducing the hot compaction process and the self-reinforced composite materials produced by this manufacturing route.

The hot compaction technology was initiated as a research project in the IRC in Polymer Science & Technology at Leeds University in 1990. This led to a series of research programmes during the period up to 1995 when the commercial possibilities were recognised and a small spin-off company Vantage Polymers Ltd was set up, under the auspices of Leeds University. Professor I M Ward was Managing Director, Mr. Derek Riley Marketing Director and Dr. Peter Hine was a key member of the technical team. During the period 1995–2000, manufacture, postforming to produce real parts and detailed property measurements were undertaken, with the focus on hot compacted polypropylene which showed the greatest commercial potential. In 2000, full scale commercialisation was commenced by Amoco Fabrics GmbH (now Propex Fabrics GmbH). In this chapter, the achievements of all these three phases will be described, with particular emphasis on the requirements for successful forming of parts from hot compacted polypropylene sheet. The products are marketed under the trade name CURV®.

9.2 The hot compaction process

In a traditional polymer composite, a polymeric matrix is typically reinforced by a stiff ceramic fibre, more often than not glass or carbon. Recently, significant research has been carried out into the manufacture and preparation of a new class of polymer composites where *both* phases are polymeric. The reinforcing phase is usually an oriented polymer, with enhanced properties due to preferred molecular orientation and the matrix phase is usually an isotropic polymer.

The production of all polymer composites, has been the subject of investigation by several researchers, using, in general, two discrete and different polymeric components. In a recent development, Peijs and his colleagues[1] have produced polypropylene composites from oriented tapes coated with a thin skin of lower melting point polymer, which can be melted and resolidified on cooling to form the matrix phase. This is not dissimilar from a very much earlier study, where Porter and his team[2] combined highly drawn polyethylene fibres with polyethylene polymer as the matrix, using the higher melting point of the fibres to ensure that the isotropic polymer could act as the matrix without melting the fibres. Other methods include film stacking,[3-5] powder impregnation,[6] solvent impregnation[7] and the use of a sudden reduction of pressure to induce melting in the matrix phase.[8]

The approach taken at Leeds, termed hot compaction, is different from those noted above in that it starts with only one component, an array of oriented fibres or tapes, most conveniently in the form of a woven fabric. Research over a number of years[9-11] showed that if this assembly is heated to a critical temperature, a thin skin of material on the surface of each oriented fibre or tape is melted, which on cooling recrystallises to form the matrix of the material.

The essential feature of the hot compacted self-reinforced composites is that they consist of two phases, both of identical chemical composition, an oriented fibre phase consisting of the central parts of the original fibre or tapes which have not melted and a matrix phase of melted and recrystallised material. This is illustrated very clearly in Fig. 9.1, which shows the DSC melting endotherm for

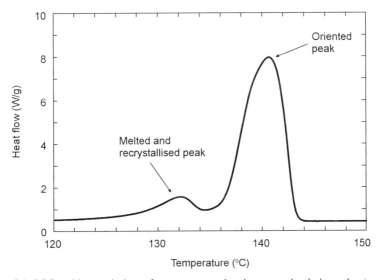

9.1 DSC melting endotherm for a compacted melt spun polyethylene sheet.

222 Composites forming technologies

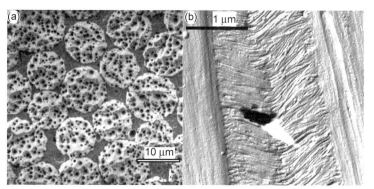

9.2 Etched micrographs from unidirectional PE fibres: (a) SEM picture of a transverse section of compacted fibres, (b) TEM picture of an interstitial lamellar region and its junction with adjacent fibres.

a compacted sheet produced from a parallel array of melt spun high modulus polyethylene fibres.

It can be seen that the melting peaks for the oriented fibre and the recrystallised isotropic matrix phases are quite distinct. Similar results have been obtained for other polymers, notably polypropylene, where the commercial exploitation has been concentrated, and polyethylene terephthalate. Our understanding of the structure of the hot compacted materials has benefited greatly from electron microscopic studies of permanganate etched samples, undertaken by Professor David Bassett and his colleagues at Reading University in a very valuable collaboration with the Leeds team. Figure 9.2 shows typical electron micrographs of etched samples of compacted unidirectional PE fibres. Figure 9.2a shows a SEM picture of a transverse section, and clearly shows the oriented fibres surrounded by a matrix of melted and recrystallised polyethylene. More detail of the recrystallised polymer is seen in Fig. 9.2b, which shows a TEM picture of a region between two polyethylene fibres. The picture shows that the 'matrix' polyethylene recrystallises epitaxially onto the remaining fraction of the oriented fibres.

The key variable in the production of these self-reinforced composites is the hot compaction temperature, which determines the proportion of melted material. Extensive experience has shown that to obtain satisfactory consolidation for compaction of fabrics, this should be about 25%, although for a unidirectional array this can be as little as 10%.

While the original research was carried out on unidirectionally arranged arrays of oriented fibres, the commercial materials are based around woven assemblies of oriented tapes, leading to more balanced in-plane properties. The mechanical stiffness of these compacted woven sheets can be calculated by assuming that these are two phase systems with continuity of strain between the oriented tape or fibre phase and the matrix phase. For a woven structure the composite modulus, $E_{composite}$ is then given by

$$E_{composite} = \frac{V_f E_L}{2} + \frac{V_f E_T}{2} + (1 - V_f)E_m \qquad 9.1$$

where V_f is the volume fraction of the oriented phase, E_L, E_T are the longitudinal and transverse moduli of the oriented phase respectively, and E_m is the modulus of the matrix phase.

To a good approximation it can be assumed that $E_T = E_m$ giving

$$E_{composite} = \frac{V_f}{2}E_L + \frac{(2 - V_f)}{2}E_m \qquad 9.2$$

Table 9.1 shows the results of these calculations for polyethylene, polypropylene and polyethylene terephthalate. It is important to note that whereas for the high modulus polyethylene fibre sheet the matrix stiffness makes a very small contribution to the overall stiffness, this is not the case for polypropylene and PET, where the modulus of the oriented phase is more modest. It is also of interest to note that for the woven PET multifilaments the measured modulus is lower than that predicted by the model. This is attributed to the effect of crimp that is present in the woven multifilament cloth, but which is much reduced when using woven flat tapes.

The matrix phase is also important in terms of its ductility which is relevant for large strain properties such as impact and thermoforming. In practical applications the use of woven fabric is essential, with the result that the matrix properties are very important as this dominates the strength between the layers (which can be measured using a peel test). Clearly such issues are also of paramount importance in considering successful postforming, as will be further discussed later.

Because of commercial importance, the hot compaction behaviour of woven polypropylene fabric has been studied in detail using a combination of mechanical tests, DSC and morphological measurements.[12-15] In the optimum compaction conditions there is epitaxial recrystallisation of the melted phase (similar to that seen with polyethylene – Fig. 9.2a), with the lamellae growing

Table 9.1 Predictions of in-plane compacted sheet modulus (based on the phase properties) in comparison with experimental measurements

	Polyethylene	Polypropylene	PET
Oriented element shape	Tape	Tape	Multifilament
Original oriented element modulus (GPa)	88.0	11.0	14.0
Melted matrix modulus (GPa)	0.51	1.2	2.8
Predicted compacted sheet modulus (GPa)	29.4	4.9	7.0
Measured compacted sheet modulus (GPa)	30.0	5.0	5.8

Using a volume fraction of $30 \pm 5\%$.

out from the surfaces of neighbouring oriented tapes and any gaps on the structure are filled by row structures. With regard to the mechanical behaviour it is essential that there is sufficient ductility in the matrix phase for this to remain coherent up to the comparatively high failure strain (~15%) of the oriented PP tapes. This means that the molecular weight of the polymer has to be reasonably high (Mw > 300,000).

In two recent publications[14,15] the important parameters which determine the hot compaction behaviour of woven oriented polypropylene were identified by studies on compacted sheets produced from five commercial woven cloths based on four different polypropylene polymers. These cloths were chosen to define the boundaries of the key parameters which include the polymer molecular weight, the orientation of the tape, and comparisons of reinforcement with fibres or tapes and coarse and fine weave fabrics. A key result was the importance of the ductility of the matrix phase which requires that the molecular weight of the polymer should be moderately high, i.e. Mw > 300,000. There is also a requirement that for the best performance the stiffness and strength of the reinforcing phase should be high, which means choosing a polymer which can be drawn to a satisfactorily high draw ratio. Although low molecular weight polymers are often used for ease of draw, during the compaction process low molecular weight polymer will recrystallise to give a brittle material. Finally, with regard to the reinforcement shape (tape or fibre) and the weave style, the best combination of properties was found with flat tapes (avoiding the crimp usually present with woven fibre cloths) and a balanced weave style (plain weave). These geometry effects were monitored by determining interlayer adhesion using a T-Peel test (ASTM D1876) where samples 10 mm wide and 100 mm long were tested in an Instron at a cross-head speed of 100 mm/min, comparing both weft and warp directions. The morphological studies showed that the low molecular weight polymers tended to recrystallise as opposing transcrystalline layers, the junctions of which provided a low resistance path for peeling.

9.3 Commercial exploitation

To establish a baseline for commercial exploitation and evaluation of the properties of hot compacted PP, large sheets (~2 m × 1 m) were produced in an autoclave.[16] Following this, trials were undertaken to develop a continuous process using a double belt press at Hymmen GmbH, Bielefeld, Germany.[16] These trials led to the commercial operation of Propex Fabrics GmbH and the present CURV® products. Table 9.2 shows a comparison of the properties of self-reinforced PP sheet, isotropic PP sheet, random short glass fibre filled PP and PP sheet reinforced with continuous glass fibre, and indicates the strengths of self-reinforced polymers in general. It can be seen that self-reinforced PP sheet shows a very interesting combination of properties with a density

Table 9.2 A comparison of the mechanical properties of polypropylene based materials

		Hot compacted PP sheet (Curv®)	Isotropic PP homo-polymer	Random mat short glass/PP 40wt% fibre	Continuous sheet glass/PP 60wt% fibre
Density	kg/m³	920	900	1185	1490
Notched Izod impact strength	J/m	4750 (20°C) 7500 (−40°C)	200	672	1600
Tensile strength	MPa	180	27	99	340
Tensile modulus	GPa	5.0	1.12	3.5–5.8	13
Heat deflection temperature	°C/455 kPa °C/1820 kPa	160 102	110 68	154	155
Thermal expansion (@20°C)	10⁻⁶/°C	41	96	27	21

Comparison data for other materials taken from www.matweb.com
Quoted values are averages of all commercially available grades
Compacted PP made on a pilot plant – www.curvonline.com

comparable to isotropic PP, tensile strength, tensile modulus and thermal expansion intermediate between the random short glass fibre and the continuous glass fibre PPs. Moreover, self-reinforced PP shows outstanding impact performance and it is particularly important that this is retained and even greater at −40°C where isotropic PP is brittle.

9.4 Postforming studies

Research to date indicates that the thermoformability of a self-reinforced polymer sheet lies between the two extremes of an isotropic polymer and a continuous glass fibre reinforced polymer. An isotropic polymer sheet can usually be easily formed by raising it to an elevated temperature above its glass transition, and stretching it into a mould of the required shape, often with the aid of a vacuum.[17] At the other extreme, a continuous glass fibre reinforced polymer composite sheet, where the fibre is inextensible, usually requires matched metal moulds, a significant forming pressure, and a temperature such that the matrix phase is melted and the fibres can rotate as required.[18] A self-reinforced polymer sheet falls between these two extremes because it is composed of oriented polymer elements and an isotropic polymer matrix. The oriented elements, while not inextensible like a glass fibre, are more resistant to stretching compared to an isotropic polymer, although the extensibility increases with increasing temperature.

In light of its commercial importance, the majority of the detailed thermoforming studies have been carried out on self-reinforced polypropylene sheet.

Three major programmes of work have been carried in this area: the first,[19] carried out at Leeds, used a combination of tensile tests and a spring-loaded hemispherical matched metal mould to assess the thermoformability over a range of temperatures and strain rates: in the second, which was part of an EPSRC sponsored programme 'DEPART', the group at Nottingham utilised a combination of picture frame tests and bias extension tests to examine thermoformability:[20] finally in the third, forming trials were carried out using a commercial glass mat thermoplastic (GMT) manufacturing line: the major findings in each of these three programmes will now be discussed.

9.4.1 Tensile tests and matched hemispherical mould experiments

The first programme of work, described in detail in reference 19, combined tensile tests at elevated temperatures with model thermoforming experiments carried out using a hemispherical, sprung loaded, matched metal tool installed inside a temperature-controlled cabinet. As was described above, the oriented elements have a significantly higher melting point compared to the isotropic matrix phase (and this separation is increased if the sheet is constrained), suggesting two potential thermoforming scenarios. Either thermoform at a temperature below the melting point of the matrix phase (for PP this is around 160°C), where the sheet will behave as a homogeneous polymer, or thermoform at a temperature above the melting point of the matrix phase, where the fibres can then rotate as appropriate. The aim was to establish under what conditions of temperature, pressure and part design, the above two scenarios for forming could apply.

Hot tensile tests were carried out on samples cut from hot compacted PP sheets. Tests were carried out over a range of temperatures (140 to 170°C) and a range of crosshead speeds (5 to 500 mm/min): temperature was found to be the most significant factor. The measured stress-strain curves all showed a similar shape, shown schematically in Fig. 9.3: an initial linear region up to a yield point at a strain of around 5%, followed by a strain hardening region of lower slope (termed the post yield modulus) up to failure. The post yield modulus is considered an important parameter as this indicates the resistance of the material to the degree of plastic deformation normally required for thermoforming. With increasing test temperature the post yield modulus was found to fall from 75 MPa at 140°C to 40 MPa at 170°C while the failure strain increased from 60% at 140°C to 90% at 170°C.

Thermoforming tests were carried out over the same temperature range using a spring-loaded hemispherical matched metal mould. The whole assembly was placed into a temperature controlled oven installed on an Instron tensile test machine, allowing the temperature of the mould and sheet to be accurately controlled and allowing the force during forming to be measured.

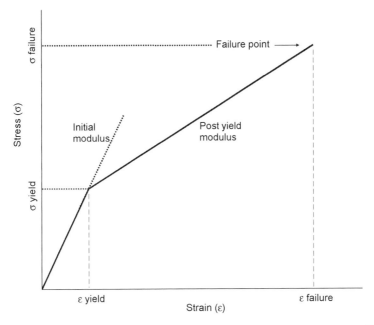

9.3 A schematic diagram of a typical stress-strain curve for hot compacted PP sheet.

For the first set of tests a fixed displacement of 10 mm was used, which resulted in partial forming of the hemispherical shape. The maximum forming force was found to decrease linearly with temperature: the maximum forming force was also found to show a linear relationship with the force measured at a similar strain in the hot tensile tests: this confirmed that a major component of the forming force is the resistance of the oriented elements to tensile stretching.

Sections taken from the thermoformed samples made at the different temperatures were examined for any signs of delamination, and a very interesting observation was made. At the lower temperatures of 140 and 150°C no delamination was seen: at 160°C a small amount of interlayer damage was seen, while at 170°C no damage was seen. These results can be understood by looking at a typical DSC melting endotherm for hot compacted, self reinforced PP sheet shown in Fig. 9.4.

As was seen in the trace for polyethylene (Fig. 9.1) there are two peaks, a lower one associated with the melted and recrystallised material and a higher one associated with the oriented elements. At 140 and 150°C, which is below the melting point of the matrix phase, the material behaves as a homogeneous sheet. At 170°C, which is above the melting point of the matrix phase, the material will form more like a traditional fibre composite with a molten matrix: however, on cooling the matrix phase will recrystallise and bond the layers together again. At 160°C, the matrix phase is not quite molten, but will be quite soft being so close

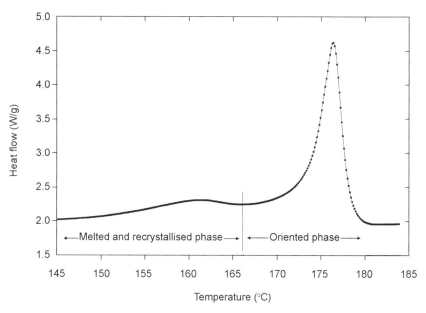

9.4 DSC melting endotherm for self reinforced PP sheet.

to the melting point. Significant interlayer shear could therefore break the matrix at this temperature, but because the matrix is not molten there is no mechanism for rebonding on cooling.

In a second set of experiments, the tool was completely closed, leading to a fully formed hemisphere: these tests were carried out both with a clamped sheet, and allowing the sheet to flow into the mould under a light pressure to restrict any wrinkling. Sections taken from the formed hemispheres indicated that if the sheet was clamped, then delamination often occurred. Conversely if the sheet was allowed to flow into the mould, then a well formed part with no delamination was the result.

From these observations it can be concluded that 160°C is the worst choice for thermoforming self-reinforced PP sheet: for successful forming two alternative strategies can be proposed. If the strain is moderate, then the material should be formed at the highest temperature possible before melting of the matrix commences, 150°C, and the sheet will behave as a homogeneous material. The material can be clamped if required. If a deeper draw is required, then the material should be formed at 170°C, when the matrix is molten, allowing the oriented elements to rotate (and also stretch to some extent). Both approaches require matched metal tooling (or a pressure assist diaphragm) and in the second case the material should be under pressure during cooling to aid consolidation. Both forming regimes are also aided if the sheet is allowed to flow into the mould under the action of a small pressure to restrain wrinkling. However, as we will see in the next section, at 170°C the sheet also has to be clamped to restrain shrinkage, so these two aspects have to be designed into the holding fixture.

9.4.2 Picture frame and bias extension tests

This second study built on the results of the experiments described in the previous section. Here the emphasis was on in-plane measurements of thermoformability using the picture frame and bias extension geometries (for a description and comparison of these two test geometries see refs 21 and 22). The picture frame consists of four equal length rails clamped over the sheet in a square configuration and hinged at the four corners.[23] If two opposing corners are then pulled apart, the square deforms into a rhomboid, forcing the clamped sheet to likewise deform. The force required to pull the frame is measured and the shear force can be determined. This test is normally used with either unconsolidated glass fibre based textile fabrics, or glass fibre reinforced composites where the matrix is molten, and is used to examine the effects of the weave style on formability (for example, the determination of the locking angle where the fibre tows can no longer rotate). The second geometry used in these experiments, which provided further information on the deformation mechanisms, was the bias extension test. Here a sample is cut from the sheet such that the main directions of the woven cloth are at ±45° to the direction of the applied force.

Following the work described above, three temperatures were used in these tests, 150, 160 and 170°C. First the picture frame results. At 150°C, the material was found to deform homogeneously in shear, with significant associated stretching of the oriented elements. At 160°C the deformation was found to be more like a continuous fibre reinforced composite prepreg, while at 170°C the deformation was very like a textile-type material (i.e., rotation of the oriented elements but no stretching). Two different weave styles were examined and there were significant differences in their formability, indicating that this is an important aspect which must be taken into account when assessing thermoformability. Another important result was the degree of shrinkage seen in the sheets at the different temperatures. At 150°C, this was found to less that 1% so the sheet could be heated in an oven if required with no loss of properties or shape. However, at 170°C, once the matrix was molten, the shrinkage could be as high as 30% (depending again on the weave style), indicating that for forming at this highest temperature the sheet has to be clamped. As would be expected, the measured shear force decreased as the temperature was increased, in line with the previous tensile tests. Also, minimal rate dependence was measured, also in line with the previous results.

The bias extension tests (carried out at 150 and 170°C) showed the same trends as the picture frame tests, and if a normalisation approach was used[24] then the quantitative agreement was good between the two test geometries. In this test, the oriented elements were found to have extended at both temperatures, though more so at the lower temperature (6% at 150°C and 4% at 170°C). The locking angle was also found to depend on the temperature: at 150C this was 35°, whereas at 170°C the locking angle was 65°. This confirms that for forming a shape with a deep draw or a complicated section requiring significant intra-ply shear, then a temperature of 170°C will double the allowable shear deformation.

To summarise these results, the new ingredient in the forming of self-reinforced PP sheets, when compared to traditional composite forming, is that the reinforcing elements can be extended under certain conditions. Self-reinforced PP sheet has been shown to be a very low rate dependent material that is sensitive to forming temperature. The mode of material deformation depends both on the boundary conditions applied to the edge of the sheet during deformation and also on the relationship between the forming temperature and the melting temperature of the 'matrix' phase. If the sheet is relatively unconstrained, as in a bias extension test, the material will tend to deform through both pure shear and tow extension, irrespective of temperature. However, if the material is constrained, as in the picture frame test, below the matrix melt temperature the material will deform as a plastic sheet and undergo both shear and significant tow extension, whereas above the matrix melt temperature the material will deform more like a typical continuous fibre reinforced composite, i.e. no tow extension.

9.4.3 Forming trials using a commercial GMT production line

The aim of the third study was to transfer the results described above, generated under carefully controlled laboratory conditions, to a practical forming situation using a commercial GMT line. The line was composed of a conveyor belt fed set of heating ovens, placed in front of a compression press. Installed on the press were a pair of GMT metal moulds for producing an undertray part, similar to the one shown in Fig. 9.5a below. While the mould halves were not matched for the forming of a self-reinforced PP sheet, they did allow a series of thermoforming trials to be carried out. In these tests, the decision was made to form below the melting point of the matrix phase, as the sheet to be formed had to pass through the heating ovens unclamped. From the results described in the previous two sections it can be seen that the ideal temperature for the sheet to be formed at is 150°C. However, there will be a delay between the heated sheet exiting the ovens and the time to place the sheet into the mould plus the mould closing time. It is therefore necessary to heat the sheet to a temperature above 150°C to allow for this delay. The results above also showed that when the temperature approached 170°C, and the sheet was unclamped, significant shrinkage occured. This behaviour therefore sets the boundaries for the processing parameters: heat the sheet as quickly as possible to a temperature not higher than 165°C and then quickly transfer it to a mould, close the mould and thermoform when the sheet temperature had dropped to close to 150°C.

To investigate the temperature profile of the sheet throughout the process, a data-logger was used which was able to travel through the oven with the sheet enclosed in an insulated box: data could subsequently be downloaded onto a PC. For the first experiments, the three individually controlled oven zones were set to a typical GMT profile of 222, 205 and 208°C. The first zone is set to a higher temperature to get the sheet heated quickly, while the next two stages allow the temperature to approach the optimum value without overshoot, which is very important to stop shrinkage and loss of molecular orientation in a self-reinforced PP sheet. After a few scouting experiments, a dwell time of 30 seconds was used in each heating zone. On exiting the oven, the sheet temperature was found to be ~150°C, cooling to 110°C after 30 seconds. Raising the dwell time to 40 seconds increased the temperature on exit to around 160°C.

For a more detailed experiment, six thermocouples were placed over the surface of a one metre square, 1.3 mm thick, sheet of self-reinforced PP. On exiting from the oven the average temperature was measured as 164 ± 2.5°C: the sheet remained flat at this temperature suggesting that this is a good compromise between heating and shrinkage. After 45 seconds the temperatures were read again and gave an average of 116 ± 2.5°C. This gave a fall of 48°C, and a cooling rate of close to 1°C/second for this sheet thickness. This suggests that a time of 15 seconds between exiting the oven and final tool closure would give the optimum forming temperature of 150°C (in this process this time was 20 seconds with the results that the sheet would be at 145°C for forming).

To assess the temperature that would be attained in the centre of a sheet, two 1.3 mm sheets were placed together and sent through the oven. Thermocouples were placed on the top and bottom surfaces and in between the sheets. The dwell time in each oven zone was increased to 50 seconds. On exit the temperatures were 152 ± 3°C on top, 145 ± 3°C in the middle and 148 ± 1°C on the bottom (the oven heats from both sides). This suggests that for a 2.4 mm sheet, there would be a slightly lower temperature in the centre, although this could be better if the sheet was a single thickness of 2.6 mm, rather than two 1.3 mm sheets taped together. A longer dwell time could also be beneficial in allowing the centre to catch up.

In the final set of tests, the heated sheet was carried through to forming. This was carried out on a different forming line, using a four zone heating oven. In this case a dwell time of 42 seconds in each zone proved optimum (zone temperatures of 176, 264, 225 and 205°C respectively). This gave an average sheet surface temperature of 159°C on exit and an average surface temperature as the mould closed 20 seconds later of 145°C. Good parts were made using these conditions with no sign of delamination or stress whitening, which was confirmed by subsequent mechanical tests including stress-strain measurements and puncture impact.

9.4.4 Final summary

The results from the three research programmes indicate two possible thermoforming strategies depending on the degree of deformation and complexity of part required. For reasonably simple shapes and low degrees of deformation (10% tensile extension, 30% shear deformation) then the best temperature for the sheet at the point of forming is 150°C. At this temperature the shrinkage of the sheet is low so it can be heated unclamped in an oven if required. Matched tools are necessary for forming at this temperature. To obtain a sheet temperature of 150°C at the point of forming requires heating the sheet to a higher temperature in an oven: a temperature of 165°C is the highest that should be used before the onset of shrinkage: for a ~2 mm sheet this gives around 15 seconds for transfer of the sheet and mould closure. This thermoforming strategy is the most commonly used commercially.

For more complicated shapes then the best temperature is 170°C, and matched tooling is again required. At this higher temperature, the tools must be closely matched to the final dimensions of the finished part taking into account the respective degrees of thinning in areas of higher deformation. This can make tool design challenging, but one solution is to have at least one half of the mould 'flexible' which is then able to conform to the thickness changes of the sheet, while maintaining forming pressure and consolidation after forming. At the higher forming temperature the material must be clamped to restrain shrinkage, although it is an advantage to allow it to flow into the mould during forming as this reduces the forming force and reduces inter-ply shear and delamination.

For both approaches, only modest thermoforming pressures are required.

Examples of commercial products, thermoformed using these two different strategies, are presented in the next section.

9.5 Key examples of commercial products

As discussed, self-reinforced polypropylene sheet may be postformed into parts over a reasonably wide temperature range at relatively low pressures. Taken in combination with the portfolio of mechanical properties and light weight this makes the material ideal for a number of manufacturing processes in a wide range of industries.

9.5.1 Automotive parts

One of the most interesting aspects of the behaviour of self-reinforced PP sheet is the high resistance to impact and abrasion, especially in view of the fact that these high values are retained at low temperatures. This notable property makes self-reinforced PP sheet an ideal material for automotive exterior components, especially under body shields. Such parts are subjected to very high levels of impact and abrasion, yet need to be lightweight and easily removed for maintenance.

Abrasion tests have been carried out by Daimler Chrysler,[25] comparing Curv® hot compacted PP sheet with three other materials – a conventional GMT with 20% glass fibre content, a 'heavy duty' GMT with 35% glass fibre and additional PET reinforcing fabric to improve its abrasion resistance and a natural fibre reinforced PP. The results indicate that hot compacted PP has significantly higher resistance to abrasion by grit and small stones than the glass fibre and natural fibre reinforced alternatives. In addition to ongoing vehicle trials, mechanical tests have been carried out to compare the mechanical properties of self-reinforced PP and GMT undershields. Such testing has shown that not only does self-reinforced PP exhibit higher impact and abrasion resistance than glass reinforced materials but the lower density allows weight savings as high as 50% in the finished part.

Figure 9.5a shows a hot compacted PP undershield installed on a vehicle. This part, thermoformed from a hot compacted sheet using matched aluminium tools, is surrounded by a number of similar GMT components, indicating the potential wide usage of replacement materials.

Under body components require no decoration and are left in their 'natural state'. However, visible exterior body panels often require painting to a Class A finish to match the rest of the vehicle. In common with other forms of polypropylene, self-reinforced PP has very low surface energy – typically 30 dynes/cm as produced. Nevertheless, standard flame treatment and plasma treatment methods are sufficient to raise the surface activity to enable the material to be painted using conventional painting systems. Flame treatment raises the surface activity to around 40 dynes/cm whereas two passes from a

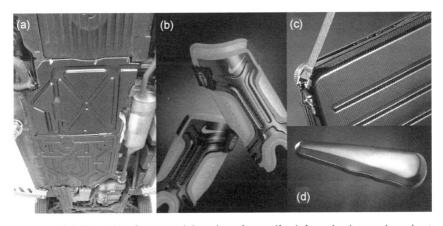

9.5 Examples of commercial products from self-reinforced polypropylene sheet (Curv®): (a) automotive undershield (centre of picture) with original GMT components to top and bottom; (b) Nike (BPS Contour) soccer shin guards; (c) laptop computer case; (d) violin case.

hand held plasma gun can raise the surface activity to around 70 dynes/cm – a level which is maintained for more than seven days. Painting with an automotive primer and top coat following such surface preparation provides a Class A automotive finish which meets industry standards for temperature and humidity cycling, stone chip and adhesion.

In the case of vehicle interiors the competition from lower cost materials such as wood-filled PP is more intense. However, self-reinforced PP is already used in safety-critical areas such as air bag release mechanisms and impact-resistant seat backs where guaranteed performance over a wide range of temperatures is required. Furthermore, in combination with other materials, such as all-polypropylene laminates, self-reinforced PP is seeing applications in load floors and rear parcel shelves where stiffness at moderately high temperature tends to preclude the use of unreinforced polypropylene.

9.5.2 Sports protection

The combination of lightweight and outstanding impact performance makes hot compacted PP a candidate for sports protection. The first application to make use of this material was for a range of soccer shin guards launched onto the market in early 2003. The finished items are shown in Fig. 9.5b. These shin guards combine Curv® with a foam to produce a component which outperforms other candidate combinations. Manufacture of the self-reinforced polypropylene protective insert is achieved using unheated matched metal tools in a simple hydraulic press. Multiple sheets are heated prior to moulding in an air convection oven giving an overall cycle time per piece of less than 30 seconds.

9.5.3 Protective cases and luggage

Protective transportation cases for electronic devices, musical instruments, etc. have been manufactured in a wide variety of materials ranging from metals to wood coated with fabrics. Self-reinforced thermoplastic composites offer an ideal combination of light weight with high levels of impact resistance, moisture resistance and ease of manufacture. The material is easy to work and features such as hinges and fasteners can be included using conventional processing.

Figures 9.5c and 9.5d show two different types of case produced by low pressure compression forming. In both these applications a relatively high depth of draw is required which would be difficult to achieve by simple compression moulding of the sheet material. To overcome this these components are produced with the aid of a sprung clamp frame which allows the sheet to slip into the tool in a well controlled manner, eliminating creases and creating good definition with a relatively tight radius in the corners.

9.5.4 Safety shoe components

Besides having very high levels of impact resistance, self-reinforced polypropylene also exhibits exceptionally high compressive strength, in excess of 500 MPa, which results in a material with very high penetration resistance. The material does not possess high speed ballistic resistance but it is remarkably resistant to slow speed puncture as required for insoles in safety footwear. The only cost-effective alternative to self-reinforced polypropylene in this application is steel. Advantages of the thermoplastic composite solution include much lower thermal conductivity for greater comfort, non-metallic to prevent false alarms in metal detectors in airports and other places of work and, for certain applications, non-conductive (a special requirement in fire fighters' boots). Combined with an all-PP toe cap, capable of withstanding 200 joule impacts (an international standard) for the first time allows the manufacture of safety footwear without metals or glass.

9.5.5 Audio equipment

There are three key requirements of materials used to make audio cones for loudspeaker systems: the material must be resistant to moisture, it must have high stiffness/weight ratio and to prevent resonance it must have self-damping properties. Many different materials have been used over the years but all fail in at least one area. For example, paper cones can be made with high stiffness/weight but they are affected by moisture unless coated with a lacquer which then affects the damping qualities; carbon fibre and Kevlar cones are light weight, high stiffness, impervious to moisture but they have poor damping; injection moulded PP has excellent damping but it is not stiff enough to give a clear sound. Self-reinforced polypropylene is a unique material in that it meets all three requirements and for this reason is used in some of the highest quality hi-fi systems produced.[26]

9.6 Future developments

Hot compacted PP sheet was a natural choice for the initial exploitation of the hot compaction technology. However, in both the Leeds researches and in studies elsewhere, other polymers have been identified for possible commercialisation. As described above, the earliest researches were undertaken on polyethylene and in collaboration with Westland Helicopters it was confirmed that the very low absorption of electromagnetic radiation of this polymer is valuable with regard to radome covers.

Table 9.3 shows a comparison of the room temperature properties of hot compacted sheets produced from a very wide range of polymers. Polyethylene (PE) is exceptional in showing very high values for modulus and strength,

Table 9.3 A comparison of the properties of the optimum compacted sheet for a range of different polymer types

		PE Tensylon	PP Curv®	PET	PEN	POM
Oriented phase type		Tapes	Tapes	Fibres	Fibres	Tapes
Oriented phase arrangement		Woven	Woven	Woven	0/90	0/90
Compaction Temperature (°C)		153	191	258	271	182
Oriented phase modulus (GPa)	E_L	88	11	14	22	22
Matrix phase modulus (GPa)	E_M	0.5	1.2	2.8	3.3	3.2
Initial compacted sheet modulus (GPa)		30	5	5.8	9.6	10
Compacted sheet failure strength (MPa)		400	182	130	207	280
Compacted sheet failure strain		2	15	10	6	6
Density (kg/m^3)		980	910	1400	1410	1420

reflecting the outstanding mechanical properties of the highly oriented fibres incorporated in the sheets. Polyethylene naphthalate (PEN) and polyoxymethylene (POM) also show significantly higher values of sheet stiffness and strength than PP, but it is in respect of the temperature performance of all these polymers that exciting prospects emerge. Considering Table 9.3 it is clear that PEN and POM are outstanding candidates for a new generation of hot compacted materials. Taking into account the likely cost of the compacted sheets, PET (polyethylene terephthalate) can also be seen as a very interesting possible polymer for the technology. However, a major point is that the postformability of all these materials has to be explored before it can be concluded that they show an adequate overall portfolio of properties for extensive commercialisation.

9.7 Acknowledgements

The hot compaction project has been very much a team effort and we wish to acknowledge the research and development input from our colleagues at the Universities of Leeds (K.E. Norris, W. Prosser and M.J. Bonner), Reading (D.C. Bassett and R.H. Olley) and Nottingham (A.C. Long, M.J. Clifford and P. Harrison). We also wish to acknowledge the contributions of our many colleagues from Propex Fabrics, Roger Whenmouth from the Ford Motor Company and Robert Bjekovic from Daimler Chrysler. Finally we wish to acknowledge the financial support from amongst others, EPSRC, BTG, Hoechst Celanese, Ford Motor Company and BP Amoco Fabrics GmbH.

9.8 References

1. Loos, J, Schimanski, T, *et al.*, 'Morphological investigations of polypropylene single-fibre reinforced polypropylene model composites', *Polymer*, 2001 **42**(8) 3827–3834.

2. Capiati, N J and Porter, R S, 'The concept of one polymer composites modelled with high density polyethylene', *Journal of Materials Science*, 1975 **10** 1671–1677.
3. Marais, C and Feillard, P, 'Manufacturing and mechanical characterization of unidirectional polyethylene-fiber polyethylene-matrix composites', *Composites Science and Technology*, 1992 **45**(3) 247–255.
4. Teishev, A, Incardona, S, *et al.*, 'Polyethylene fibers-polyethylene matrix composites: Preparation and physical properties', *Journal of Applied Polymer Science*, 1993 **50** 503–512.
5. Ogawa, T, Mukai, H, *et al.*, 'Mechanical properties of ultrahigh-molecular-weight polyethylene fiber-reinforced PE composites', *Journal of Applied Polymer Science*, 1998 **68**(9) 1431–1439.
6. Hinrichsen, G, Kreuzberger, S, *et al.*, 'Production and characterization of UHMWPE fibers LDPE composites', *Mechanics of Composite Materials*, 1996 **32**(6) 497–503.
7. Lacroix, F, Lu, H Q, *et al.*, 'Wet powder impregnation for polyethylene composites: preparation and mechanical properties', *Composites Part A – Applied Science and Manufacturing*, 1999 **30**(3) 369–373.
8. Rein, D M, Vaykhansky, L, *et al.*, 'Controlling the properties of single-polymer composites by surface melting of the reinforcing fibers', *Polymers for Advanced Technologies*, 2002 **13**(10–12) 1046–1054.
9. Hine, P J, Ward, I M, *et al.*, 'The Hot Compaction of High Modulus Melt-Spun Polyethylene Fibers', *Journal of Materials Science*, 1993 **28**(2) 316–324.
10. Olley, R H, Bassett, D C, *et al.*, 'Morphology of Compacted Polyethylene Fibers', *Journal of Materials Science*, 1993 **28**(4) 1107–1112.
11. Ward, I M, Hine, P J, *et al.* (March 1992). Polymeric Materials, British Patent Office GB2253420.
12. Hine, P J, Ward, I M, *et al.*, 'The hot compaction of woven polypropylene tapes', *Journal of Materials Science*, 1998 **33**(11) 2725–2733.
13. Teckoe, J, Olley, R H, *et al.*, 'The morphology of woven polypropylene tapes compacted at temperatures above and below optimum', *Journal of Materials Science*, 1999 **34**(9) 2065–2073.
14. Hine, P J, Ward, I M, *et al.*, 'The hot compaction behaviour of woven oriented polypropylene fibres and tapes. I. Mechanical properties', *Polymer*, 2003 **44** 1117–1131.
15. Jordan, N D, Bassett, D C, *et al.*, 'The hot compaction of woven oriented polypropylene fibres and tapes. II Morphology of cloths before and after compaction', *Polymer*, 2003 **44** 1133–1143.
16. Hine, P J, Bonner, M, *et al.*, 'Hot compacted polypropylene sheet', *Plastics Rubber and Composites Processing and Applications*, 1998 **27**(4) 167–171.
17. Macauley, N, Harkin-Jones, E, *et al.*, 'Thermoforming of polypropylene', *Plastics Engineering*, 1996 **52**(7) 33–38.
18. Hou, M, 'Stamp forming of continuous glass fibre reinforced polypropylene', *Composites Part A – Applied Science and Manufacturing*, 1997 **28**(8) 695–702.
19. Prosser, W, Hine, P J, *et al.*, 'Investigation into thermoformability of hot compacted polypropylene sheet', *Plastics Rubber and Composites*, 2000 **29**(8) 401–410.
20. Harrison, P, Long, A C, *et al.*, Investigation of thermoformability and molecular structure of Curv TM. *Automotive Composites and Plastics*, 3–4 December 2002, Basildon, Essex, UK.
21. Lebrun, G, Bureau, M N, *et al.*, 'Evaluation of bias-extension and picture-frame test

methods for the measurement of intraply shear properties of PP/glass commingled fabrics', *Composite Structures*, 2003 **61**(4) 341–352.
22. Harrison, P, Clifford, M J, *et al.*, 'Shear characterisation of viscous woven textile composites: a comparison between picture frame and bias extension experiments', *Composites Science and Technology*, 2004 **64**(10–11) 1453–1465.
23. Peng, X Q, Cao, J, *et al.*, 'Experimental and numerical analysis on normalization of picture frame tests for composite materials', *Composites Science and Technology*, 2004 **64**(1) 11–21.
24. Harrison, P, Clifford, M J, *et al.*, Shear characterisation of woven textile composites. *10th European Conference on Composite Materials*, 3–7 June 2002, Bruges, Belgium.
25. Bjekovic, R and Klimke, J, *Kunstoffe Trends*, 2002.
26. Howard, K, 'Hot compacted polypropylene – a new diaphragm material', *Gramophone*, 1998 114–115.

10
Forming technology for thermoset composites

R PATON, Cooperative Research Centre for Advanced Composite Structures Ltd, Australia

10.1 Introduction

Pre-impregnated continuous fibre reinforcement (called prepreg) is a form of reinforcement very commonly used in the manufacture of 'advanced composite' structures. Prepreg technology is strongly associated with the aircraft industry, but is also used extensively in the manufacture of sporting equipment, racing and performance cars, wind-turbine blades, and racing yachts. Although continuous fibre impregnated with thermoplastic resin, and short fibre impregnated with thermoset resins, are sometimes also called prepreg, in this chapter the term will describe continuous fibre reinforcement impregnated with a partially-cured thermosetting resin, typically epoxy.

Currently most continuous-fibre-reinforced thermoset composite parts for aircraft are formed by successive lamination and shaping of single plies of prepreg on a shaped mould surface. Only one ply is normally handled at a time, and the shaping of the ply is done at the same time as the lamination. Where the parts are made from fabric prepreg, the shaping and lamination of the single plies is almost always done by hand. This is very appropriate and efficient for small production runs, and complex components, but the hand-laminated components are less competitive for high-volume production.

To automate production of advanced (thermoset) composite parts from prepreg, the aircraft industry has progressively developed machine tools called Automated Tape Laying (ATL) and Automated Tow Placement (ATP) machines. These numerically controlled machines resemble large six-axis gantry milling machines, but incorporate a very expensive prepreg tape dispenser as the end-effector. They lay down unidirectional prepreg tape along a pre-determined numerically controlled path to create laminates of unidirectional tape prepreg which are then vacuum-bagged and cured in an autoclave.

Many carbon-epoxy parts, however, such as ribs, spars or brackets, are too small or too complex in shape to be efficiently manufactured using such machines. To bring the benefits of automation to the manufacture of some of these parts, some aerospace industry engineers have started to reorganise the layup process into two stages. A flat stack of plies, typically 1 mm to 5 mm thick, with the correct shape and layup, is first created by hand layup, by ATL machine, or by a dedicated flat laminating machine. This flat stack is then transferred to the mould tool, where it is formed (shaped) as one unit around the mould tool, before being cured in the normal way.

This chapter is thus concerned with the forming of stacks of prepreg which will subsequently be cured under heat and compaction to form a laminated thermoset composite component. The process of forming prepreg is not as widely known, or well developed, as the corresponding processes for the forming of dry reinforcement stacks (generally known as preforming) or the forming of preconsolidated thermoplastic composite laminates (often known as thermoforming). However, the process is very efficient, and is likely to become much more popular.

10.2 Practicalities of forming thermoset prepreg stacks

There are many common types of thermoset prepreg. The fibre reinforcement is normally a type of glass fibre or carbon fibre, although other fibres such as boron, aramid, basalt, quartz or even steel are sometimes used. The fibre is impregnated with a thermosetting resin mixture such as an epoxy or bismaleimide. This resin is distributed evenly through the reinforcement and usually partially cured, to prevent resin run-off and to give the prepreg the right handling qualities. The aircraft industry mostly uses carbon-epoxy prepreg designed to be cured at around 175°C to 180°C, and this material should be assumed for the rest of this chapter.

In most cases prepregs have been designed to be laid up by hand. (Prepreg tape may be optimised for ATL, with more accurate control of the resin amount, and lower tack. However, the author is unaware of any prepreg which has been specifically optimised for ease of forming.) The degree of cure of the resin, the resin content, and the tackiness of the resin, have been adjusted to suit a manual operator, often working with gloves, who wants to be able to handle large plies of prepreg easily, and shape it without excessive force. The operator needs the tackiness (tack) to be high enough to hold the ply in place after layup, but not so high that the ply cannot be safely removed if it is laid down incorrectly. The prepreg is normally delivered covered with two plies of 'backing paper' to keep it clean until layup. Unidirectional tape prepreg is usually from 0.1 to 0.2 mm thick: fabric prepreg thickness typically ranges from 0.2 to 0.4 mm.

Prepreg is perishable. The resin gradually stiffens, hardens, and loses its tack if exposed to the air and to room temperature. If not being used, prepreg is normally sealed and stored at $-18°C$, and needs to be warmed to room temperature before use. Therefore the deformation characteristics depend on the time that the prepreg has been exposed to room temperature, otherwise known as the 'age' or 'out-time' of the prepreg. A maximum allowed 'out-life' of between 10 days and a month is typical for prepreg designed for hand layup. The forming characteristics of the prepreg can change considerably over this period.

The viscosity and tack of the epoxy resin is quite sensitive to temperature and humidity as well as age. It is also somewhat sensitive to shear history. Although process specifications allow some heating of prepreg in order to facilitate hand shaping, heating to temperatures higher than is safe for skin contact is usually not allowed during hand layup.

Composite laminates can be designed with quite anisotropic structures. However, in most cases, carbon-epoxy parts found in aircraft include reinforcement fibres oriented towards each of the four points of the compass: i.e. fibres oriented at 0°, 45°, 90°, and 135° (−45°) to a given reference direction. Many have a layup which is 'quasi-isotropic', with an equal number of plies in each direction. Design guidelines discourage grouping of plies with the same orientation together. This means that in parts made with fabric prepreg, most 0°/90° plies are in contact with two ±45° plies, and in parts made with tape prepreg, a 0° ply might typically be in contact with a 45° and a 135° ply. It is a normal requirement that the stack layup is symmetrical about its centre ply to avoid warping of the laminate following cure and subsequent cool down.

10.3 Deformation mechanisms in woven fabric prepreg

Woven fabric prepreg stacks have the same deformation modes as dry fabric stacks. The most important modes are interply slip and intraply shear, while out-of-plane bending of the fabric and tows, and in-plane bending of tows, are also necessary. Slippage between the tows does not occur so readily as with dry fabric. Refer to Chapter 1 for a detailed description of stack and fabric deformation mechanisms and associated characterisation techniques.

10.3.1 Interply slip

In the forming of thermoset prepreg stacks, interply slip (also known as interply shear) is the most important deformation mechanism. Interply slip is the relative movement between individual prepreg plies during forming. Interply slip is necessary in virtually all forming of prepreg stacks, and can accommodate two types of stack deformation. Firstly, during the bending of a flat ply stack over a radius (single-curvature forming) each ply must slide over the other, as otherwise high tensile strains would occur in the outermost plies and compressive strains in the innermost plies. Secondly, during forming of a stack over a compound curve, the necessary local intraply shear will be different in adjacent plies with different orientations. Interply slip must occur to allow this.

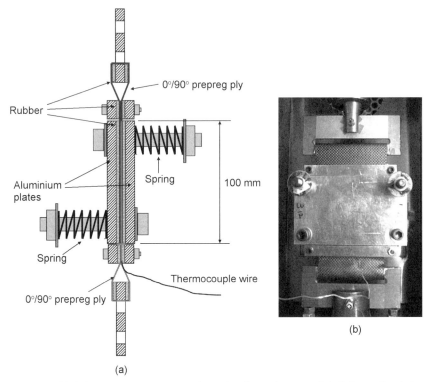

10.1 Apparatus to measure interply slip resistance. (a) Schematic of the apparatus. (b) Photo of the apparatus during a test.

As plies are relatively thin, and are in contact with plies of different orientation, a considerable amount of work may be required for relative movement of the large area of ply interfaces. Resistance to interply slip can be quite high, and is dependent on resin viscosity and the compaction pressure on the stack. Higher slip resistance is experienced under higher compaction pressures. Interply slip resistance is also strongly dependent on temperature, as the viscosity of the resin is a major controlling factor.

To characterise the resistance to interply slippage between prepreg plies, several fixtures including that shown in Fig. 10.1 (Young and Paton 2001) have been used. In the test, the upper plate grips the outer prepreg specimen and the lower stationary plate holds the inner prepreg specimen. The resistance to slip between the inner and outer prepreg specimens is measured and recorded by a tensile test machine. The normal (compaction) pressure on the two specimens may be adjusted using the four calibrated springs through the side plates. The normal pressure remains relatively constant despite small changes in spring length caused by 'bedding in' of the inner and outer specimens.

Figure 10.2 shows typical curves of the interply slip resistance against displacement (Phung *et al.* 2003). In the tests shown here, the prepreg has a high

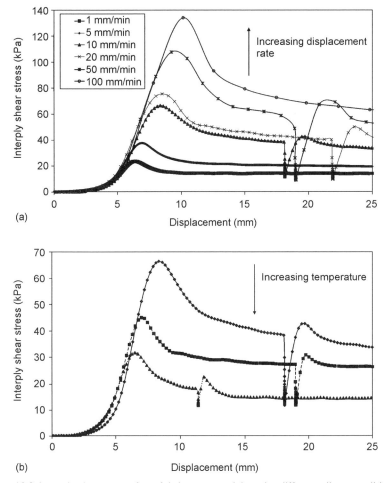

10.2 Interply shear stress for a fabric prepreg: (a) under different slip rates, (b) at different temperatures.

initial static resistance followed by a decrease to a more constant level of resistance. The load oscillations to the right of the plots record intentional temporary stoppages in the test to measure force relaxation.

During forming, it may be necessary for prepreg plies to slip across tooling or diaphragm surfaces. It has been found that slip resistance against these surfaces is similar to that against another prepreg surface.

10.3.2 Intraply shear

Intraply shear, (also called trellising, or extension along the bias of the weave) is the major deformation mechanism that allows an individual prepreg ply to conform to a compound curvature. The resistance to intraply shear, and the

244   Composites forming technologies

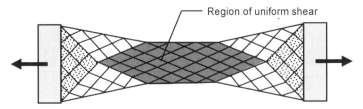

10.3 Schematic representation of a bias extension test fabric coupon, showing the different shear regions.

amount of shear possible before the fabric locks or jams, is dependent on the weave type and tow spacing, as well as resin characteristics.

Both the bias extension test and the picture frame shear test can be used to characterise the intraply shear of prepregs (Young and Paton 2001). However, the bias extension test is easier to conduct. The resin stops the fabric from fraying, almost eliminates inter-tow slip, and keeps the fabric together. Conversely, the picture frame test requires very careful set-up with carbon fibre fabrics, as the fibre is so stiff that tiny fabric misalignments lead to the fibre being loaded in tension during the test, increasing the apparent shear load on the fabric. Thus the bias extension test is preferred. The principle of the test is depicted in Fig. 10.3 (Phung 2004).

Results from bias extension tests on fabric prepregs typically look like the curve shown in Fig. 10.4 (Phung 2004), with three distinct stages in the force-displacement curve. As load is first applied there is some fibre straightening and slippage between the tows, as well as fabric shear. In some cases there also appears to be a 'yielding' effect in the resin as it starts to shear. In the second

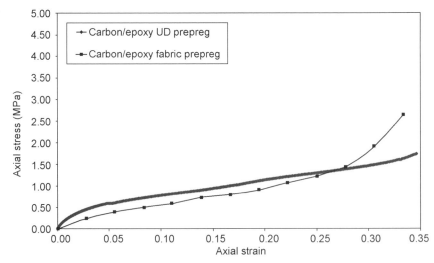

10.4 Stress–strain curve for bias extension tests on one ply of ±45 prepreg fabric and two plies [+45, −45] of prepreg tape.

region of the curve, shear occurs at a steadily increasing load. In the third region there is an accelerating increase in load caused by the locking (jamming) of tows with the closure of spaces between tows preventing any further easy rotation of tows towards the loading direction.

It should be noted that the phenomenon of tow locking is easily observed in tests and receives a lot of attention from researchers. However in practical manufacturing situations with prepreg stacks, intraply shear strains rarely reach the level required for locking to occur.

Correlation of forming experiments with simulations suggest that lateral compaction pressure on the prepreg fabric during shearing strongly affects the shear force required. Unfortunately this effect is difficult to measure directly, and a successful apparatus for this is not known to the author.

10.3.3 Fabric bending

A third required deformation mode is out-of-plane ply bending. As the plies are usually quite thin, the bending stiffness of woven prepreg is generally very low in comparison to its in-plane tensile strength, and the prepreg can be formed around tight radii. Ply bending resistance is not usually a significant factor in forming of prepreg stacks. However, as input is required for simulations, bending stiffness has been measured under a static load by a self-weight cantilever bend test (Young *et al.* 2001, Wang *et al.* 2006). The visco-elastic nature of the resin makes the deflection time-dependent. The deflection increases rapidly at first and then slowly approaches an equilibrium asymptote value.

10.3.4 Tow bending, buckling and crimping

In-plane tow bending is required where tows pass between regions of different intraply shear strain, such as shown in Fig. 10.3. Resistance to this leads to intertow slip, followed by tow separation or squeezing in these areas.

Tow bending resistance, however, is very important for a different reason. Tow buckling is a critical deformation mode, but an unwanted one. The most frequent defect in parts made by forming prepreg stacks is a wrinkle (a line of buckled tows). Although small amounts of tow buckling, or tow buckling in a lightly loaded area, may be tolerated, obvious tow buckling normally leads to rejection of an aerospace carbon-epoxy part. Even if the contour of the part is not affected (i.e. the buckles are in-plane rather than out-of-plane) the local failure load of the laminate, especially in compression, will be severely degraded. This makes tow buckling a significant problem during forming. The key problem in prepreg forming could be summed up as 'how to encourage all other modes of deformation while avoiding tow buckling'.

The buckling resistance of the tow is complex to measure, and is imperfectly understood. It is directly dependent on fibre type and size, and tow geometry (especially the degree of crimp), and has a complex relationship to resin viscosity. More work is needed to characterise tow buckling resistance.

10.3.5 Intertow slip

Intertow slip is the relative slippage of tows in the prepreg at their cross-over point. Generally intertow slip is a minor deformation mechanism in prepregs, making only a small contribution to the forming capability of a fabric. Intertow slip is most obvious at prepreg edges where tows can sometimes be separated from their neighbour due to high traction forces. However, one of the benefits of using prepreg is that the fabric edges are, in general, much more stable than in dry fabrics.

10.3.6 Fibre stretching

Generally, carbon reinforcing fibres only stretch at most 1% to 2% before failure. Therefore fibre stretch during forming is minimal and has little influence on the formability of the prepregs. However, woven fabric prepregs may stretch a little in the fibre direction (less than 0.5%) by straightening of the weave crimp, and may be compressed a similar amount in the fibre direction by an even distribution of increased weave crimp, without the strain being easily detected or being considered as a laminate defect.

10.3.7 The influence of temperature and thermal history

As each of the deformation mechanisms are dependent to some degree on the viscosity of the pre-impregnated resin, and the resin viscosity itself is dependent on time and temperature history, prepreg forming properties are time/temperature dependent. Interply slip resistance, intra-ply shear stiffness, and bending stiffness are dependent on viscosity, and thus are initially reduced by increasing temperature in the prepreg. (However, prolonged exposure to elevated temperature will increase resin viscosity as the resin cures.) For this reason it may sometimes be useful to carry out forming at an elevated temperature. However, it should be noted that tow buckling resistance is also reduced at elevated temperature.

Deformation mechanisms in prepreg are substantially viscoelastic. Formed shapes typically revert at least partly towards their original flat shape if removed from the tooling and not restrained. Prepreg deformation is also sensitive to strain rate. A thousand percent increase in strain rate may lead to a hundred percent increase in interply slip or intraply shear load. The time dependency and shape recovery potential is very dependent on the prepreg resin and thermal history.

10.3.8 Stack forming behaviour

When a fabric is required to conform to a complex shape such as the rib shape in Fig. 10.5, the orientation of the weave controls the local intra-ply shear required. Figure 10.7 shows the shape of a rib for an aircraft control surface leading edge with the drape pattern of a 0°/90° ply (left), and a ±45° ply (right), predicted using the DRAPE simulation software (Bergsma 2000). It can readily be seen that the required shear pattern is different for the two fabric orientations.

These shear patterns predicted using drape simulation software may only be realised by forming single plies under ideal conditions. In multi-ply stacks the resistance to interply slip imposes constraints on local intraply shear, and it is observed that the shear pattern of surface plies in such stacks often does not match that predicted by draping theory. Intraply shear deformation is usually less than predicted. The consequent residual compression stress in the plies may be dissipated by the formation of local wrinkles, or by increased local crimp in the fabric.

10.4 Tape prepreg

The deformation mechanisms for tape prepreg are in many ways similar to those of fabric prepreg. Interply slip behaviour in particular is similar (Phung *et al.* 2003). Out-of-plane bending resistance is higher in the fibre direction, and much lower in the transverse direction.

The important difference is in intraply shear behaviour. Intraply shear resistance is thought to be considerably lower than for fabric prepregs, but is somewhat difficult to measure directly. Bias extension tests on two ply [+45, −45] specimens give similar results to those on a single ±45 fabric prepreg (see Fig. 10.4) and to some extent produce the intraply shear behaviour expected of a pin-jointed net (Potter 2002a), although there is no tow locking effect, and both intertow slip and interply slip become more obvious at higher extensions. However, much of the apparent intraply shear resistance in such bias extension

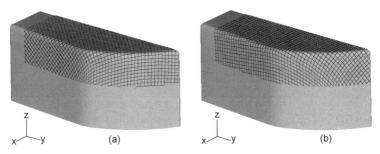

10.5 Solid model of the leading edge rib tool 'draped' with a fabric at 0/90° and ±45° orientations using simulation software. Locations of maximum shear can be seen for each case.

tests stems in fact from the interply slip between the two unidirectional plies (Phung 2004).

Tape prepreg has an additional deformation mode, transverse stretching or bunching, which can assist the forming of stacks into or over double curved tooling (Potter 2002b). However, the same mechanism can lead to unstable local deformation such as splitting of the surface plies in the forming of tape prepreg stacks: for best results surface plies should not be oriented with fibre direction parallel to any stack edge subject to significant traction. Local microbuckling of tows can also be seen in some cases (Phung 2004, Potter 2002a).

Although there seems to have been less research into the forming of tape prepreg than fabric prepreg, there are good reasons to believe that stack forming technology should be much better suited to tape prepreg than fabric prepreg (Potter 2002b). Certainly the advantages of automated forming methods over hand layup should be greater for prepreg tape than prepreg fabric.

10.5 Forming processes

Forming can be done by a variety of processes. Diaphragm processes and rubber presses are common. Matched dies and folding machinery are also sometimes used. Rollers and matched form dies may be used to form continuous single plies or ply stacks in an automated continuous process forming constant section profiles.

Diaphragm forming may involve a single diaphragm or two diaphragms which encapsulate the preform stack.

10.5.1 Single-diaphragm forming

Single-diaphragm forming is most suitable for simply curved shapes such as ribs with straight flanges. The diaphragm is only capable of effectively transmitting forces normal to its surface, making it difficult, using this method, to form shapes with significant double curvature. A forming box arrangement similar to that shown in Fig. 10.6 (Young and Paton 2001) can be used. Air is pumped out of the cavity and the diaphragm deforms under the action of atmospheric pressure.

10.5.2 Double-diaphragm forming

In double-diaphragm forming a common arrangement is similar to that shown in Fig. 10.7 (Young and Paton 2001). The upper and lower diaphragms are sealed together at their edges, and air is removed from the resulting cavity to supply a clamping force to the prepreg stack. This enables a range of forces to be transferred into the stack. Once vacuum is established between the diaphragms, the cavity between the mould and the lower diaphragm is evacuated. The

Forming technology for thermoset composites 249

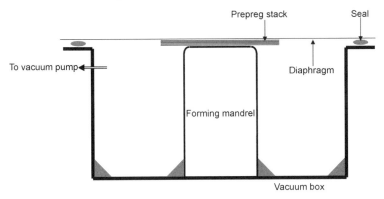

10.6 Basic tooling arrangement for single-diaphragm forming.

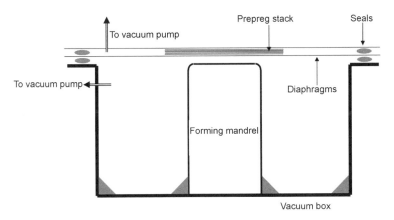

10.7 Basic tooling arrangement for double-diaphragm forming.

diaphragms are then stretched and formed into the box cavity, under the action of atmospheric pressure, taking the prepreg stack along with them.

10.5.3 Rubber press forming

A rubber press is also commonly used to form prepreg stacks. The tooling usually consists of a rubber die and a matching metal die. The rubber tool may be either the male or female tool. A metal male can be used with a rubber female to make simple rib type shapes similar to those shown in Fig. 10.11.

A metal female tool similar to that in Fig. 10.8 can be used to make more complex parts (Piegsa 2003), such as the flanged near-hemisphere shown in Fig. 10.11. The blank holder(s) may be quite complex devices, and must be carefully designed to apply sufficient 'friction' in the right places so that the plies are held in tension throughout the forming process, even as the plies slip under the blank holder. Excessive load on the blank holders will lead to excessive punch loads or

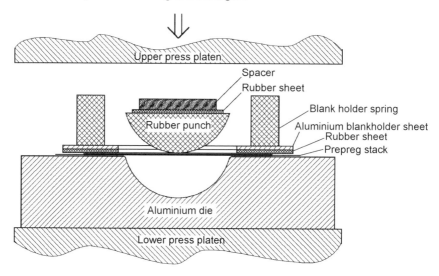

10.8 Rubber male forming punch and metal female die with blank holder.

even rupture of the plies. Quite complex parts such as stiffened wing ribs can be made using this sort of process.

10.6 Tooling and equipment

10.6.1 Vacuum pump or compressor

Most diaphragm forming is done using a vacuum source, as it is generally sufficient to use atmospheric pressure to drive down the diaphragm(s). For double-diaphragm forming, there are two cavities to be evacuated, and for best results the pressure in these cavities should be controlled separately. The forming box cavity must be evacuated to pull down the diaphragm(s) and the rate of evacuation naturally controls the forming rate. Unless very rapid forming is required for economic reasons, the cavity is normally evacuated in a few minutes. High vacuum is not necessary. The cavity between the diaphragms may require higher vacuum before and during forming, in order to ensure continuous clamping pressure on the stack, and also to remove air, moisture and volatiles from the prepreg stack. Where more force is required, it is practical to use an inflatable bladder to pressure a single diaphragm against a stack and hard mandrel. A rubber or rubber-faced punch may also be used.

10.6.2 Heat sources

For many applications the prepreg needs to be heated to ensure it is at the optimum temperature for forming. Many heat sources can be used. Radiant heating devices such as infra-red lamps are particularly effective, as carbon

Forming technology for thermoset composites

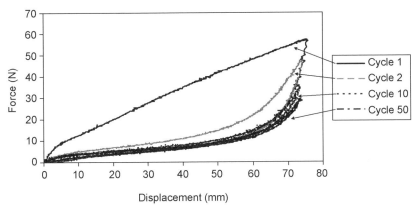

10.9 Force-displacement behaviour of silicone rubber at up to 300% strain after different numbers of cycles.

prepreg has a matt black surface, and reasonable heat conductivity. Where stacks are thick, radiant heating may be ineffective or insufficient, and it may be necessary to heat the stack from both sides, or heat the forming mandrel itself.

10.6.3 Diaphragm materials

For forming prepreg stacks, two types of diaphragm are common. Silicone rubber sheet, a hyperelastic material, is perhaps the most popular. High-temperature polymer films, which behave in an elastic-plastic manner during deformation, are also used. The choice of diaphragm depends on both circumstances and preferences.

Silicone rubber diaphragms are much more expensive, but can be re-used many times which saves costs in the long term. This makes them especially suitable for manufacturing smaller preforms, and higher manufacturing rates. Silicone rubbers also exhibit some 'permanent set' due to the Mullins effect. Figure 10.9 (Bibo and Paton 1999) shows the typical stress-strain behaviour of a silicone rubber used for diaphragm forming, and the Mullins effect.

Expendable high-temperature polymer diaphragms can only be used once. They can be prone to leaks though small pin-holes. The quite different stress-strain behaviour of these two diaphragm materials means that the design of the forming process can be dependent on the type of diaphragm.

10.7 Diaphragm forming tooling

The diaphragm forming process is most efficient when many parts can be formed at once. Sets of forming mandrels may be permanently placed within a dedicated forming box, especially if the parts are complex. In this case the cavities and sealing rim may be carefully designed for optimum efficiency. The box must, of course, be vacuum tight, and able to withstand normal loads of one atmosphere without deformation or collapse.

Alternatively, a larger generic forming box may be used, into which forming mandrels are placed and grouped as required. This approach may reduce tooling cost, but is more suited to simpler parts and lower throughput, as it is more difficult to optimise the forming of individual parts when all the tooling has to be removed and replaced frequently.

Whichever type of forming box is used, the box has vacuum connections and quick seal provisions on its flanges. It may have a folding lid or frame with integral diaphragm. Where higher temperatures are used for forming, the part mandrels need to be dimensionally stable and built from the usual tooling materials – steel, nickel, Invar or carbon-epoxy. Internal tooling which is solely used to shape the cavity or support the diaphragms may be made of less stable materials.

10.8 Potential problems

When forming carbon-epoxy prepreg stacks, the most common defects are various types of wrinkles. One type can be found in the inner plies of a radius, even in simple-curved shapes, where they are caused by the bending of the stack placing the inner plies under a compression force normal to the axis of the corner.

A second common type is an out-of-plane wrinkle involving all the plies in the stack, as seen in Fig. 10.10. These wrinkles are caused by the resistance of the plies in the stack to the shearing necessary to accommodate a shortened 'path length'. (After the forming operation shown in Figs 10.5 and 10.10, the 'path length' along the curved portion of the rib flange is less than the 'path length' of the section of flat stack from which it was formed.) This may be because the many interfaces in a quasi-isotropic layup have, in total, too much resistance to

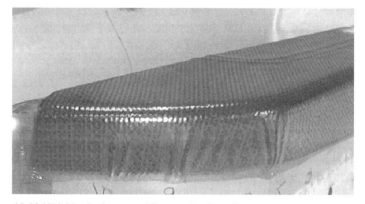

10.10 Wrinkles in the curved flange of a rib perform.

slip to allow the individual intraply shear required, or because the required in-plane shear cannot be accommodated by the local fabric orientation. Such wrinkles are likely to result in the part being rejected, because the local part thickness is too great, or the fibre orientation is locally incorrect.

Where matching hard tooling is used for forming, or a hard tool and soft tool combination, it is possible to break fibres, leading to rejection of the part. In diaphragm forming, this is rare – more common is a situation where the strength and stiffness of the plies, and the geometry of the tooling, result in bridging of the plies or incomplete forming of the stack around the tooling.

10.9 Process capabilities

Diaphragm forming is very successful for forming parts such as ribs and spars over male tools. These parts typically have a 'C'-shaped cross-section. Straight flanges are the easiest to form, and may often be formed using a single diaphragm, or in matched moulds. However, convex-shaped flanges with a radius as tight as 50 mm can be successfully formed from flat prepreg stacks by double-diaphragm forming. Parts with a mild concave flange may also be formed in the right conditions. In this case the limiting factor will usually be fibre bridging, so the orientation of the fibres with respect to the tooling is absolutely critical.

The double-diaphragm forming process as used by the Cooperative Research Centre for Advanced Composite Structures is capable of forming most of the typical rib and spar shapes found in the substructure of aircraft control surfaces. The shapes formed have included those in Fig. 10.11. Spar shapes with joggles in the flanges can also be formed. The critical factors are normally the length of the flange, the thickness of the stack, and the curvature of the flange, if any. As these increase, forming becomes more difficult.

The part at lower left in Fig. 10.11 is a flanged dome shape formed using a rubber male punch and a metal female tool with blank holder.

10.10 Future trends

Forming of flat, prepreg stacks into shaped preforms can be very successful if the deformation requirements can be understood. Although the fabric stack is in practical terms inextensible in any fibre direction, the intraply shear, interply slip, and ply bending deformation modes enable many shapes to be successfully formed. Some aircraft control surface rib shapes are actually easier to form in prepreg than in aluminium. The forming process is also easier, with lower tooling costs than the thermoforming of thermoplastic composites, as the temperatures and loads required are substantially less.

The technology of forming reinforcement stacks is likely to become much more important as efforts continue to lower the cost of continuous-fibre

254　Composites forming technologies

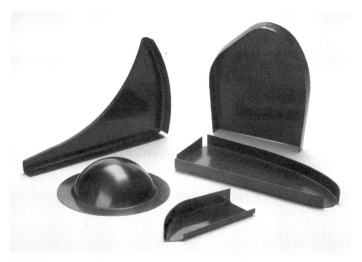

10.11 Prepreg preform shapes formed at CRC-ACS (clockwise from top left): a DDF rudder leading-edge rib with a concave flange; a DDF section of a rudder closure rib; a DDF flap leading-edge rib; a DDF rudder leading edge rib; a flanged dome shape formed using a rubber male punch and a metal female die with blank holder.

reinforced polymer composites. Widespread introduction of carbon-epoxy composite into the automotive industry, for example, would appear to require robust forming technology. Whether this forming technology involves dry fibre fabrics, thermoset prepregs or thermoplastic laminates, will depend heavily on the economics of the different material forms, and recyclability requirements, as well as the forming technology.

The lack of any plastic deformation mode for the continuous reinforcement fibres is, however, a significant hindrance. It seems likely that some low cost and robust form of long discontinuous fibre prepreg will emerge in the near future which will allow a certain degree of effective 'fibre extension' to occur during forming. This would significantly increase the number of shapes that could be formed, and give a tremendous boost to composites forming technology.

10.11 References

Bergsma, O, 'Drape Software Beta Release 1', Technical University Delft, The Netherlands, 2000.

Bibo, G, and Paton, R, 'Mechanical properties of diaphragm materials' *20th SAMPE Europe Conference,* Paris, 1999.

Phung, T, *Deformation mechanisms of composite prepregs during forming.* PhD Thesis, RMIT, 2004.

Phung, T, Paton, R, and Mouritz, A P, 'Characterisation of the interply shearing resistance of carbon-epoxy unidirectional tape and fabric prepregs', *6th*

International ESAFORM Conference on Material Forming, Nuova Ipse Editore, 2003.

Piegsa, R, *An investigation into the forming of carbon-epoxy prepreg stacks into female moulds*. Diplomarbeit Thesis, University of Stuttgart, 2003.

Potter, K, 'Bias extension measurements on cross-plied unidirectional prepreg' *Composites Part A*, 33, 2002a.

Potter, K, 'Beyond the pin-jointed net: maximising the deformability of aligned continuous fibre reinforcements' *Composites Part A*, 33, 2002b.

Wang, J, Lin, H, Long, A C, Clifford, M J, and Harrison. P, 'Predictive modelling and experimental measurement of the bending behaviour of viscous textile composites', *Proc. 9th Int. European Scientific Association for Material Forming (ESAFORM) Conference*, Glasgow, April 2006.

Young, M, Cartwright, B, Paton, R, Yu, X, Zhang, L, and Mai, Y-W, 'Material characterisation tests for finite element simulation of the diaphragm forming process', *4th International ESAFORM Conference on Material Forming*, University of Liege, 2001.

Young, M, and Paton R, 'Diaphragm forming of resin pre-impregnated woven carbon fibre materials', *33rd International SAMPE Technical Conference*, SAMPE 2001.

11
Forming technology for thermoplastic composites

R BROOKS, University of Nottingham, UK

11.1 Introduction

This chapter gives an overview of the technologies currently used for forming thermoplastic composites (TPCs). Unlike thermosets, which undergo a chemical curing reaction for geometry stabilisation during forming, the forming process for TPCs is relatively simple and involves purely physical changes to the material. Essentially, the matrix is melted by heating, the part shape is formed and, finally, the geometry is stabilised by cooling. There are, of course, other factors to consider, such as the relatively high viscosity of the thermoplastic melt and the degree of consolidation which can be achieved within a specific processing window of time, temperature and pressure. Part quality will clearly depend on the latter. How these more fundamental issues are addressed by the forming technologies currently used by industry is the main subject of this chapter.

The specific forming technology used depends to a large extent on the TPC material form. Is it supplied as partially or fully consolidated? Is the reinforcement in a fabric or aligned form? Section 11.2 describes the various intermediate forms of TPC available and how, in many cases, matrix and reinforcement have been combined to aid the subsequent forming and consolidation processes. Before the various forming technologies are described, the basic principles of isothermal and non-isothermal forming are explained in Section 11.3. The main forming technologies are then covered in some detail in Section 11.4 including compression moulding of flow materials and more aligned structural TPCs, vacuum forming, diaphragm forming, bladder moulding and roll forming. The aim is to give an overview of each technology and address the factors which dictate its application and scope. The chapter concludes with a brief consideration of some recent developments.

11.2 Thermoplastic composite materials (TPCs) for forming

Thermoplastic composites are available commercially in a number of different intermediate forms, each of which determines the appropriate forming

Forming technology for thermoplastic composites 257

technologies required for making parts from the material. The material forms can be categorised according to their reinforcement fibre architecture and their matrix distribution or level of consolidation in the intermediate form. Thus, fibre architecture can be random, a textile form or fully aligned while matrix distribution can be partially consolidated, fully consolidated, commingled or powder impregnated. Various combinations of these attributes result in the material forms available commercially. Table 11.1 illustrates the broad range, their form (fibre architecture and matrix distribution) and some key properties

Table 11.1 Range of thermoplastic composites, their form and key mechanical properties (stiffness and strength)

Supplier	Thermoplastic composite	vf (%)	Tensile modulus (GPa)	Tensile strength (MPa)
GMT				
Quadrant	GMT 40wt% glass/PP random	24	5.8	81
Quadrant	GMtex 40 wt% glass/PP 1:1	24	4.9	120
Quadrant	GMtext 40 wt% glass/PP 4:1	24	10.5	248
Glass fabrics				
Comfil	Commingled glass/PET twill	44	22	260
St-Gobain Vetrotex	Twintex glass/PP 2:2 twill	35	13	300
Ten Cate	CETEX glass/PPS fabric	50	23	324
Plytron	Glass/PP 0:90	35	16	360
St-Gobain Vetrotex	Twintex glass/PP 4:1	35	24	400
Bond laminates	TEPEX 22 101 glass/PA66 twill	47	22	400
Bond laminates	TEPEX 22 102 glass/PA6 twill	47	22.4	404
St-Gobain Vetrotex	Twintex glass/PP 1:1	50	25	440
Ten Cate	CETEX glass/PEI fabric	50	26	484
Bond laminates	TEPEX 22 109 glass/PA46 twill	46	23.5	550
Glass UD				
St-Gobain Vetrotex	Glass/PP UD	50	38	700
Plytron	GN tape glass	35	28	720
Carbon fabrics				
Bond laminates	TEPEX 22 206 carbon/PA12 twill	50	55	610
Ten Cate	CETEX carbon/PPS fabric	50	56	617
Ten Cate	CETEX carbon/PEI fabric	50	56	656
Bond laminates	TEPEX 22 209 carbon/PA46 twill	50	55	780
Bond laminates	TEPEX 22 201 carbon/PA66 twill	48	53	785
Schappe Tech.	Stretch broken carbon/PA12	56	61	801
Hexcel	Towflex 22 carbon/nylon twill	51	66	821
Carbon UD				
Hexcel	Towflex 22 carbon/nylon UD	51	116	1665
Ten Cate	Carbon/PEI UD	60	128	1890
Schappe Tech.	TPFL carbon/PEEK slivers UD	52	130	2150
Cytec	APC-2 carbon/PEEK UD	63	151	2400
Gurit	SUPreM carbon/PEEK UD	60	142	2800

when formed into laminates or parts. Although not a comprehensive list, the table includes examples from most of the main suppliers and focuses on glass and carbon fibre reinforced materials. Data has been obtained from a number of sources including materials suppliers and information published by the CORONET European Network in Thermoplastic Composites.[1] Other reinforcements, levels of reinforcement volume fraction and a significantly wider range of resins are also available and suppliers are willing to provide data. To better understand the range of materials, strength versus stiffness data from Table 11.1 is plotted in Figs 11.1(a) and 11.1(b) for glass and carbon reinforced systems respectively. At low stiffness and strength, flow materials, e.g. glass mat thermoplastic (GMT), compete with short fibre moulding compounds (not included here) in semi-structural applications. These materials can be moulded

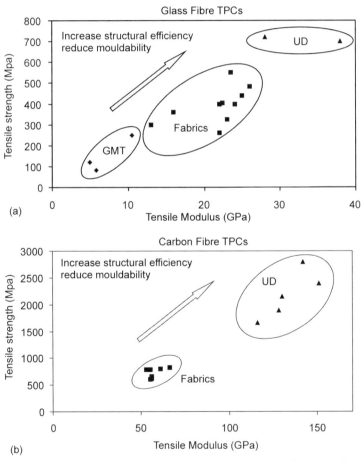

11.1 Mechanical properties (stiffness and strength) of thermoplastic composites: (a) glass fibre-reinforced TPCs and (b) carbon fibre-reinforced TPCs.

into very complex shapes but have limited structural performance. For structural applications, continuous fibre reinforcement is required. As shown in the figures, fabric forms of reinforcement provide a level of structural performance suitable for many engineering applications. They also retain a capability to be formed into reasonably complex shapes including double curvature and concave surfaces. For maximum structural performance, however, unidirectional (UD) forms of material are required. These are often available as fully impregnated systems with high volume fraction of reinforcement and as a result are difficult to form into anything other than simple shapes. In general, as structural efficiency of the material increases (e.g., through alignment or increase in the volume fraction of the reinforcement), the mouldability, i.e. ability to form into shape, reduces. This reduction is often accompanied by an increase in cycle time and a reduction in production volume. Comparing Fig. 11.1(a) with Fig. 11.1(b) also shows, as expected, that significantly greater structural performance can be achieved with carbon systems. This clearly comes at a cost penalty.

To facilitate a better understanding of the materials a brief description of the main material forms now follows.

11.2.1 Partially consolidated – random reinforcement

This form of TPC is commonly called glass mat thermoplastic or GMT for short and consists of several layers of glass fibre mat (chopped strand mat, random continuous strand mat or unidirectional mat combined with either of these two) in a thermoplastic matrix, most commonly polypropylene. The material is used widely in semi-structural automotive applications, e.g. seat backs, front ends and bumpers, and can be manufactured at very high rates using compression moulding (see Section 11.4.1) giving economic high volume production. The raw material is supplied as partially consolidated rigid sheets manufactured by one of two processes: slurry deposition (also called wet laying) or melt impregnation. Slurry deposition involves mixing discontinuous fibres and polymer powder with water to form an aqueous slurry. The solids are then filtered from a head box containing the slurry by a moving screen. A randomly dispersed mat is formed which is then dried and partially consolidated in a heated double belt press. Melt impregnation is the more common method of production and involves passing a glass fibre mat through a preheating oven and combining directly with extruded polymer sheet. Up to six layers of reinforcement, interleaved with polymer melt, may be combined to form the composite sheet. The glass mat is bonded together using thousands of felt needles. This laminate then passes through calendering rolls and a heated double belt press for consolidation. GMT is a flow material and forms by flow moulding (see Section 11.4.1). Thus, volume fractions are typically 20vf% or less and hence parts are generally used in semi-structural applications only. Key suppliers of GMT are Quadrant Plastic Composites and Azdel Inc.

11.2.2 Commingled fabrics

The production of commingled tows involves mixing continuous filaments of reinforcement and matrix to produce a tow bundle which can subsequently be woven into fabric possessing good drape properties. A number of techniques for commingling exist. The most common (developed by St-Gobain Vetrotex for their product Twintex – commingled glass/PP or PET) involves extruding continuous polymer fibres through a spinneret and combining with a drawn tow of glass fibres. Typical reinforcement volume fractions for commercial fabric made from this type of commingled tow is 35 vf%, although levels up to 50 vf% and higher are achievable. An alternative process (developed by Schappe Techniques) involves the stretch breaking of carbon fibre tows and their combination with discontinuous polymer fibres into a commingled yarn of discontinuous (but long) fibres. This process produces materials with volume fractions of 50–60 vf%. The fabrics produced from commingled tows/yarns can be formed at significantly reduced pressures and times, a consequence of the reduced flow distance for the resin in the commingled structure. Textile weaving techniques for producing woven fabrics from commingled yarns have had to be developed. Handling of a combination of stiff reinforcement and flexible matrix fibres in the same yarn is difficult. The sizing requirements are particularly important as damage to the reinforcement must be minimised, yet good dispersion between reinforcement and matrix must be maintained during weaving. Recent developments have resulted in the production of three-dimensional woven products from commingled yarns. Near net shape preforms can be made this way which have good interlaminar properties improving impact performance.

11.2.3 Powder impregnated fabrics

Powder impregnated tows are produced by impregnating the fibre bundle with fine particles of polymer using either a fluidised bed or water slurry. The process is particularly relevant to resins with high viscosity although a broad range of materials can be used, such as PEEK, PES, PPS, PEI in addition to PP and PA. Powders with diameters similar to reinforcement fibre sizes would be ideal although economics normally prevents this and powders in the range $15\,\mu$m to $150\,\mu$m are typical. For good dispersion of powder, pins or rollers are often used to open the fibre tow or alternatively air knives and venturis have been proposed. Both mechanical and electrostatic forces control the dispersion process. After powder dispersion, tows are usually passed directly through infra-red ovens (rapid heating) to sinter the powder to the reinforcement fibres. Advantages of powder impregnation include the fact that flow lengths are short and that flow during impregnation takes place predominantly along, rather than transverse to the fibres. This results in very much lower impregnation times. Powder impregnated tows have to be handled carefully in subsequent textile processes such as weaving into fabric or braiding. The FIT (Fibre Imprégnée de Thermoplastique) process has been developed to keep the powder in place during such textile processes. It involves coating the powder impregnated tow with a thin outer coating of polymer to keep the powder in place. Rapid impregnation can then take place during part forming. One drawback of the FIT process, however, is the debulking which takes place during the forming process which can lead to distortion and fibre buckling. A disadvantage of the powder impregnation method is that the unidirectional prepreg very often has low transverse strength and requires a high content of polymeric binding agent which remains in the system. This, however, is not a problem for fabric forms. Typical reinforcement volume fractions for powder impregnated fabrics are 50–60 vf%.

11.2.4 Solvent impregnated forms

Amorphous resins such as polyethersulphone (PES) and polyetherimide (PEI) can be dissolved in organic solvents and the solution used to form prepregs in a similar way to thermosets. Because of the low viscosity of the solution, impregnation of high reinforcement volume fractions can be attained in UD or fabric form. The materials can also be formed using common thermoset forming processes. TenCate's Cetex® materials are a typical example. However, the nature of these resins is such that the final part can have poor resistance to solvent attack and during processing volatiles can be released.

11.2.5 Fully consolidated – UD tape

A number of different fully consolidated thermoplastic prepregs exist, the longest standing of which is Cytec's 'Aromatic Polymer Composite' (APC-2). APC-2 is a fully impregnated unidirectional carbon/PEEK tape manufactured by a modified pultrusion process. Its high structural properties have ensured its use in high performance aerospace applications. Low molecular weight (and hence low viscosity) PEEK is used to impregnate the carbon fibre tow in an extrusion die. The molecular weight is then increased as a result of a reagent previously applied to the fibre surface. Chain growth occurs initially in the fibre-matrix interphase and moves out to the remainder of the matrix. The process can operate at high speed and produce material with high volume fraction (60vf% or higher). Full impregnation allows the tapes to be fabricated into components at low pressure for a short time (typically 1 MPa for 10 mins). Sheet, or wide tape forms are also available. Because the material is fully impregnated, wide tapes do not possess the required drape properties. To overcome this limitation, narrow tapes are woven into fabric form. Alternatively, tapes can be used in other fabrication processes such as tape laying and filament winding. Similar products exist using other resins, e.g.

PEKK, PEI, PPS and with glass reinforcement, e.g. glass/PP (Plytron® is an example) and glass/PA systems.

11.3 Basic principles of TPC forming technologies

Shape forming of TPCs is not done as an isolated process but involves a number of simultaneous and/or sequential operations including heating, forming, consolidation and cooling of the material. Although TPCs do not have to undergo a curing reaction, as do thermosets, there are specific properties of a TPC which require careful control of the process cycle during forming to achieve all of the above operations and produce a quality part. In particular, the thermoplastic matrix, at melt processing temperatures, has a high viscosity ($\sim 10^3$ Pa s) by comparison with a thermoset (~ 1 Pa s). For forming, the material must firstly be heated above the matrix melting temperature where it will be sufficiently flexible to partially conform to a shaped mould. Sufficient pressure and time, at the melt temperature, is then needed to achieve full mould conformation, impregnation of the fibre reinforcement by the matrix and consequent full consolidation of the TPC. The required time at pressure depends on the matrix viscosity, the heating regime and the TPC material form. For the latter, as described in the previous section, a number of forms have been developed specifically to reduce the impregnation/consolidation times. Fully preconsolidated, commingled and powder impregnated forms all reduce the flow distance for the matrix allowing lower pressure processing and/or shorter cycle times.

To achieve the above, both 'isothermal' and 'non-isothermal' processes are used in practice. Figures 11.2(a) and 11.2(b) illustrate temperature-pressure-time profiles for each of these approaches.

In the isothermal process, shown in Fig. 11.2(a), the preform (and tool) are heated to above the matrix melt temperature, i.e. the forming/consolidation temperature, and pressure is applied whilst the temperature is held constant at this level. Pressure is then maintained during cooling to below the recrystallisation temperature to ensure the consolidation achieved during the isothermal part of the cycle is maintained and the resulting part has low void content ($<1\%$). Quenching (rapid cooling) can take place beyond this time to speed up the process. Whereas isothermal moulding like this can achieve high quality parts with excellent consolidation, as required by the aerospace industry, for instance, the cycle time is compromised by having to cycle the tool temperature during the process.

By contrast, a shorter cycle time can be achieved using the non-isothermal process shown in Fig. 11.2(b). Here, the preform is preheated to the forming/ consolidation temperature usually separately in an oven. The preheated charge is then rapidly transferred to a forming tool located in a pressing station (e.g., press or vacuum, etc.). Pressure is then applied quickly at the same time the charge is

Forming technology for thermoplastic composites 263

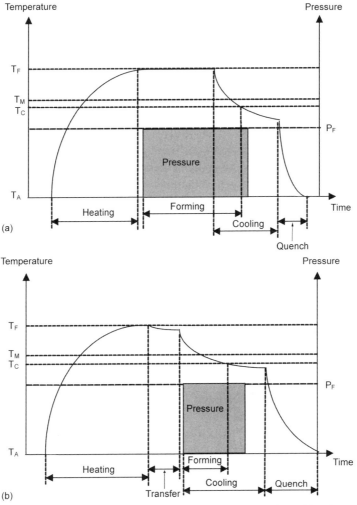

11.2 Temperature and pressure vs. time profiles for: (a) isothermal forming and (b) non-isothermal forming of thermoplastic composites.

cooling on contact with the tool, i.e. non-isothermal forming. Pressure can then be removed when the temperature has reduced to below the matrix recrystallisation temperature and quenching can follow if required. The non-isothermal process requires no temperature cycling of the tool and consequently heating and cooling times can be significantly shorter than the isothermal process. Process cycle times are therefore shorter. However, there is a limited time and temperature window during which forming and consolidation can take place. Consequently, the method is often applied to materials with lower reinforcement volume fraction and/or with forms requiring lower consolidation pressures and times, e.g. commingled and powder impregnated materials. The non-isothermal process is also attractive where medium to high production volumes, i.e. short cycle times, are required, and higher void content levels are acceptable, such as in the automotive industry.

11.4 Forming methods

11.4.1 Compression moulding

Compression moulding is used here to denote processes where the forces required to form a part are provided by a press (usually high pressure), containing a matched metal (male and female) or metal-rubber (i.e. deformable) tool. In the case of TPCs, compression moulding was originally developed for the forming of flow materials, e.g. GMT, however the process has more recently been used for both textile forms and fully aligned forms of TPC. In these cases, the process is often referred to as 'stamp forming'. The compression moulding (stamp forming) process is non-isothermal and capable of high production rates with cycle times typically under one minute. Although closely related, it is appropriate to consider the forming of flow materials and aligned materials separately.

Flow moulding of GMT

Glass mat thermoplastic (GMT) is supplied in the form of partially consolidated sheets of random glass mat and polymer (typically PP) manufactured by either a slurry deposition process or a melt impregnation process. The use of random reinforcing fibres (e.g., chopped strands or continuous swirl mat) at relatively low volume fractions (20–40 wt%) results in a material, in melt form, capable of flowing in the mould. The GMT sheet is fabricated into parts by cutting into blanks and stamping or flow moulding in a matched metal tool in a press. Net shape parts are produced. Detailed guidelines for the forming/moulding of GMT can be obtained from materials suppliers and the following overview draws on information supplied by Quadrant Plastic Composites.[2]

The forming process, schematically shown in Fig. 11.3, involves the following stages:

1. Several blanks are preheated in an infra-red or hot circulating air oven to the processing temperature (200–220°C for PP based GMT). Conveyor ovens are normally used for a continuous process. Blanks loft considerably during heating (two to three times original thickness) as the frozen-in forces created during GMT sheet manufacture are released. This lofting has consequences for material heating at the later stages.
2. Blanks are transferred by hand or robot to a matched metal mould, the latter being at a temperature of 20–80°C.
3. Hydraulic pressure is applied by the press. High closing speeds of 600 mm/s are necessary to retain material temperature and are followed by compression

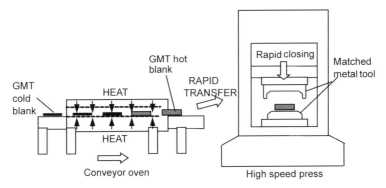

11.3 Schematic diagram of the compression flow moulding process for GMT.

speeds of ~30 mm/s. Full pressures of 10–20 MPa (100–200 bar) are held for less than six seconds. This allows the GMT to flow in the mould. During flow, the material is cooling against the cool mould surface (i.e. a non-isothermal process) and therefore the material has a limited flow distance to part thickness ratio.

4. Cooling takes place in the mould at a typical rate of 4 to 5 seconds per mm of part wall thickness. Shrinkage is ~0.3–0.5%.

The whole process is high speed with cycle times of approximately 30–60 seconds. Processing times for GMT compare favourably with SMC (sheet moulding compound) which has a moulding cycle of 1.5–6 minutes. Other advantages the material possesses over SMC is that pressure is only required briefly at the bottom of the press stroke and not continually. Because the material flows, complex parts containing ribs, undercuts, inserts and varying wall thickness can all be made.

Quality of the final part depends on careful attention to a number of factors, the following in particular:

- *Blank configuration* – must be cut to the mould size and configured for part geometry; stacked in centre of mould to allow regular flow; should not be folded; overlapping to help knit weld lines; designed for flow, e.g. no bridging of ribs; no pinching by mould side walls.
- *Heating* – do not heat blank surface any greater than 230°C for PP based GMT (indicated by smoking of the material); heat blank internal temperature to between 190°C and 215°C. Transfer quickly to mould to prevent pre-cooling.
- *Pressing* – important to maintain a high compression rate (30 mm/s), otherwise incomplete parts will result.

Compression moulds used for GMT must have close tolerance telescoping shear edges (see Fig. 11.4(a)) to seal off the mould cavity at its outer shape before the GMT reaches the edges. At the same time air is allowed to escape

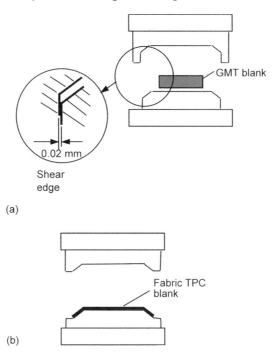

11.4 Schematic diagram of (a) a GMT matched metal mould with shear edge and (b) a hot stamp forming tool.

without causing flashing. Clearance of these edges should be 0.02 mm to 0.03 mm and the surfaces at these edges should be hardened (flame hardening is suitable). A heavy duty guidance system should be used to prevent wear of the sealing edges, and safety stop blocks are necessary to prevent the two halves of the mould closing on each other. The stop blocks are not used to determine part thickness but are normally set at 0.5 mm below minimum part thickness. Mould temperatures should be held constant at 40°C to 70°C using a carefully designed cooling system. The aim is to maintain a uniform temperature distribution across the mould surface.

Ejector pins should be used to eject the part evenly, in some cases from both halves of the mould. Venting is important, particularly at ribs or bosses where overheating and scorching can otherwise occur. For ease of flow and ejection the mould surface should be highly polished. GMT causes very little wear, so mould surface hardening is not necessary except near shear edges as already discussed.

Production moulds are generally made from tempered steel with a medium hardness. Apart from high volume, hardening is not necessary. Prototype tooling can be made from a variety of materials such as Kirksite or aluminium. For these, cooling channels can be omitted, as can the ejector system and mould surface polishing may not be necessary.

The forming of GMT into parts using flow moulding is a mature technology principally developed for the high rate production of semi-structural automotive parts.

Hot stamping of commingled fabrics and aligned prepregs

For higher performance parts, fabric forms or fully aligned TPCs, with higher volume fractions, are used (see earlier section on 'Material forms'). Such materials can also be formed into shape using compression moulding with matched metal tooling. Shear edge tools are not necessary in this case as the materials do not flow in the mould (see Fig. 11.4(b)). Commingled and powder impregnated TPC fabrics exhibit the better draping properties, although, as for all materials, preheating to above the matrix melt temperature takes place prior to forming and consequently, even fully aligned materials are sufficiently flexible to conform to shape. Fabric forms can, however, generally be formed to more complex double curvature shapes and to greater draw depths.

As with GMT, the intermediate material, e.g. fabric, is preheated, most commonly in an infra-red oven to melt the matrix. The material is then transferred rapidly (to minimise cool down) to a cool matched metal tool which closes quickly and forms the part into shape whilst the material cools (i.e. a non-isothermal process). Unlike GMT, the material does not flow in the mould and hence it is more of a 'hot stamping' process. As this is not a flow process and because the intermediate material already has a high degree of impregnation, forming can be undertaken at significantly lower pressures (1–2 MPa) compared to GMT moulding (10–20 MPa). Forming pressures do vary over the part depending on the local draw angle and part thickness. Horizontal surfaces, for instance, support high pressures and can achieve optimum consolidation levels with minimum void content. More vertical surfaces on the part, however, support lower pressures and can suffer from poor consolidation. Soft tooling techniques (described below) can alleviate this problem. Other factors important when forming fabrics are the maintenance of fibre direction and architecture during heating and transfer to the tool and the avoidance of wrinkling in regions of high fabric shear when forming. Both of these problems can be avoided by the use of a blank holder. The fabric is stretched under light load and held by springs biaxially in the fabric plane during heating. The blank holder thus restrains any shrinkage during heating and holds the fabric in place during transfer to the tool. During forming, the blank holder also offers a relatively low level of membrane pressure whilst the material draws in. This pressure has been shown to increase significantly the maximum attainable fabric shear angle before wrinkling occurs.

Matched metal tooling results in a pressure distribution across the part which depends on the part geometry, in particular, the draw angle and often results in variable consolidation, as discussed above. The variation in pressure across the tool can be reduced or, in some cases, eliminated by using a compliant material for one half of the tool, e.g. the female half.[3] Elastomers such as silicon rubber or polyurethane are commonly used. The pressure distribution with such a tool is generally an improvement over matched metal tools and can result in near hydrostatic pressure across the tool and consequently more consistency in consolidation. Disadvantages of rubber tooling include increased cycle times due to reduced cooling rates caused by the low thermal diffusivity of the rubber and reduced production volumes achievable before tool damage and wear occurs.

11.4.2 Vacuum forming

Vacuum forming is a common technique for forming TPCs similar to the method used for thermosets. Useful information on the process including tooling and part design can be found in the Twintex® vacuum moulding manual supplied by St-Gobain Vetrotex.[4] Figure 11.5 shows the basic isothermal process. The TPC laminate is laid up on a single-sided mould (metal, metal shell or composite) and covered with a release film (i.e. peel ply) and breather cloth. The former prevents the breather cloth bonding to the laminate and the latter ensures all air is extracted under vacuum, and even distribution of pressure. A vacuum bag is then positioned over the arrangement and sealed around its edges against the mould. Typical bag materials include disposable nylon film or reusable silicon rubber. The air is pumped out until vacuum pressure up to 0.9–0.95 bar is achieved. The whole arrangement is then placed in a circulating air oven and the temperature raised to above the matrix melting temperature for forming and consolidation to take place. Consolidation time depends on part size, material thickness and degree of void content allowable in the final part. Generally 10–15 minutes consolidation time will result in void content below 2%. Longer times are necessary for better quality. It is also important that

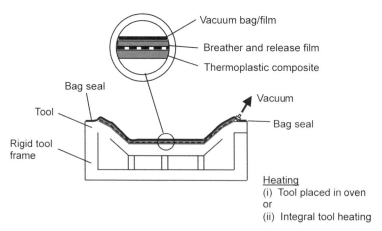

11.5 Schematic diagram of the isothermal vacuum forming process.

sufficient heating time has been allowed to ensure the tool reaches the processing temperature and the laminate is heated through its full thickness. Large tools (for large parts) can result in significant heating times (hours) making the technique, in most cases, a low volume process. To speed up the process, shell tools with lower thermal inertia can be used but this will depend on the economics of the process. Following consolidation, the tool, etc., is removed from the oven and allowed to cool before the part can be removed. Vacuum is maintained during cooling. Again, depending on the size and construction of the tool, cooling can be a slow process. An alternative to oven heating, is to heat the tool directly using oil or electrical methods. This can, in some cases, speed up the process, but its viability depends very much on the part size and shape. More complex shaped parts prohibit uniform heating by this method.

Non-isothermal vacuum forming enables faster cycle times to be achieved. In this case the TPC laminate is preheated in an oven, transferred and laid up rapidly on a cool tool following which a vacuum is applied immediately. This can be achieved by the use of a vacuum table as shown in Fig. 11.6. The table comprises a reusable silicone rubber bag mounted in a frame capable of being closed and sealed against the table. Thus, when the preheated laminate is transferred to the tool, the bag can be closed rapidly and a vacuum pump activated to remove air and apply the vacuum. Time is critical in the non-isothermal technique with transfer time being the most important, which should be less than 20 seconds to ensure material temperature is retained. The transfer process can be improved by using transfer plates, i.e. preheated metal plates, also to assist temperature retention during transfer. Tool temperature can also be raised to slow down the cooling process, however, this will clearly affect cycle times and can result in unacceptable surface finish to the part. Non-isothermal

11.6 A vacuum table.

vacuum moulding is clearly a rapid process but is restricted to the size of part mouldable. Above about 1 m² projected area, preheated charge transfer becomes problematic.

11.4.3 Diaphragm forming

Both single and double diaphragm forming processes are used commercially for forming TPCs.[5] Figure 11.7 shows a schematic representation of the double diaphragm forming process in an autoclave. The TPC laminate is placed between two flexible diaphragms (membranes) which are then clamped around their edges. The diaphragm-laminate stack is then clamped across the mould and the whole is placed in the autoclave. The space between the diaphragms is then evacuated. This serves to hold the laminate in place during heating and forming and prevents wrinkling. The autoclave temperature is then raised above the matrix melt temperature and pressure (typically up to 7 bar) is applied. The laminate itself is not edge clamped and is thus free to conform to the mould shape (along with the flexible diaphragms) by friction slip against the diaphragms and inter-laminar slip within the laminate. Relatively deep draw and complex curvatures can be achieved in diaphragm formed parts. The diaphragms must be flexible at temperature and a number of different materials are used, depending on the laminate melt temperature. Thin superplastic aluminium diaphragms and high temperature polyimide (PI) films, e.g. Upilex-R®, are commonly used for high performance laminates such as Carbon Fibre/PEEK (APC-2) while silicone rubber sheets have been used for lower performance/lower temperature materials such as Glass Fibre/PP (Twintex®). The process is isothermal and, because of mould temperature cycling, has a relatively long cycle time although this can be shortened by using shell type moulds. As a result of the high pressure and isothermal moulding, very high quality laminates with low void content can be achieved, however, this is countered by the relatively long cycle times.

To speed up the cycle time, diaphragm forming can be undertaken non-isothermally using vacuum forming and press methods. In this process, the

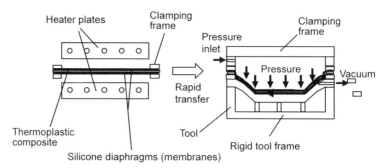

11.7 Schematic diagram of the double diaphragm forming process.

laminate contained between a double diaphragm, is preheated between heaters and rapidly transferred using a conveyor mechanism to the forming station. Pressure of 5 bar is applied above the diaphragms using inert nitrogen gas and a vacuum is applied below through the single-sided tool. This ensures rapid forming of the charge onto the cool tool surface.

11.4.4 Bladder moulding

Bladder moulding is used to form hollow TPC vessels, e.g. pressure vessels, pipe sections etc. Both isothermal and non-isothermal variants of the process have been developed giving cycle times of 10–15 minutes and 2–3 minutes respectively.[6] The technique utilises commingled tows braided into hollow sleeves which are formed and consolidated on the internal surface of a hollow mould using a pressurised bladder. Figure 11.8 illustrates the isothermal process. Firstly, the braided TPC sleeve is assembled over a bladder. Typical bladder materials include compliant silicone rubber, capable of withstanding melt temperatures, or rigid net shape extrusion blow moulded polymer, e.g. nylon, which will soften but not melt when raised to the TPC melt temperature. Both TPC sleeve and bladder are then pinched between the two halves of a hollow mould. An initial low pressure (~2 bar) is applied to prevent the collapse of the bladder-sleeve arrangement during heating, following which the tool temperature is raised above the TPC matrix melt temperature. Pressure is then increased (to ~10 bar) to ensure rapid bladder expansion and forming and consolidation of the TPC braid against the inner tool surface. Direct tool heating or oven heating methods are both used in practice. Following an appropriate consolidation time, the tool is cooled and the hollow part removed.

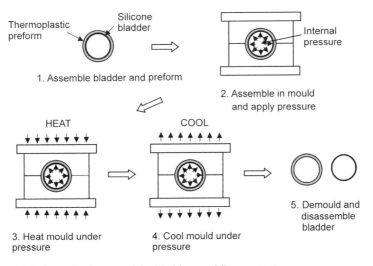

11.8 Schematic diagram of the bladder moulding process.

11.4.5 Roll forming

Roll forming is an established metal processing route for producing constant cross-section shaped profiles. The basic principle is to form the material progressively through a series of pairs of forming rolls until the desired cross-section is achieved. Only in the last ten years or so has the process been applied on a commercial scale to the forming of continuous fibre reinforced thermoplastic composite sheet/strips. Figure 11.9 shows the basic stages required to roll form TPC strips as reported in an investigation of the process by Henninger[7] and summarised as follows. In this process, strips (800 mm × 90 mm) of glass fibre/PP and glass fibre/PA, both fully consolidated, 0/90 twill weave, are firstly fed on a conveyor belt into an infra-red heater. As the heater only heats the surface layers of the strip, it must be of sufficient length to allow time for the heat to conduct and melt the matrix through the strip thickness. Overheating of the surface must also be avoided if degradation of the matrix is not to occur. Following heating, the strip temperature is allowed to cool back towards (but above) the matrix recrystallisation temperature in a dwell zone before it is fed into the forming rolls. The entry temperature to the rolls is a critical process parameter and temperatures 10–20°C (depending on production speed) below the matrix melting temperature have been found to be optimum. Too low an inlet temperature results in high void content in the formed strips i.e. poor consolidation. Too high a temperature results in spring back or spring forward depending on throughput speed. Forming takes place in a number of stages and must be completed before the material cools to the recrystallisation temperature. Following forming to the final shape, further shape consolidation is achieved by passing through further rolls. However, to avoid spring back/forward it is important that the material cools to below the recrystallisation temperature before passing out of the final rolls. To achieve this, forced cooling may be necessary in the shape consolidation zone.

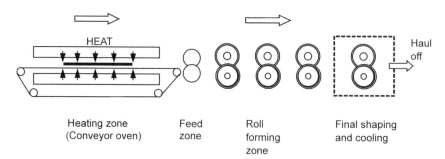

11.9 Schematic diagram of the roll forming process.

11.5 Some recent developments

11.5.1 Co-moulding

Combining material forms to achieve optimum part design and mouldability has been the subject of much activity in recent years. Wakeman et al.[8,9] report on the development of 'over-injection moulding' of thermoplastic composites, both UD and fabric forms. The technology involves robotic placement of UD tape or tows or the placement of pre-stamped fabrics into an injection tool. Preheating of the composite forms to just below the matrix melting temperature ensures good final consolidation and fusion with the over moulded long fibre thermoplastic (LFT) injected into the mould. The continuous TPC forms provide structural efficiency where required and the LFT enables ribs and complex part features to be moulded. Cycle times are low, in line with LFT injection moulding. Combination of GMT and fabric TPC in a compression moulding process,[10] to form a prototype automotive door cassette, has been shown to achieve similar results. Figure 11.10 shows the charge placement for this part resulting in a highly structural waste rail (top of the cassette) and a relatively complex shaped main

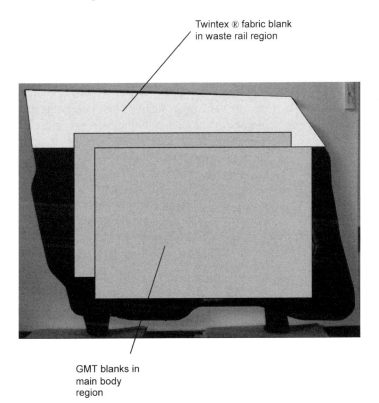

11.10 Blank configuration for a co-moulded automotive door cassette (after Wakeman et al.[10]).

body with lower mechanical properties. A similar process development, termed E-LFT (Endlessfibre reinforced Long Fibre Thermoplastic) has been reported by Ruegg et al.[11] and combines the local placement of continuous fibre TPC with LFT in a compression moulding process. Commercial material combinations, using the above principles, are now available for parts moulding, e.g. Quadrant Plastic Composites' GMText® combines two skin layers of commingled TPC, Twintex®, with an intermediate layer of GMT. The flowability of the latter aids the forming process for the former by allowing for thickness variations and evening out pressure variations across the mould.

Many of the above processes have moved into mainstream manufacture of production automotive parts at high volumes and the materials combinations and co-moulding processes are expanding the application areas for both GMT and LFT.

11.5.2 Sandwich moulding

TPC sandwich structures generally possess significantly higher stiffness and strength to weight ratios than monomer TPC structures and can offer lightweight solutions in many applications. The forming of flat TPC sandwich structures/ panels and beams has previously entailed pre-consolidation of the TPC skin in a press forming process and subsequent reheating (often the skin surface layers only) before bonding to a thermoplastic core, typically a low density polymer foam or honeycomb. This bonding process can be achieved continuously by feeding reheated skins over a cold core in a double belt laminating process or in a batch process using compression moulding. In the latter process, tool stops are required to prevent over crushing of the core during bonding. Recent work has aimed at developing the single shot moulding of these structures, i.e. consolidation of the TPC skin at the same time as the sandwich panel is formed. By removing the skin pre-consolidation process, this will save on manufacturing time and cost. The problem is to avoid over crush of the core during the process while still attaining good skin consolidation. This requires correct choice of forming pressure relevant to the crush strength of the core material. Brooks et al. have demonstrated that good quality glass/PP skinned sandwich structures with PP foam cores can be made using a low cost single step vacuum forming process.[12] Heated transfer plates are used to retain skin temperature during transfer from the heater to the mould.

11.5.3 Reaction moulding

Recent material developments have resulted in a class of low molecular weight cyclic thermoplastic polyesters (CBTs) which can be melted to a very low viscosity resin in the temperature range 120–160°C.[13] With the addition of a catalyst, the resin polymerises into high molecular weight thermoplastic PBT.

Because of the initial low viscosity, these resins can be combined with high volume fraction fibre reinforcement in a similar way to thermoset resins. Thus, many moulding/forming processes used for thermosets are also applicable to composites manufactured with CBT resins such as resin transfer moulding and resin film infusion.[14,15] Cyclics offer the property advantages of TPCs with the processing advantages of thermoset composites, e.g. impregnation of high volume fraction reinforcement. However, the fact that the resins do undergo polymerisation in the mould precludes them from the very high rate (low cycle times) associated with most TPC forming processes.

11.6 Conclusions

Forming methods for thermoplastic composites are developing at a rapid rate. This chapter has provided an overview of the processes currently available for forming TPCs into shape in addition to some recent developments. Whereas isothermal processes can produce high quality parts, cycle times are compromised by the need to thermally cycle the tool during processing. Conversely, non-isothermal processes reduce significantly the cycle time but generally at a cost of part quality. The challenge therefore is to form high quality, structurally efficient parts at high production rates. Significant innovation in both materials forms and the forming technologies themselves in recent years are addressing this challenge and it is likely that further developments will continue into the foreseeable future.

11.7 References

1. CORONET European Network in Thermoplastic Composites, *www.coronet.com*.
2. 'GMT Part Design and Moulding Guidelines', Quadrant Plastic Composites, *www.quadrantplastics.com*.
3. Mallon P J and Obradaigh C M, 'Compliant Mold Techniques for Thermoplastic Composite', in *Comprehensive Composite Materials*, Ed. A Kelly and C Zweben, Volume 2, Ch. 2.26, 874–912.
4. Twintex® Vacuum Moulding Manual, St-Gobain Vetrotex, Febraury 2005, *www.twintex.com*.
5. Pantelakis S G and Baxevani E A, 'Optimization of the diaphragm forming process with regard to product quality and cost', *Composites: Part A*, 33, 2002, 459–470.
6. Wakeman M D, Vuilliomenet P and Manson J-A E, 'Isothermal and non-isothermal bladder inflation moulding of composite pressure vessels: a process/cost study', *Proc. 24th SAMPE Europe Conference,* 1–3 April 2003, Paris, France.
7. Henninger F and Friedrich K, 'Production of textile reinforced thermoplastic profiles by roll forming', *Composites: Part A*, 35, 2004, 573–583.
8. Wakeman M D, Beyeler P, Eble E, Hagstrand P-O, Hermann T, and Manson J-A E, 'Hybrid thermoplastic composite beam structures integrating UD tows, stamped fabrics and injection/compression moulding', *ECCM-11, Proc. of 11th European Conference on Composite Materials*, 31 May–3 June, 2004, Rhodes, Greece.

9. Wakeman M D, Hagstrand P-O, Ecabert B, Eble E, and Manson J-A E, 'Robotically placed continuous fibre tow as tailored structural inserts for over injection moulding', *Proc. 25th SAMPE Europe Conference,* 30 Mar–1 April 2004, Paris, France.
10. Wakeman M D, Cain T A, Rudd C D, Brooks R and Long A C, 'Compression moulding of glass and polypropylene composites for optimised macro- and micromechanical properties. Part 4: Technology demonstrator – a door cassette structure, *Composites Science and Technology*, 2000, 60, 1901–1918.
11. Ruegg A, Stotzner N, Jaggi D and Ziegler S, 'E-LFT process – a new mass production process for structural lightweight', *Proc. 25th SAMPE Europe Conference*, 30 Mar–1 April 2004, Paris, France, 318–324.
12. Brooks R, Kulandaivel P and Rudd C D, 'Vacuum moulding of thermoplastic composite sandwich beams', *Proc. 25th SAMPE Europe Conference,* 30 Mar–1 April 2004, Paris, France, 494–500.
13. Brunelle D, Bradt J E, Serth-Guzzo J, Takekoshi T, Evans T L, Pearce E J and Wilson P R, 'Semicrystalline polymers via ring-opening polymerization: Preparation and polymerization of alkylene phthalate cyclic oligomers, *Macromolecules*, 31, 1998, 4782–4790.
14. Rosch M, 'Processing of advanced thermoplastic composites using a novel cyclic thermoplastic polyester one part system designed for resin transfer moulding', *Proc. 24th SAMPE Europe Conference*, 1–3 April 2003, Paris, France, 305–310.
15. Coll S M, Murtagh A M and O'braidaigh C M, 'Resin film infusion of cyclic PBT composites: A fundamental study', *Proc. 24th SAMPE Europe Conference*, 1–3 April 2003, Paris, France, 311–317.

12
The use of draping simulation in composite design

J W KLINTWORTH, MSC Software Ltd, UK and
A C LONG, University of Nottingham, UK

12.1 Introduction

Composites forming is an important part of the overall development process for components and structures made of composites materials. Although composite materials are used extensively for applications that are not strength-critical, this review will concentrate on structures needing high mechanical performance.

The military aerospace market has used high-performance composites for many years, but civil applications are now growing in importance with a huge increase in the use of CFRP (carbon fibre reinforced plastics) for full-sized civil aircraft.

While large composite structures are now manufactured using tape or tow placement, the majority of structures are laid up of plies of sheet materials that have to be applied to 3D mould surfaces. The application methods remain predominantly manual, assisted by laser projection systems. Robotic application systems are being tested to improve the quality of layup and these are being introduced. Automated forming processes are also growing in popularity, including diaphragm forming for thermosets (Chapter 10) and stamping for thermoplastics (Chapter 11).

In this chapter the use of draping and forming simulations within the composites design and development process will be discussed. This will focus on the use of modelling techniques to inform component design and to assist with component manufacture. In particular the use of forming simulations within the composites design and optimization environment will be described.

12.2 Zone and ply descriptions

During the design and analysis tasks of the composites development process, engineers use both zone and ply descriptions of a laminate as shown in Fig. 12.1.

Analysts have traditionally used finite element analysis techniques to determine the performance of composite structures using a mesh-based approximation of the model. From the perspective of the finite element analysis code, the

278   Composites forming technologies

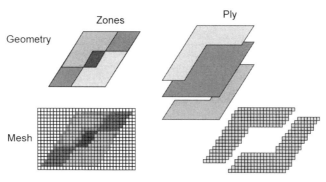

12.1 Alternate descriptions of the laminated composites model.

manufacturing requirements are irrelevant and only the local properties of the laminate on each element are significant. This has lead to the use of a 'Zone' description of the laminate, where a 'laminate material' consisting of a stack of layers of homogeneous material is defined and assigned to the underlying mesh. This description is quick to generate for simple structures, but impossible to use on complex structures where the fibre orientation varies continuously.

Designers have traditionally defined the layup using curves defining ply boundaries on an underlying surface, i.e. a 'Ply' description. This description mirrors the final manufactured configuration closely and allows the designer to draw in details such as ply drop-offs and insert placement. However, drawing this detailed description is relatively time-consuming.

These traditional descriptions are not compatible and the interaction between analyst and designer is severely impeded by the different descriptions. To improve the efficiency of the composites development process, analysts now often use a ply description on the mesh, as pioneered in Laminate Modeler from MSC.Software (Klintworth and MacMillan, 1992). This allows rapid definition and modification of the composites structure and ready incorporation of the results of draping and forming simulations in the finite element model.

Recently, as more work has become centralised in PLM (product lifecycle management) systems, the zone description used in preliminary work is now supported by mainstream design-centric systems. The most sophisticated systems offer some capabilities for transferring zone descriptions to ply descriptions in a semi-automated fashion.

12.3   Composites development process

The composites development process spans a number of tasks and functions as shown in Fig. 12.2. In the majority of aerospace or automotive applications, the basic geometrical shape is fixed by aerodynamic and packaging constraints defined earlier in the project cycle. This data is typically generated and stored using an enterprise-wide CAD system such as CATIA.

The use of draping simulation in composite design 279

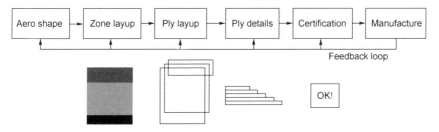

12.2 The composites development process.

After the initial shape has been determined, preliminary structural analysis is undertaken in order to quantify the thickness and orientation of reinforcing materials in the structure required to carry the imposed loads. The finite element analysis technique is typically used to quantify the mechanical and thermal performance of the structure. In the aerospace industry, this work is typically undertaken using software packages such as MSC.Patran and MSC.Nastran®. At this stage, the composite model is defined using laminate materials (e.g., MSC.Nastran PCOMP cards) or ply descriptions based on the finite element model.

Once preliminary sizing is completed, the composites model is transferred to the CAD system where the manufacturing details of the model are added. In particular, ply drop-off and splicing details are specified to ensure that no ambiguity is possible during the manufacture of the structure. These processes are typically undertaken using a composites design tool like CATIA Composites Design, which defines all composite data in terms of geometrical entities. Increasingly advanced simulation tools are becoming available for simulating fibre orientations, even on complex surface geometries as shown in Fig. 12.3. These simulations have brought techniques previously restricted to specialists into the domain of the design engineer. Accounting for fabric behaviour earlier leads to fewer problems in the downstream process.

For composite structures, it is likely that the detailed design process will introduce a number of changes in the structure. Therefore, it is wise to re-evaluate the final performance of the structure just before manufacturing. Certification analysis has been greatly improved recently by the introduction of ply tracking in the commercial finite element analysis codes to allow identification and interpretation of laminate results. For example, the new PCOMPG card in MSC.Nastran 2004 will automatically sort layer results on the basis of global plies as illustrated in Fig. 12.4.

Once design requirements have been satisfied, the manufacturing information must be generated in the form required by the manufacturing process. A feature of the past few years has been the advance in techniques for simulating composites manufacture. For example, nonlinear FEA solvers such as MSC.Marc and ESI PAM-FORM can now undertake detailed forming analysis

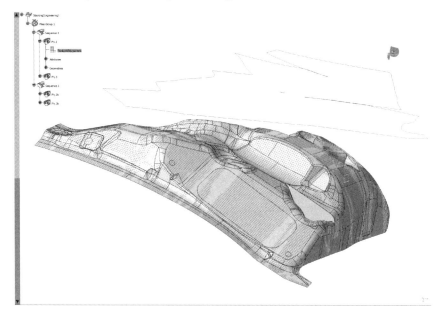

12.3 Advanced fibre simulation in the design environment.

of fabrics and composites to predict regions liable to wrinkling. These codes now also have the capability of simulating curing, involving the coupled analysis of mechanical, thermal and resin curing problems. These problems are particularly important for thick open sections typically used on spars or frames. The use of FEA for forming simulation is discussed in more detail in Chapter 3, whilst curing simulation and implications for component distortion are described in Chapter 7.

Clearly, the composites development process is multidisciplinary in nature and makes use of a wide range of tools and skills. Most fundamentally, the way in which composites data are defined by analysts and designers is fundamentally different from traditional materials. This has introduced substantial inefficiencies in the development process and delayed the use of composite materials in cost-sensitive industries.

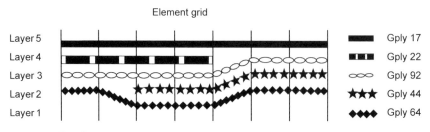

12.4 Global ply modelling in FEA codes.

12.4 Composites data exchange

Design environments are becoming increasingly sophisticated and permit baseline mechanical analysis using finite element methods. However, the design environment remains unsuitable for advanced analysis, particularly when abstraction needs to be changed, as the design environment usually assumes that the finite element mesh is tied to the geometry. This approach cannot allow for typical analysis techniques like the use of multi-point constraints to model hinges. Therefore, specialised analysis environments retain their usefulness and a key requirement during the development process is to transfer models between the design and analysis environments as shown in Fig. 12.5.

Such data transfer was traditionally effected using file transfer. Unfortunately, file formats are not rich enough to transfer full data incorporating both geometry and associated data. Happily, modern design systems incorporate rich interfaces so it is possible to transfer all required data in a seamless operation. Figure 12.6 shows the transfer of a complete ply model from design to analysis environments, complete with layup tree and associated data.

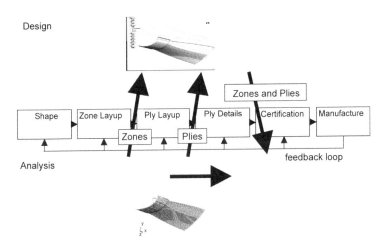

12.5 Data transfer between design and analysis environments.

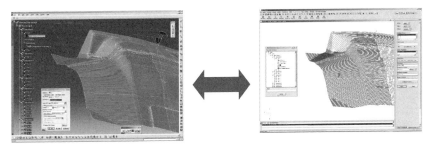

12.6 Example of data transfer.

12.5 Draping and forming simulation

During the composites development process, it is important to simulate the draping or forming of the plies to predict problems, estimate material requirements, generate manufacturing data and provide accurate fibre angles for subsequent finite element analysis. Most tools in commercial use implement kinematic draping algorithms that use a geometrical approach to provide accurate simulations for hand-layup processes. Where friction is significant, such as if matched die tooling or diaphragm forming techniques are used, full finite element techniques can be utilized.

12.5.1 Kinematic draping simulation

To form flat material to a 3D surface, the material must shear. Kinematic draping simulation begins with an assumption of the material shearing behaviour which depends on the material structure. Typical material models are for scissor and slide deformation, related to biaxial woven fabric and unidirectional material respectively (Fig. 12.7). Based on this assumption, the incremental draping over a surface can be calculated. (i.e. the local draping problem). Results can be used directly to provide fibre orientation data for structural analysis, to determine ply net-shapes for fabric cutting, and to approximate manufacturing difficulties such as wrinkling when shear deformation exceeds a user-defined limit ('locking angle').

To define the overall fibre pattern, the way in which the fabric extends from an initial point must be assumed. A unique draped pattern can be obtained by specifying two intersecting fibre paths (generators) on the surface of the component or forming tool. The remaining fibres are positioned using a

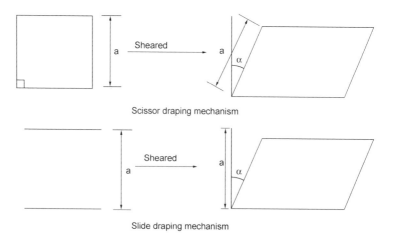

12.7 Scissor and slide draping mechanisms for woven and unidirectional materials.

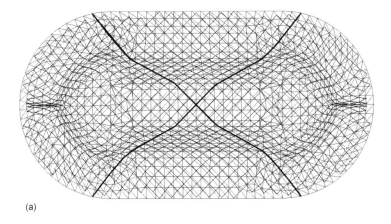

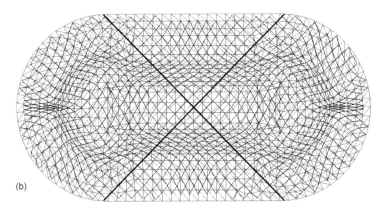

12.8 Kinematic drape simulations for a double dome geometry, based on geodesic (top) and projected (bottom) generators.

mapping approach, which involves solving geometric equations to determine the intersection of the surface with possible crossover points for the fibre segments. Several strategies are available for specifying the generator paths, and their correct specification is critical as this will determine the positions of all remaining fibres. This is illustrated in Fig. 12.8, which compares predicted fibre patterns obtained using a scissor model with geodesic and projected generator paths. A number of different approaches have been developed, tested and incorporated into commercial codes. For example, the MSC.Patran Laminate Modeler allows the user to use geodesic, minimum energy, or minimum shear extension methods.

The user may also be able to apply constraints to the simulation by defining initial warp or bias directions using lines. These lines in turn can be drawn on 2D shapes and projected onto the 3D mould surface to allow very accurate fibre

284   Composites forming technologies

placement on the shop floor. The most advanced kinematic simulations allow the user to define the order in which the draping proceeds. This concept is very familiar in the metal forming industry where multi-stage stamping is common, and provides excellent accuracy for typical manufacturing situations.

12.5.2 Finite element forming simulation

Finite element techniques are used in situations where friction effects are important, or where material behaviour cannot be approximated using one of the simple deformation modes outlined above (e.g., for non-isothermal stamping of thermoplastic composites). Such techniques also allow ply wrinkling to be predicted directly via determination of compressive loads within plies. In general, because detailed material characterization is required, extensive testing is necessary before simulation begins (see Chapter 1). Nevertheless, finite element techniques can be useful for extending and checking kinematic simulations for specific simulations. The basis of such techniques is described in Chapter 3.

12.5.3 Comparison of kinematic and finite element forming simulations

Kinematic simulations are well proven and are usually completed in less than one second. By comparison, finite element techniques take between about a minute for an implicit single-ply simulation to hours or days for a multi-ply explicit simulation. Therefore, finite element techniques are currently primarily a research topic for all except the most specialised applications. However, the use of such techniques may become more widespread as automated forming processes are adopted.

12.6 Linking forming simulation to component design analysis

12.6.1 Initial zone sizing

During the initial sizing phase of composites product development, ubiquitous optimization techniques can be applied to define the initial layup. The general optimization problem can be written as:

$$\begin{aligned}
&\text{minimize:} & &F(X) & &\text{objective function} \\
&\text{subject to:} & &g_j(X) \leq 0 & &j = 1, 2, \ldots n_g & &\text{inequality constraints} \\
& & &h_k(X) = 0 & &k = 1, 2, n_h & &\text{equality constraints} \\
& & &x_i^l \leq x_i \leq x_i^u & &i = 1, 2, \ldots n & &\text{side constraints} \\
&\text{where:} & &X = \{x_1, x_2, \ldots x_n\} & & & &\text{design variables}
\end{aligned}$$

Many different methods have been developed to solve this mathematical problem, and the numerical methods required are relatively mature.

12.6.2 Optimization in FEA codes

The numerical optimization techniques can be used with finite element analysis using the latter as a 'black box' for calculating results for a given set of design variables. However, the usefulness of numerical optimization has been increased enormously by embedding numerical optimization algorithms into the finite element analysis codes themselves (Klintworth, 2001).

By doing this, multiple constraints can be applied to a model simultaneously. For example, users can limit stresses, displacements or even derive results like failure indices simultaneously. Furthermore, constraints can be imposed for multiple analysis types at the same time. For example, users can specify a minimum natural frequency from a normal modes analysis together with a maximum stress from a linear static solution.

Optimization techniques based on gradient techniques require the calculation of the sensitivities of the design constraints to changes in the design variables. This was traditionally calculated via finite difference techniques, but can be calculated analytically within the FEA code. Together with the use of double precision variables, this improves the speed and numerical accuracy of the optimization algorithm by orders of magnitude.

In addition to gradient techniques, other numerical methods such as topology optimization algorithms and fully stressed design techniques can be integrated into a FEA solver with commensurate speed and accuracy gains (Klintworth, 2005).

12.6.3 Zone description

The zone description in a FEA code consists of a laminate record, which is referenced by a set of elements, typically shells. For example, the MSC.Nastran PCOMP card has the format shown in Table 12.1.

The LAM option allows explicit specification of individual layers of the laminate, as well as new smeared laminate options that are of particular relevance to optimization problems. By initially neglecting the stacking

Table 12.1 Format of the MSC.Nastran PCOMPG card

1	2	3	4	5	6	7	8	9	10
PCOMP	PID	Z0	NSM	SB	FT	TREF	GE	LAM	
	MID1	T1	THETA1	SOUT1	MID2	T2	THETA2	SOUT2	
	MID3	T3	THETA3	SOUT3	etc.				

Table 12.2 Effect of laminate options in MSC.Nastran

LAM option	New 2001	Membrane [A]	Bending [B]	Coupling [D]	Ply results	Comments
BLANK		Y	Y	Y	Y	Default
SYM		Y	Y		Y	
MEM	Y	Y			Y	Wing skins
BEND	Y		Y		Y	
SMEAR	Y	Y	Y	Y		Smeared
SMCORE	Y	Y (core N)	Y			Smeared with core

sequence of the laminate, extraneous modes resulting from membrane-bending coupling are suppressed and the algorithm converges more rapidly. The laminate options consider the effects shown in Table 12.2.

The SMEAR and SMCORE options model 'percentage ply laminates', where the overall percentage of fibres in various directions are known, but the stacking sequence is not. This description is exactly what is required at the beginning of the design process.

12.6.4 Design variables for thickness and orientation

Within MSC.Nastran, the design variables are defined using the DESVAR card. These can be linked with any property variable (e.g., PCOMP) using the DVPREL1 and DVPREL2. This allows enormous flexibility in setting up the composites optimization model.

In theory, the user can treat every thickness and every orientation on every PCOMP card as a design variable. However, the use of orientation as a design variable is limited by two factors. First, the principal stiffness of a layer of material rotated through a small angle Θ varies approximately proportional to $\cos^2\Theta$. This means that small rotations can have very little effect on the constraints, which in turn means that it is numerically difficult to find an optimum solution and convergence is slow. In addition, it is also very easy to achieve a solution where layers in adjacent zones are at different angles, which cannot be achieved sensibly during manufacture.

It is therefore common practice to define design variables as the thickness of layers that have a fixed orientation with respect to some basis, e.g. a projection of a vector onto the shell surface. For the case of unidirectional plies, the nominal orientations of $-45°/0°/45°/90°$ are usually used, with the core oriented at zero degrees. This gives a total of five design variables of thickness per PCOMP card. For a typical automotive composite model having 100 PCOMP regions, this yields 500 design variables, which can be processed satisfactorily using desktop computers.

12.6.5 Discrete optimization

Materials supplied in sheet form and used to manufacture composite structures are supplied in particular thicknesses reflecting the chosen reinforcement. For example, carbon fabric-based prepreg materials typically have a thickness of 0.25 mm.

Therefore, it is useful to constrain the design variables to some multiple of this thickness in order to achieve a result that can be manufactured. A new feature with MSC.Nastran 2001 allows users to define allowable values in a table. For the optimization calculation, continuous variables will be used for reasons of numerical stability. Then, after a user-defined design cycle (the default is the final cycle), the variables will be forced to the most appropriate discrete value allowing for the model constraints.

12.6.6 Zone orientation

It was noted above that one common technique of orienting the zone laminate orientation is through the use of a projected coordinate system. While this is simple for a relatively flat component like a wing skin, problems arise with curved shapes. In these cases, a kinematic draping algorithm can be used to simulate the drape of a fabric ply over the surface. From this, the orientations of the warp fibre on every element can be calculated and stored in a field within MSC.Patran, which can then be used to orient the zones. This provides a much more realistic orientation of the zones in a way that follows the orientation of manufactured plies. By accounting for manufacturing at this early stage of the development process, the user can be sure that plies indicated by the zone optimization process are actually producible.

12.6.7 Ply-based analysis

Once the initial sizing is completed, the analyst can define the initial layup using plies defined on the shell element mesh. This technique allows rapid modification of the model by simply adding and removing plies, or changing their coverage. At the same time, possible manufacturing problems can be identified using the draping simulation found in suitable tools.

Once a ply layup has been defined, the model needs to be translated to the zone-based laminate material description supported by commercial finite element codes. Then after completion of the analysis run, the results (on a zone layer basis) need to be sorted in the format of the layup plies to allow for effective interpretation. This translation between ply layup and zone layer descriptions is shown in Fig. 12.9.

The translation from ply layup to zone description is generally specific to the analysis code used. First, the orientation system supported by the analysis code is chosen by the user. For example, MSC.Nastran users can orientate laminate

288   Composites forming technologies

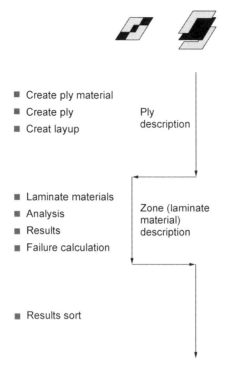

- Create ply material
- Create ply
- Creat layup

Ply description

- Laminate materials
- Analysis
- Results
- Failure calculation

Zone (laminate material) description

- Results sort

12.9 Relationship between ply and zone modelling.

materials using a coordinate system or an angle from the first edge of each element. Second, the laminate materials on each element can be formulated based on the draped directions and the orientation system. Finally, sorting procedures are used to minimise the number of laminate materials generated within a user defined tolerance. Typical values are 5° of layer orientation and 5% of layer thickness.

Following the finite element calculations, the results can be sorted back on the basis of plies. Historically, this sorting was done as a separate postprocessing operation by the users. However, modern FEA analysis codes can track the global ply identifiers so they can automatically sort results themselves. This capability (MSC, 2005) makes the use of ply descriptions seamless in the analysis environment. This in turn allows designers and analysts to communicate easily.

12.6.8 Accounting for material shear

When fabric shears during the forming process, it changes thickness and the fibre orientations change. For a typical woven fabric with warp and weft fibres initially at right angles, this changes the mechanical properties markedly. For dry fabrics to be impregnated by resin transfer moulding, this also changes the permeability and hence the resulting resin flow pattern (Long *et al.*, 1998).

Calculating the change is dependent on the material system used. Predictive techniques based on micromechanics to predict local elastic constants and layered shell elements to represent the fibres within each ply can be used for simple materials such as non-crimp fabrics or UD prepreg (Crookston *et al.*, 2002). However for woven materials the accuracy of such an approach is questionable. Hence commercial laminate modelling systems allow the user to create material property sets for particular shear states (MSC, 2003). These sets can be defined theoretically or experimentally as required. Then the system will use the appropriate property set for the particular shear state on an individual element. This clearly increases the number of laminate materials required for the analysis.

12.6.9 Failure analysis

Failure analysis is used on an everyday basis for failure calculation in FEA codes. The most common failure analysis is a simple empirical criterion within the plane of each layer (Hill, Hoffman, Tsai-Wu or Maximum Strain are common), with a simple maximum stress criterion for out-of-plane calculations. This allows analysts to quickly identify lowest Margins of Safety on whole models, elements within models, and even plies on each element.

Specialised postprocessors are available to examine multiple failure criteria over multiple loadcases (Anaglyph, 2003). This capability is especially important where a structure has many potentially critical results cases. A particular instance of this occurs when structures are weakly nonlinear so loadcases cannot be generated by linear superposition.

For strongly nonlinear problems where loading may continue after failure, it is necessary to implement failure criteria as user subroutines within nonlinear implicit and explicit analysis codes. This allows users to track progressive failure in components under ultimate loading. This is especially important for crushing and crash analysis. This level of detailed analysis is seldom completed on a routine basis due to its complexity.

12.6.10 Stochastic analysis

Conventional gradient optimization techniques can be very useful for initial sizing, but their usefulness is limited further down the design process. The main problem is numerical instability, but the zone composites description also excludes manufacturing constraints. Therefore the resulting optimized model cannot be manufactured.

However, if the user generates a ply layup model, design improvement reflecting manufacturing constraints can be effected naturally as the plies are already manufacturable. Consider the wing section where a rib intersects the skin as shown in Fig. 12.10. Here, it is essential that plies lie in the wing skin, on

290 Composites forming technologies

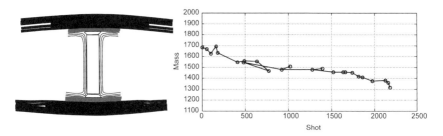

12.10 Stochastic design improvement (SDI) of a section.

pads between rib and skin, in the rib and on joins between rib and skin. Using a ply layup, it is possible to set up a model that reflects the manufacturing requirements.

This ply model can then be improved using different techniques such as stochastic methods (Monte Carlo) or Genetic Algorithm (GA) approaches. An example of a stochastic design improvement of an aerofoil is shown in Fig. 12.10. This required several thousand runs of MSC.Nastran taking several days to run. However, with the increasing power of computers, this sort of improvement will soon be all in a day's work.

12.6.11 Solid element generation

Composite structures have traditionally been analysed using shell elements. For the linear regime, these elements typically use classical laminate theory to calculate effective in-plane and bending stiffnesses in the shell. Equilibrium methods are typically used to calculate approximate out-of-plane intralaminar stresses (i.e. bond stresses between plies). These methods continue to serve the industry successfully for relatively thin structures.

However, this approach does not yield sufficiently detailed recovery of out-of-plane shear and normal stresses. Consequently, specialized finite elements such as P/COMPOSITE from PDA Engineering were developed in the 1980s. These incorporated high-order shape functions and allowed the definition of plies at an arbitrary orientation in space at the Gauss points. These powerful elements, however, needed detailed models that were not economic to build, and use of these very advanced technologies has tapered off.

Over the past decade, general-purpose solid laminated composite elements have been incorporated in many mainstream finite element codes. These elements generally work under the following assumptions:

- The laminate material referenced by the element is sandwiched between opposite faces of the element. For example, in an 8-noded hexahedral element, the layers are parallel to the 1,2,3,4 and 5,6,7,8 faces of the element.
- If the sum of the layer thicknesses does not match the physical spacing of the

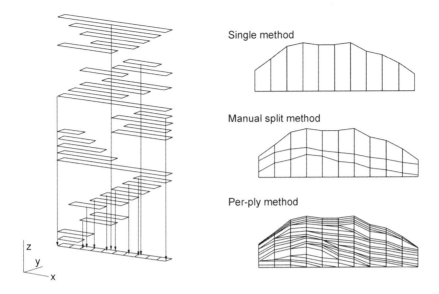

12.11 Solid generation from ply layup.

appropriate faces, the laminate material is effectively scaled to fill the gap or compress the excess material.
- The layers are assumed to cover the entire area of the appropriate faces of the elements, i.e. ply drop-off within the element is not supported.

These assumptions, particularly the requirement that the element faces are approximately parallel to the laminate layers, restrict the extent to which solid composites elements can be used effectively. However, this means that extrusion of the elements from a shell model is a highly appropriate method of model generation.

Recently, composites modelling tools have been extended to automatically generate solid composite elements by extruding shell models through an appropriate thickness. Sophisticated controls are provided, including the use of 'split' plies, to generate the required solid mesh as shown in Fig. 12.11.

12.7 Conclusions

The composites development process is complex and multidisciplinary, and can be facilitated by analysis tools at several key stages. Draping or forming simulation tools allow fibre orientations to be predicted across the component, and also allow determination of possible manufacturing problems. Coupled with modern techniques for definition of zones and plies, this information can be linked directly to the lay-up procedure. This data also informs structural analysis by ensuring that the correct material orientations and hence mechanical properties are specified in the different zones. Given increases in processing speeds, these techniques can be incorporated within design iteration and optimization schemes, ultimately leading to more efficient and effective composites design.

12.8 References

Anaglyph (2003) Laminate Tools 2.0 User Guide, Anaglyph Limited, London, UK.

Crookston J, Long A C and Jones I A (2002) Modelling effects of reinforcement deformation during manufacturing on elastic properties of textile composites, *Plastics Rubber and Composites* 31(2) 58–65.

Klintworth J W (2001) Multidisciplinary Optimization of Composites Structures, *International SAMPE Technical Conference, Seattle*, November 2001.

Klintworth J W (2005) Detailed Simulation of Primary Composite Structures, *SAMPE Long Beach*, May 2005.

Klintworth J W and MacMillan S (1992) Effective Analysis of Laminated Composite Structures, *Benchmark*, December, 20–22.

Long A C, Blanchard P J, Rudd C D and Smith P (1998) The development of an integrated process model for liquid composite moulding, *Composites Part A* 29(7) 847–854.

MSC (2003) MSC.Patran 2003 Release Guide, MSC.Software Corporation, Los Angeles, California, USA.

MSC (2005) MSC.Patran 2005 Release Guide, MSC.Software Corporation, Los Angeles, California, USA.

13
Benchmarking of composite forming modelling techniques

J L GORCZYCA-COLE and J CHEN, University of Massachusetts Lowell, USA and J CAO, Northwestern University, USA

13.1 Introduction

Woven-fabric reinforced composites (hereafter referred to as woven composites) have attracted a significant amount of attention from both industry and academia, due to their high specific strength and stiffness as well as their supreme formability characteristics. However, applications of these materials have been hampered by a lack of low-cost fabrication methods, as well as robust simulation methods. Designing low-cost manufacturing processes requires accurate material modeling and process simulation tools. Recognizing these requirements, a group of international researchers gathered at the University of Massachusetts Lowell for the NSF Workshop on Composite Sheet Forming in September 2001. The main objectives of that workshop were to better understand the state-of-the-art and existing challenges in both materials characterization and numerical methods required for robust simulations of forming processes. One direct outcome of the workshop and the effort to move towards standardization of material characterization methods was a web-based forum exclusively for research on forming of woven composites, established in September 2003, at http://nwbenchmark.gtwebsolutions.com/. Other outcomes of the workshop are in the form of publications, such as this one, highlighting recommended practices for experimental techniques and modeling methods.

Material property characterization and material forming characterization were two main areas related to material testing identified at the 2001 NSF Composite Sheet Forming Workshop. Standard material testing methods are necessary for researchers to understand the formability of the material, the effect of process variables on formability, and to provide input data and validation data for numerical simulations. Thus, the researchers embarked on a benchmarking project to study, understand and report the results of material testing efforts currently in use around the world for woven composites to make recommendations for best practices.

294   Composites forming technologies

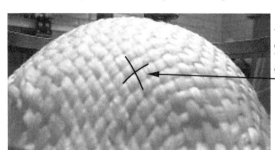

13.1 Fabric shearing.

Three different commingled fiberglass-polypropylene woven-composite materials were used for this research. The materials were donated by Vetrotex Saint-Gobain in May 2003 and were distributed in July 2003 to the following research groups: Hong Kong University of Science and Technology (HKUST) in Hong Kong, Katholieke Universiteit Leuven (KUL) in Belgium, Laboratoire de Mécanique des Systèmes et des Procédés (LMSP) in France, Northwestern University (NU) in the USA, University of Massachusetts Lowell (UML) in the USA and University of Twente (UT) in the Netherlands.

As intra-ply shear is the most dominant deformation mode in woven composite forming (Fig. 13.1), the trellis-frame (picture-frame) test (Fig. 13.2) and the bias-extension test (Fig. 13.3) were identified for further study related to material shear-property characterization. This paper focuses on the shear property determined from the results of the trellis-frame-test. Five of the six research groups listed submitted data for the experimental trellising-shear part of the benchmark project. A summary and comparison of the trellis-frame test methods and the findings from all participating research groups is presented. Future publications will focus on the bias-extension results.

A summary of the properties of the materials used in this study is presented in Section 13.2. More detailed information about the fabric is also listed on the forum website. Trellis-frame-test results along with the experimental techniques

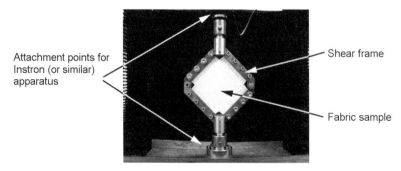

13.2 Trellising-shear test apparatus.

Benchmarking of composite forming modelling techniques 295

13.3 Bias-extension test apparatus.

are the focus of Section 13.3. Also, all the data reported are available for download at the forum website. Section 13.4 discusses how the data from these tests can be used to advance the benchmarking effort related to the numerical modeling of the benchmark fabrics in thermostamping simulations. Conclusions and future work are presented in Section 13.5.

13.2 Forming process and fabric properties

As stated in the introduction, the three types of woven fabrics used in this study were donated by Vetrotex Saint-Gobain (Fig. 13.4). The fabric properties, as reported by the material supplier and benchmark participants, are listed in Table 13.1. Each fabric is comprised of yarns with continuous commingled glass and polypropylene (PP) fibers. These fabrics were chosen due to their ability to be formed using the thermostamping method.

(a) Plain weave   (b) Balanced twill weave   (c) Unbalanced twill weave

13.4 Woven fabrics tested.

Table 13.1 Fabric parameters (as reported by the material supplier unless specified otherwise)

	TPEET22XXX	TPEET44XXX	TPECU53XXX
Manufacturer's style			
Weave type	Plain	Balanced twill	Unbalanced twill
Yarns	Glass/PP	Glass/PP	Glass/PP
Weave	Plain	Twill 2/2	Twill 2/2
Areal density, g/m^2	743	1485	1816
Yarn linear density, tex	1870	1870	2400
Thickness*, mm	1.2 (NU)	2.0 (NU)	3.3 (NU)
Yarn count, picks/cm or ends/cm			
Warp	1.91 (KUL) 1.93 (HKUST) 1.95 (NU)	5.56 (KUL)	3.39 (KUL)
Weft	1.90 (KUL) 1.93 (HKUST) 1.95 (NU)	3.75 (KUL)	1.52 (KUL)
Yarn width in the fabric, mm (**standard deviation)			
Warp	4.18 ± 0.140** (KUL) 4.20 (HKUST) 4.27 (NU)	1.62 ± 0.107** (KUL)	2.72 ± 0.38** (KUL)
Weft	4.22 ± 0.150** (KUL) 4.20 (HKUST) 4.27 (NU)	2.32 ± 0.401** (KUL)	3.58 ± 0.21** (KUL)

* ASTM Standard D1777 (Applied pressure = 4.14 kPa)

For commingled woven fabrics, the thermostamping process is a rapid manufacturing method similar to the stamping method used to form metal parts. The main difference between the thermostamping process and stamping process is the addition of heat. An oven is present at the start of the forming process to heat the fabric blank. The tools are also heated in the thermostamping process so that the fabric blank does not cool before it is fully formed. Recall from the introduction that the objective of the thermostamping process is the reorientation of the yarns through rotation or shearing to produce the desired shape from a fabric blank (Fig. 13.1). Any wrinkling present in the final part would indicate a defect. In the oven, the fabric is heated above 165°C to melt the PP fibers. When the fabric blank leaves the oven it is placed beneath a punch. A binder is rapidly placed over the fabric blank to apply in-plane tension to the yarns. This tension aids in the prevention of wrinkling during the stamping process by causing the yarns in the fabric to rotate and take on the shape of the die. The punch then presses the fabric into a die. The metal tools are heated to slow the rate at which

Benchmarking of composite forming modelling techniques 297

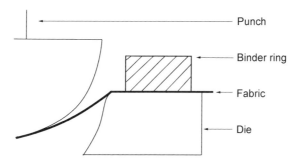

13.5 Schematic of thermostamping components.

the PP cools and solidifies (Fig. 13.5). In the final formed part, the glass fibers act as the reinforcement to the PP matrix. A schematic of the process is shown in Fig. 13.5.

13.3 Experimental

A trellis frame or picture frame is a fixture used to perform a shear test for woven fabrics. As stated in the introduction, the test procedures used by the participating researchers were not identical. However, the tests were equivalent in principle. As recommendations for best practices and standardized test procedures were one of the desired outcomes of this research, researchers did not limit all groups to perform the test using the exact same procedure. By using varying procedures researchers could study and recommend methods for data comparison.

Before discussing the test procedure in detail, it should be noted that some groups chose not to submit results for all of the fabrics included in this study. Table 13.2 shows the fabrics tested by each participant.

All researchers reported load histories and global-shear-angle data for picture-frame tests conducted at room temperature. The researchers decided that even though temperature effects were an important part of the process, initial

Table 13.2 Tested fabrics by participating researchers

Group	Plain weave	Balanced twill weave	Unbalanced twill weave
HKUST	Y	N	N
KUL	Y	Y	Y
LMSP	Y	Y	Y
UML	Y	Y	Y
UT	Y	Y	N

Note: (Y = data reported; N = data not reported)

298 Composites forming technologies

comparisons among results obtained using non-standard test procedures should be conducted without varying the temperature. Once the differences were understood at room temperature, the additional complexity of comparing results at elevated temperatures would be incorporated into the study.

One main difference in the procedure used by the five groups who submitted results can be observed in the type of frame used. Although the frames used in this study are not identical (Fig. 13.6, Table 13.2), all of them have common features. For example, the corners of each frame are pinned. When the fabric is loaded into the frame, it is clamped on all edges to prevent slippage. The corners are cut out of the sample to allow the tows to rotate without wrinkling the fabric. Thus, it appears that each sample has four flanges (Fig. 13.6). The

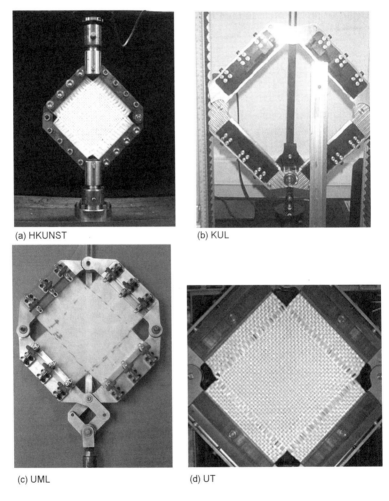

13.6 Picture frames designed, fabricated and used by the research groups.

Benchmarking of composite forming modelling techniques 299

sample size is considered to be the area of the fabric without the flanges as that area represents the amount of fabric that is deformed during the test. It is the area that encompasses the tows which must rotate at the crossover points during the test.

One may also note different clamping mechanisms in Fig. 13.6. It was assumed that all clamping mechanisms held the fabric rigidly in the frame and there was no slippage. Thus, differences in clamping mechanisms are not taken into account in the analysis of the results. With the fabric properly aligned and tightly clamped in the frame, the distance between two opposing corners is increased with the aid of an Instron testing machine (or similar tensile test apparatus). Figure 13.7 shows a fabric sample loaded in a picture frame in the starting position and in the deformed position. Using this test method uniform shearing of the majority of the fabric specimen is obtained. Displacement and load data are recorded to aid in the characterization of pure shear behavior. The operating principles of each frame are the same in that the fabric sample is initially square (Table 13.3) and the tows are oriented in the 0/90 position to start the test (Figs 13.6 and 13.7). Also, after the test begins and the crosshead displacement increases, pulling on the frame, the tows begin to reorient themselves as they shear (Fig. 13.7). However, the mechanism by which the fabric deforms is aided by linkages in the frames used by KUL and UML (Fig.

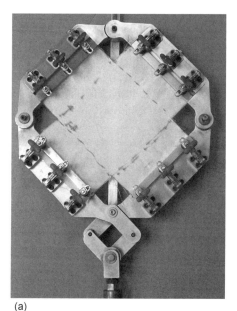
(a)

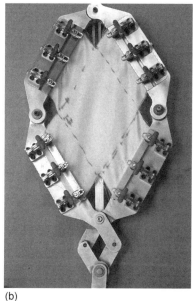
(b)

13.7 UML shear frame (a) starting position (b) deformed position. (Note that the top hinge has traveled from the bottom of the slot in the undeformed position (a) to the top of the slot in the deformed position (b).)

300 Composites forming technologies

Table 13.3 Frame size and test parameters

Group	Frame (mm)	Fabric (mm)	Speed (mm/min)	Specimen temperature
HKUST	180	140	10	Room temperature
KUL	250	180	20	
LMSP	245	240	75 to 450	
UML	216	140	120	
UT	250	180	1000	

13.6). UML's linkage was added to allow the frame to displace a greater speed than could be achieved by the Instron or tensile test machine alone. Note the inclusion of the slot in the KUL and UML frames (Figs 13.6–13.8). In addition to amplifying the distance traveled, these linkages amplify the measured force, and this amplification factor must be accounted for when the results from all the groups are analyzed and compared. When this amplification factor was removed, it was expected that the results from different groups would be comparable if a proper normalization technique was used to account for differences in sample and frame size. A detailed discussion of the amplification factor associated with the inclusion of the linkages and the various normalization techniques is included in the discussion of results. This section focuses on the similarities and differences of the test methods used by each group.

Additional differences in the test procedure among the groups were related to sample preparation. For example, to eliminate the potential force contribution from shearing of the yarns in the edge (arm) parts of the sample, HKUST removed all of the unclamped fringe yarns (Fig. 13.9). UT reported that they removed some of the yarns adjacent to the center area of the sample to prevent the material from wrinkling during testing (Fig. 13.9). In previous research by

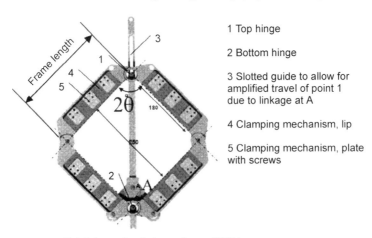

13.8 Schematic of picture frame (KUL).

Benchmarking of composite forming modelling techniques   301

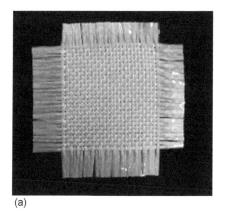 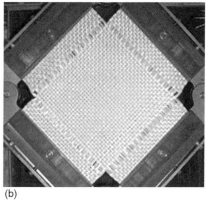

(a)   (b)

13.9 Specimens with yarns removed from arm regions (a) HKUST (b) UT.

Lussier (2000), it was reported that care must be taken not to alter the tightness of the weave or local orientation of the remaining yarns when removing some of the yarns prior to testing the fabric. This statement was further supported by HKUST who noted that theoretically in an obliquely oriented or misaligned specimen in the frame, one group of yarns would be under tension while the other would be under compression. Because a yarn cannot be compressed in the longitudinal direction, a misalignment would indicate that the yarn buckles out of the original plane and the onset of wrinkling in the fabric occurs at lower shear angles than when the specimen is properly aligned in the frame.

UT terminated their tests at the onset of wrinkling, as the shear deformation is no longer uniform once wrinkling occurs. UML noted that by 'mechanically conditioning' the specimen, i.e., by shearing the fabric in the frame several times before starting the test the variability in tension due to local deviations in orientation could be eliminated. This occurrence indicates the importance of the precise handling of both the sample and test fixture.

Figure 13.8 shows a schematic drawing of a picture frame. In this case, a displacement transducer in the tensile machine measures the vertical displacement, d, of point A (KUL, UT). Through trigonometric relations, the angle of the frame, θ, is calculated.

$$\cos\theta = \frac{\sqrt{2}L_{frame} + d}{2 \times L_{frame}} \qquad 13.1$$

where L_{frame} is the frame length indicated in Fig. 13.8. The shear angle, γ, is calculated from the geometry of the picture frame.

$$\gamma = 90° - 2\theta \qquad 13.2$$

This value, γ, is also called the global shear angle. Note that this value is taken to be an average shear value over the entire specimen. The actual shear angle at

302   Composites forming technologies

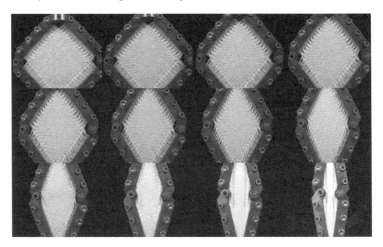

13.10 Array of images captured during the loading process (HKUST).

any point on the fabric may vary. (HKUST and UML used a similar approach to determine γ based on their displacement measurements.)

Optical methods which can aid in the determination of the shear angle at any particular point on the fabric specimen also exist. HKUST used a camera to capture arrays of images during the loading process. They then processed these images with AutoCAD, as shown in Figs 13.10 and 13.11. They found that the maximum deviation between the measured shear angle and the calculated shear angle (Eq. 13.2) is about 9.3% and that the maximum deviation typically occurs at larger shear angles. Based upon the small percent difference, the shear angle reported in this paper is the calculated shear angle. Thus, the shear angle is consistent with the method used by the other research groups.

KUL incorporated an image mapping system (Aramis) into their experiment. After photos were taken by a CCD camera, displacement and strain fields are identified by the Aramis software by analyzing the difference between two subsequent photos. Figure 13.12 shows a von Mises strain distribution over an image of a fabric sample during testing. By averaging the local shear angles

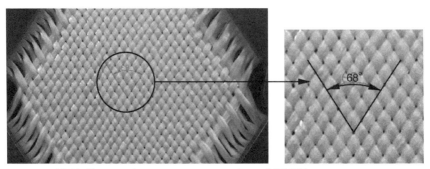

13.11 Shear angle measurement on a photo (HKUST).

Benchmarking of composite forming modelling techniques 303

13.12 Image of the fabric and the central region with the von Mises strain field (KUL).

produced by Aramis and comparing them with the global shear angles calculated from the crosshead displacement using Eqs. 13.1 and 13.2, KUL generated the graph in Fig. 13.13, which shows that the shear angle values obtained using the two methods are comparable. Again, as the difference was small between the calculated shear angle and the shear angle measured with the mapping system as well as to be consistent with the groups participating in this study, the shear angle used in this paper is the shear angle calculated from the crosshead displacement.

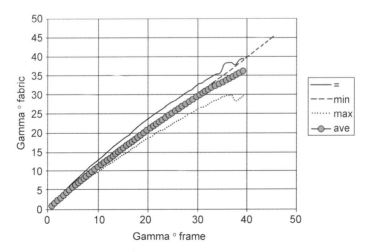

13.13 Typical relationship between the optically measured shear angle and shear angle of the frame, unbalanced twill weave (KUL).

The force needed to deform the fixture must be measured to accurately determine the actual force required to shear the fabric. HKUST and UML conducted several tests on their frames without including a fabric blank to record the force required to deform the frame, F'. This value was subtracted from the results obtained from fabric shear tests F''. The difference in these values is the force required to shear the fabric sample. Shear force, F_s, is calculated,

$$F_s = \frac{F}{2\cos\theta} = \frac{F'' - F'}{2\cos\theta} \qquad 13.3$$

where F is the load after subtracting a constant offset value from each data point to eliminate the error caused by the weight and inertia of the fixture, and θ is the angle shown in Fig. 13.8 calculated using Eq. 13.1.

KUL used a different method to measure the force required to deform the fixture, F'. Their method required a hinge (Fig. 13.14) to balance the initial weight and calibrate the results under various loading speeds.

After subtracting the offset force, F', from the measured force value, F'', the shear force, F_s, may be calculated with the aid of the frame geometry and a free-body diagram.

$$F_s = \frac{F}{2\cos\theta} = \frac{F'' - F'}{2\cos\theta} \qquad 13.4$$

13.14 Hinge for calibrating the force required to deform the frame.

Benchmarking of composite forming modelling techniques 305

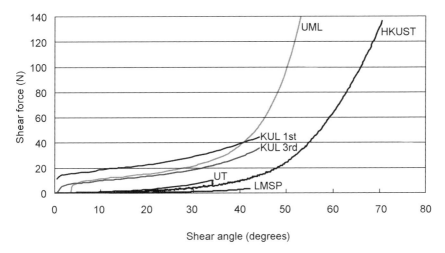

13.15 Shear force versus shear angle.

Figure 13.15 shows shear-force data as a function of the calculated shear angle (Eq. 13.2) for plain-weave fabric at room temperature. Shear forces are calculated based on the tensile machine loads according to Eqs. 13.3 and 13.4.

Based solely on Fig. 13.15, the data obtained for the shear force does not seem to be comparable. However, the differences among the frames and the sample sizes have not yet been taken into account. The following paragraphs will explain methods for comparing shear frame data obtained by different groups using different shear frames and different samples sizes. The first difference which should be accounted for is the inclusion of a linkage on the test frame. Note that KUL and UML both use shear frames with a linkage, while the other groups do not have a linkage on their shear frames. This linkage was included by UML because they needed to run the tests at a speed higher than their Instron machine could accommodate. By including the linkage and the slot, the frame could travel at a rate 4.25 times faster than was possible through the specified crosshead displacement rate. From these experiments, it has been found that the linkage introduces an amplification factor in the force calculation. Thus, to compare data when some groups have a shear frame with a linkage to the data from other groups whose shear frame does not have a linkage, the amplification factor must be removed.

Using UML's frame for the picture frame tests, the displacement and load are applied on the corner of the small amplifier frame marked 'A' instead of on the top corner of frame that clamps the fabric (Fig. 13.16). Note that point 'A' is the point at which the sliding link attaches the crosshead mount to the amplifier linkage. Therefore, a kinematic analysis of the picture frame with the amplifier is necessary for the calculation of the shear angle and shear load in the test. When performing this analysis, it should be noted that the amplifier shares two

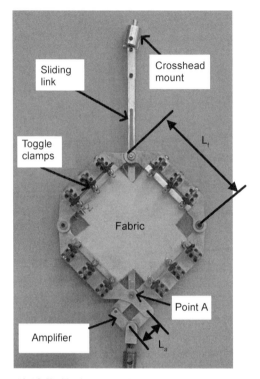

13.16 Trellis-frame test fixture (UML).

pieces of the frame with the picture frame, arms B and C. Thus, the shear angle of the amplifier equals the shear angle of the picture frame.

Beginning with the geometry of the amplifier frame, as shown in Fig. 13.17, the displacement of point 'A', δ_a, can be calculated.

$$\delta_a = 2L_a \left(\cos\left(\frac{\pi}{4} - \frac{\gamma}{2}\right) - \frac{\sqrt{2}}{2} \right) \qquad 13.5$$

where L_a is the length of one side of the amplifier frame and γ is the shear angle.

Solving Eq. 13.5 for the shear angle, γ,

$$\gamma = \frac{\pi}{2} - 2\cos^{-1}\left(\frac{\delta_a}{2L_a} + \frac{\sqrt{2}}{2}\right) \qquad 13.6$$

To calculate the shear load, the kinematics of the picture frame is studied. Figure 13.17 shows the schematic diagram of the picture frame. The free body diagrams of the side frame BC and BAF are shown in Fig. 13.18.

From Fig. 13.18, note that joint C is free for motion. Using symmetry, it can be determined that the force applied on joint C from link CD and BC is zero. Thus, performing a static analysis using the free body diagram of link BC (Fig. 13.19a),

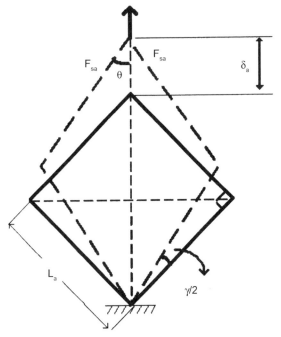

13.17 Geometry of the amplifier.

$$F_B - F_s = 0 \quad \text{or} \quad F_B = F_s \qquad 13.7$$

where F_B is the force on joint B between link BC and link BAF and F_s is the shear force the fabric sample applied to link BC.

Then, from the free body diagram of link BAF in Fig. 13.19b,

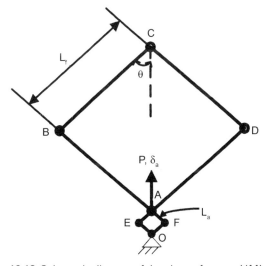

13.18 Schematic diagram of the picture frame at UML.

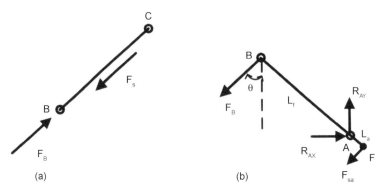

13.19 Free body diagrams of (a) link BC, and (b) link BAF.

$$M_A = 0 \quad \text{or} \quad F_B L_f \sin(2\theta) - F_{sa} L_a \sin(2\theta) = 0 \qquad 13.8$$

where M_A is the moment at point A, L_f is the length of link BC, F_{sa} is the shear force applied on the amplifier frame from the Instron machine and θ is the angle between link BC to the vertical direction as seen in Fig. 13.4. Solving Eq. 13.8 for F_B,

$$F_B = \frac{F_{sa} L_a}{L_f} \qquad 13.9$$

Defining an amplification factor as, α,

$$\alpha = \frac{L_f}{L_a} \qquad 13.10$$

From the geometry of the amplifier frame, the shear force of the amplifier, F_{sa}, can be calculated.

$$F_{sa} = \frac{P}{2\cos\theta} \qquad 13.11$$

where P is the force measured on the load cell in the crosshead.

Substituting Eqs. 13.7, 13.10 and 13.11 into Eq. 13.9,

$$F_s = \frac{P}{2\alpha \cos\theta} \qquad 13.12$$

Thus, in processing the picture frame test data at UML, Eqs. 13.6 and 13.11 are used to calculate the shear angle and the shear load, respectively. After comparing Eq. 13.12 with Eq. 13.3, it should be noted that the shear force equation is only altered through the inclusion of the amplification factor in the denominator in the left-hand side of the equation. Intuitively, this is reasonable because the amplification factor, α, for a frame with no amplification linkage would be 1 and Eq. 13.12 would then be equivalent to Eq. 13.3.

A similar analysis can be performed on the frame used by KUL. However, some differences exist because the amplification linkage in the KUL frame is

Benchmarking of composite forming modelling techniques 309

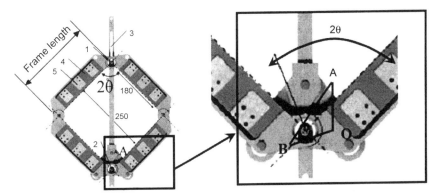

13.20 Geometry of picture frame (KUL).

inverted when compared to the amplification linkage in the UML frame (Figs 13.7 and 13.8). The geometry of the linkage in KUL's frame is shown in detail in Fig. 13.20. Note that none of the angles of the KUL amplification linkage are equal to the shear angle for the fabric as the crosshead moves in the vertical direction. Thus, the amplification factor for the KUL frame is not a constant value like it was for the UML frame. However, upon performing a kinematic analysis, an equation (as opposed to a constant value) can be determined for the amplification factor, α, and substituted into Eq. 13.12.

Figure 13.21 shows the data comparison with the amplification factor introduced by the linkages in the frames used by UML and KUL removed.

HKUST and UML mechanically conditioned the samples prior to testing, LMSP reported the results from the third repetition of the test on a single sample

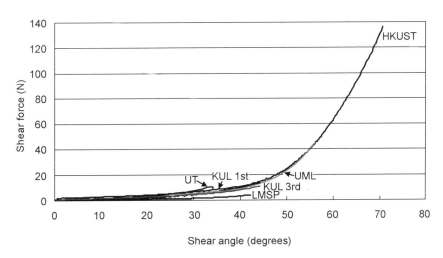

13.21 Shear force versus shear angle with linkage amplification removed from UML and KUL results.

and KUL reported data for each of three repetitions of the test on a single sample. Examining the data from the third repetition of the test on a single sample can be equated to mechanical conditioning, as the sample has deformed two times. KUL noted that the data from the second and third repetitions on a single sample were comparable with each other, but both were well below the data from the first time the sample was deformed in the shear frame. UT did not report whether their samples were mechanically conditioned prior to testing or that repeated tests were performed on the same samples at any time.

The calculations in the remainder of this report will focus on the region of the plot before 60°. It was at approximately 45° where locking began to occur for this fabric. Locking refers to the point at which the tows are no longer able to rotate and they begin to exert a compressive force on each other as the fabric is further deformed. The force required to deform the fabric begins to increase significantly as the locking angle is reached and surpassed. When the compression of the tows reaches a maximum, wrinkling begins to occur and the fabric begins to buckle out of plane. Wrinkling in a formed part is considered a defect. Thus, it is undesirable. Figure 13.22 shows the shear force vs. shear angle obtained from different groups up to 60° of shearing angle.

One proposed method for normalization was to use the frame length. Figure 13.23 presents the force results normalized by the length of the frame used by each group. As seen, the normalization brought curves closer, but noticeable deviations are still seen. Frame length could be indicative of sample size, i.e. a larger frame may indicate a larger sample size which in turn would indicate the deformation of a greater number of crossovers. However, as there is no standard ratio for the length of a test sample to the length of the frame, this method is not the best method for normalization.

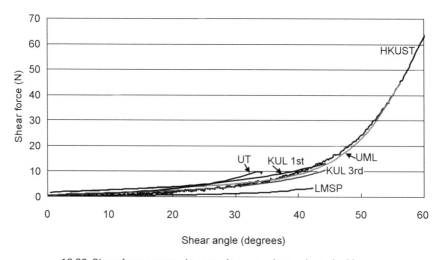

13.22 Shear force versus shear angle comparison prior to locking.

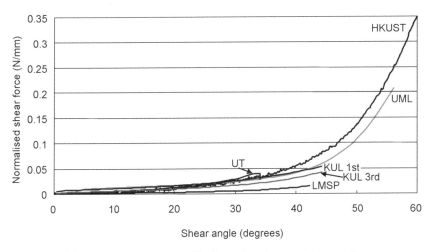

13.23 Shear force normalized by frame length versus shear angle.

The investigation continued by comparing the data when normalized by fabric area. For this research the fabric area was defined as the inner square area of the sample, i.e. the arms were neglected (Fig. 13.24). The fabric area would be related to the number of crossovers in the material. A larger sample would have more yarns resulting in more crossovers between the yarns. Figure 13.25 shows the results when the data was normalized by the fabric area. Again, this normalization technique brought the curves closer together, but researchers were interested in a more comprehensive normalization technique. A technique which took into account both the size of the sample and the size of the frame was then investigated.

Both Peng *et al.* (2004) and Harrison *et al.* (2004) have developed normalization methods using an energy method. Harrison *et al.* (2004) studied the case where the frame length is equal to the fabric length. They researched

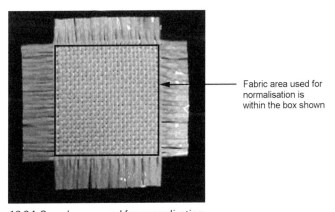

Fabric area used for normalisation is within the box shown

13.24 Sample area used for normalization.

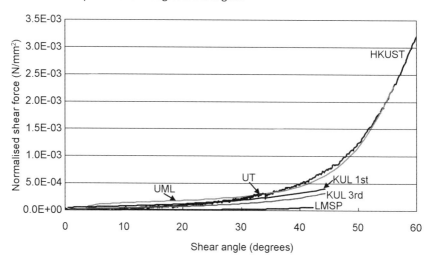

13.25 Shear force normalized by fabric area versus shear angle.

and proposed a method for comparing the force results obtained using shear frames (and as a result fabric samples) of different sizes. Peng et al. (2004) studied that case and the case where the length of the fabric sample is not equal to the length of the frame. Their equation reduces to the method proposed by Harrison et al. (2004) for the case when the fabric length is equal to the frame length. Thus, as proposed by Peng et al. (2004), to normalize the force data,

$$P_{normalized} = P_{original} \cdot \frac{L_{frame}}{L_{fabric}^2} \qquad 13.13$$

where $P_{normalized}$ is the shear force normalized according to the energy method, $P_{original}$ is the force required to shear the fabric, L_{frame} is the length of the frame and L_{fabric} is the length of the fabric.

When the length of the fabric is equal to the length of the frame, this equation becomes,

$$P_{normalized} = P_{original} \cdot \frac{1}{L_{frame}} = P_{original} \cdot \frac{1}{L_{fabric}} \qquad 13.14$$

as proposed by Harrison et al. (2004).

Figure 13.26 presents the normalized data using Eq. 13.13. Notice that the best agreement in the low shear angle region is obtained using this method when compared to normalizing the data by the frame size and the sample size independently.

In this section of the chapter, important features of the picture-frame test were presented, including: preparation of the samples, length definitions, force and shear angle measurements and calculations, repeatability, and comparison of

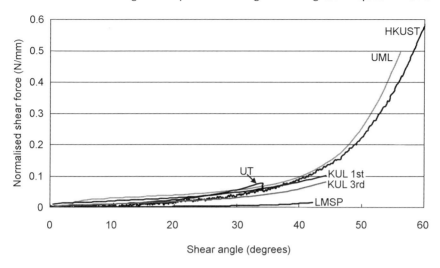

13.26 Shear force normalized using the energy method versus shear angle.

results force data with and without normalization methods. These normalization methods show that the results from different groups using different shear frames can be compared.

13.4 Numerical analyses

Researchers noted that repeatability of the experimental results was improved with tests performed on the same sample and with preconditioning. Researchers realize that fabric handling may decrease repeatability as well if the handling affects the alignment of the tows in the fabric. Thus, it may be feasible to incorporate a step into the manufacturing process that will precondition the sample or aid in the proper alignment of tows prior to the forming process. However, research has not been conducted to assess the actual impact of conditioning on the forming of a part. HKUST reported on the impact of misalignment of the fabric in the frame causing early onset of wrinkling. Again, misalignment of the fabric has not been investigated in relation to the forming of a part. One way to investigate these effects would be to stamp the actual parts and compare the results through visual inspections and experimental tests. However, realizing that this could be a costly and time-consuming process, the researchers involved with this project are continuing the benchmarking effort with a numerical forming investigation. The results from the shear tests will be used in that investigation as part of the constitutive material relations. The data from the benchmarking study of the shear behavior of the fabric can be incorporated into the numerical investigations currently underway in the benchmarking effort. The following paragraphs outline the proposed method for determining the shear modulus of the fabric from the shear force normalized using the energy method (Eqs. 13.13 and 13.14). This shear modulus can be incorporated into finite element models of the thermostamping process through the use of a user-supplied material model. As more tests are performed on the fabric, the user-supplied material models can be updated to provide a more robust analysis of the forming behavior of the materials under investigation.

By definition, the shear stress, τ, is obtained by dividing force by cross-sectional area.

$$\tau = \frac{P_{original}}{Area} = \frac{P_{original}}{L_{fabric} \cdot t_{fabric}} \qquad 13.15$$

In Eq. 13.15, by definition, the cross-sectional area, denoted by the variable *Area*, is equal to the length of the fabric, L_{fabric}, multiplied by the thickness of the fabric, t_{fabric}. Note that the denominator in Eq. 13.15 contains L_{fabric}, as does the denominator of the normalized force in Eqs. 13.13 and 13.14. Thus, to calculate the shear stress using the force normalized by fabric length, it is only necessary to divide the normalized force by the thickness of the fabric, t_{fabric}, not the area not the cross-sectional area, *Area*. Hence, the shear stress is,

$$\tau = \frac{P_{normalised}}{t_{fabric}} \qquad 13.16$$

It is then proposed that the shear modulus can be calculated from the derivative of the regression equation determined from the data points on the shear stress versus shear strain plot, if the units of shear strain are in radians. These results are shown in Fig. 13.27. Note that there is little difference in the

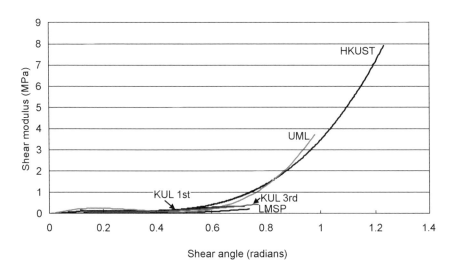

13.27 Shear modulus (MPa) versus shear angle (radians).

shear modulus among the groups until the locking angle is approached (i.e. where the shear modulus begins to rapidly increase). Thus, there should not be any significant differences in the modeling results for deformation in the region below the locking angle. However, only two groups collected data in the region where the shear modulus begins to rapidly increase as the locking angle is approached and exceeded. Thus, it is difficult to draw conclusions for this region. In future studies, such as those where the temperature effects will be considered, participating research groups should be asked to collect data over a wider range of angles so that more definite conclusions can be drawn.

13.5 Conclusions and future trends

The properties of woven fabrics are very different from conventional materials, such as bulk metals and polymers. This phenomenon lead to the interest in the woven-fabric composite material community to conduct benchmark tests. It has been shown that picture-frame tests are able to produce valuable experimental data for analytical and numerical research on woven composites. Mechanically conditioning the sample can also improve repeatability. This was shown through results from UML and KUL. UML's samples were all mechanically conditioned and appeared very repeatable. While KUL did not mechanically condition their samples, they conducted the shear test three times on each fabric blank and noted a large difference between the 1st run and the 2nd and 3rd runs. However, there was not a large difference in the results when only comparing the 2nd and 3rd runs. Based on the results from numerical studies manufacturers may want to consider the incorporation of a step into the manufacturing process to mechanically condition the fabric blank.

Test results provided by different groups show consistency but still have some deviations. Further studies are underway to help develop a standard test setup and procedure for obtaining accurate and appropriate material properties. Material responses under different speeds and temperatures will be further investigated. Calibration, sample preparation, and other important techniques to increase the accuracy of the tests will be collected and shared among the community. High temperature tests present challenges to researchers as they limit the use of optical devices and require higher sensitivity of the testing equipment. However, these optical methods showed that determining the shear angle mathematically from the crosshead displacement was a reasonable method as the shear angles obtained optically using Aramis and AutoCad did not vary significantly from the method used to calculate the shear angle from the crosshead displacement.

It was noted that a direct comparison of the region immediately preceeding the locking angle and the region following the locking angle could not be made because all groups did not take the same amount of data. However, the fact that a complete comparison cannot be made at this time, does not diminish the importance of this data. Researchers have learned that picture-frame test results obtained from different research groups using different sample and frame sizes are comparable and that the energy method for normalization appears to be the best method in the literature to date for normalizing picture-frame shear data. With the available data, the benchmarking research group can begin to use numerical methods, such as finite element analyses, to analyze the effects of fabric alignment, fabric conditioning and material type on the forming process. From the results of the numerical investigations, forming experiments can be judiciously chosen and performed to validate the numerical results.

13.6 Acknowledgements

The authors would like to thank NSF, Saint-Gobain, Inc., Hong Kong RGC (under grant HKUST6012/02E), the Netherlands Agency for Aerospace Programmes and a Marie Curie Fellowship of the EC (HPMT-CT-2000-00030).

The contributions of the following researchers are also acknowledged and sincerely appreciated: H.S. Cheng,[1] T.X. Yu,[2] B. Zhu,[2] X.M. Tao,[2a] S.V. Lomov,[3] Tz. Stoilova,[3] I.Verpoest,[3] P. Boisse,[4] J. Launay,[4] G. Hivet,[4] L. Liu,[5] E.F. de Graaf,[6] R. Akkerman[6]

1. Northwestern University – Dept. of Mech. Engng., Evanston, IL, 60208, USA.
   URL:http://www.mech.northwestern.edu/ampl/home.html
   e-mail: jcao@northwestern.edu
2. Hong Kong University of Science and Technology, Clear Water Bay, Kowloon, Hong Kong
   URL:http://www.me.ust.hk.html
   e-mail: metxyu@ust.hk
2a. Hong Kong Polytechnic University, Hung Hom, Kowloon, Hong Kong.
   e-mail: tctaoxm@polyu.edu.hk
3. Katholieke Universiteit Leuven, Belgium
   URL: http://www.mtm.kuleuven.ac.be/Research/C2/poly/index.htm
   e-mail: Stepan.Lomov@mtm.kuleuven.ac.be
4. Laboratoire de Mécanique des Systèmes et des Procédés, France
   e-mail: Philippe.Boisse@univ-orleans.fr
5. University of Massachusetts at Lowell, Dept of Mechanical Eng, Lowell, MA 01854 USA
   URL: http://m-5.eng.uml.edu/acmtrl/
   e-mail: Julie_Chen@uml.edu
6. University of Twente, Netherlands
   URL: http://www.opm.ctw.utwente.nl/en/pt/
   e-mail: R.Akkerman@ctw.utwente.nl

13.7 References and further reading

Cao, J., H.S. Cheng, T.X. Yu, B. Zhu, X.M. Tao, S.V. Lomov, Tz. Stoilova, I.Verpoest, P. Boisse, J. Launay, G. Hivet, L. Liu, J. Chen, E.F. de Graaf, R. Akkerman (2004). 'A Cooperative Benchmark Effort on Testing of Woven Composites', *ESAFORM 2004*, pp. 305–308.

Harrison, P., M.J. Clifford, A.C. Long (2004). 'Shear characterization of viscous woven textile composites: a comparison between picture frame and bias extension experiments', *Composites Science and Technology*, Vol. 64, pp. 1453–1465.

Lebrun, G., M.N. Bureau, J. Denault (2003). 'Evaluation of bias-extension and picture-frame test methods for the measurement of intraply shear properties of PP/glass commingled fabrics', *Composite Structures*, Vol. 61, pp. 341–352.

Lomov, S.V., Tz. Stoilova, I. Verpoest (2004). 'Shear of woven fabrics: Theoretical model, numerical experiments and full field strain measurements', *ESAFORM Conference 2004*.

Long, A.C., F. Robitaille, B.J. Souter (2001). 'Mechanical modeling of in-plane shear and draping for woven and non-crimp reinforcements', *Journal of Thermoplastic Composite Materials*, 14, pp. 316–326.

Lussier, D. (2000). 'Shear Characterization of Textile Composite Formability', Masters Thesis, Department of Mechanical Engineering at the University of Massachusetts Lowell.

Peng, X.Q., J. Cao, J. Chen, P. Xue, D.S. Lussier, L. Liu (2004). 'Experimental and numerical analysis on normalization of picture frame tests for composite materials', *Composites Science and Technology*, Vol. 64, pp. 11–21.

Index

abrasion resistance 233
abrasive water jet cutting 206–7
ABAQUS 51, 54, 161
affine deformation 30
Airbus Industry A380 198, 200–1, 211–12
aircraft industry 197, 240–1
  see also thermoset prepreg
aluminium 190–1
  diaphragms 270
Amoco Fabrics 220
amplification factor 305–9
amplification linkages 305–9
analytical modelling 213–15
angle classes 153, 154
anisotropic fibre distribution 153, 154
ARALL (aramid aluminium laminate) 199, 200
aramid fibres 205
Aramis 302–3
Aromatic Polymer Composite APC-2 261
aspect ratio 8
asymmetric lay-ups 181–2
audio cones for loudspeaker systems 235
autoclave pressure 192
automated tape laying (ATL) machines 239
automated tow placement (ATP) machines 239
automotive parts
  co-moulding 273–4
  compression moulding simulation 161–3, 164–5, 168–9
  hot compacted 232–4
axial loading 2, 9–10
  see also tensile behaviour
axial stresses 177–8, 179

balanced twill weave 295, 296
benchmarking 293–317

forming process and fabric properties 295–7
future trends 315–16
numerical analyses 313–15
test procedures and results 297–313
bending
  FML 208
    modelling 216–17
  mechanical work of 94, 96
  ply bending 2, 12–14, 245
  tow bending 245
bending energy 83
bending stiffness 12–13
  glass rovings 97, 98, 100
  woven prepreg 245
bias extension test 8–9, 294, 295
  constitutive modelling 41–2
  hot-compacted polypropylene 229–30
  thermoset prepreg 244–5, 247–8
  see also intra-ply shear
biaxial tension 81–2
  model for woven fabric 89–93
    comparison with finite element simulation and experiment 90–3
    outline of the algorithm 89–90, 91
    parametric description of fabric behaviour under simultaneous shear and 96–111
  tests 9–10
biaxial weave model 27
bladder moulding 271
blank holder 267
blank holding force (BHF) 128, 136–42
blank (preform) shape 131
  direct method for shape optimisation 131–4, 135
blast resistance 203
bonding technology 199
bowing 183–6

Index

breather cloth 268
Brent's method 123
broken fibres 128, 253
buckling 128–9
  test 13–14
  tow buckling 245–6
  see also wrinkles/wrinkling
bunching 248

carbon-epoxy prepreg 240
  see also thermoset prepreg
carbon fabric TPCs 257, 258–9
carbon UD TPCs 257, 258–9
Carreau-WLF model 148–9
cases, protective 233, 234
Catia Composites Design 279
cemented carbide/hard metal tool bits 206
centre of force 170–1
certification 279, 281
chemical shrinkage 179–80
  spring-in of curved parts 187–90
    shrinkage after vitrification 187–8
    shrinkage before vitrification 188–90
circumferential stresses 179
clamping mechanisms 299–300
classic laminate theory (CLT) 213–14
commingled fabrics
  benchmarking 295–316
  consolidation behaviour 18
  TPCs 260
    hot stamping 267–8
co-moulding 273–4
compaction 2, 14, 19
  compaction behaviour of reinforcements 14–17
  number of compaction cycles 16
complex parts, distortion in 192–4
composite/metal hybrids 197–219
  development of 198–201
  see also fibre metal laminates (FML)
composite modulus 222–3
composites design and development 277–92
  composites data exchange 281
  composites development process 278–80
  draping and forming simulation 282–4
  linking forming simulation to component design analysis 284–91
    accounting for material shear 288–9
    design variables for thickness and orientation 286
    discrete optimisation 287
    failure analysis 289
    initial zone sizing 284–5
    optimisation in FEA codes 285
  ply-based analysis 287–8
  solid element generation 290–1
  stochastic analysis 289–90
  zone description 285–6
  zone orientation 287
  zone and ply descriptions 277–8
compression 81–2
  fabric behaviour under simultaneous shear and tension 111
  glass rovings 97–9, 100
  model of compression of woven fabric 84–9
    outline of the algorithm 84–5
    2D laminates 86–8
    3D fabrics 88–9
    yarns in a relaxed woven fabric 83–4
compression moulding 1, 144
  TPCs 264–8
    flow moulding of GMT 264–7
    hot stamping of commingled fabrics and aligned prepregs 266, 267–8
compression moulding simulation 144–76
  examples of use 161–71
    fibre orientation simulation 164–5
    flow simulation 161–4
    quality control 169–71
    stiffness and failure analysis 166–9
  measurement of material data 172–4
  theoretical description 145–61
    calculation of fibre orientation 153–61
    flow transfer simulation 145–9
    heat transfer simulation 149–53
COMPRO 193
conduction 149
conjugate gradient (CG) method 118, 121–4
consolidation 2, 14, 181
  consolidation behaviour of prepreg 17–19
  effect on spring-in 191–2
  multi-ply forming and re-consolidation simulations 70–5
constitutive axes 52, 53
constitutive modelling 22–45
  continuum-based laminate modelling *see* continuum-based laminate modelling
  continuum models 26–9
  discrete models 23–6
  future trends 43–4
  multi-layered models 29
continuous approaches to FE analysis 49–57
  hypoelastic model 51–4
  macromechanical model used in a commercial FE code 56–7
  non-orthogonal constitutive models 54–6

continuous fibre reinforcements 48, 224–5
continuous fibres reinforced thermoplastics (CFRTP) 70–5
  prepregs 49
  simulation of CFRTP forming processes 71–2
continuum-based laminate modelling 29–43
  constitutive equation for a single layer 31–4
  kinematics 30–4
  multi-layer effects 34–5
  parameter characterisation 35–43
    bias extension 41–2
    picture frame 36–41
    pull-out 42–3
continuum drape models 26–9
  elastic models 26–7
  multi-component models 27–9
  viscous models 27
continuum theory 166, 167
control volume approach 147
convection 149
convergence speed 120–1, 122
cooling stage 127
Cooperative Research Centre for Advanced Composite Structures (CRC-ACS) 253, 254
corner thickening/thinning 181, 191–2
corrosion resistance 205
cost analysis 117
crack bridging 202–3
crack growth
  aluminium alloy 200, 202
  FML 200, 202, 214–15
crimp 60, 85, 223, 245–6
  de-crimping 9
  elementary crimp intervals 82–3
  heights 82–3, 84, 89–90
  non-crimp fabrics *see* non-crimp fabrics (NCFs)
critical fibre concentration 155
cross ply (CP) laminates 207
cure, degree of 150–1
curing, differential 181
CURV 220, 224, 233, 234, 236
  see also hot compaction
curved parts/plates 178–9
  spring-in of 186–92, 194
cutting 205–7, 218
cyclic thermoplastic polyesters (CBTs) 274–5

dampening term 156
Darcy law 47
dashboard 162, 164
data exchange 281
de-crimping 9
defects 128–30, 131
  see also buckling; wrinkles/wrinkling
deformation gradient 54
deformation energy 172
deformation mechanisms 1–21
  axial loading 2, 9–10
  compaction 2, 14–17, 19
  consolidation 2, 14, 17–19
  intra-ply shear 2, 3–9, 19
  ply bending 2, 12–14
  ply/tool and ply/ply friction 2, 10–12
  in woven fabric prepreg 241–7
degrees of freedom (DOFs) 29
delamination 227–8
density 204
dependencies 96, 103–5, 106–7
design points (parameter sets) 125–6
diaphragm forming 127–8
  thermoset prepreg 248–9
    diaphragm materials 251
    tooling 251–2
  TPCs 270–1
differencing points 150
differential curing 181
differential thermal contraction 177–9, 180, 181–2
diffusion 149
direct methods for optimisation 142
  shape optimisation 131–4, 135
discrete (mesoscopic) approach to FE analysis 50, 57–9
discrete drape models 23–6
  mapping-based schemes 23–4
  particle-based schemes 24–5
  truss-based schemes 25–6
discrete layer calculation of fibre orientation 159–60
discrete optimisation 287
distortion 177–96
  in flat parts 181–6, 194
  fundamental mechanisms 177–81
  in more complex parts 192–4
  spring-in of curved parts 186–92, 194
door panel 161–3, 164–5, 168–9
double curved laminates 212–13
double-diaphragm forming 248–9, 270–1
doublers 212
downhill simplex (DS) method 118–21, 122
drape modelling 22–3, 46–7
  constitutive modelling *see* constitutive modelling

Index   321

drape simulation 246, 282–4
  comparison with forming simulation 284
  see also composites design and development
dressed virtual humans 25
drilling 207
dry fabrics 1
ductility 223, 224

E-LFT (endless fibre reinforced long fibre thermoplastic) 274
edge milling 206–7
elastic energy 208
elastic material continuum models 26–7
elastic stress 31–2
electrical properties 204
elementary crimp intervals 82–3
empirical models 16–17
enclosed fibre angle 31
energy-based mapping method 24
energy method for normalisation 311–12, 313
energy transport *see* heat transfer simulation
equilibrium, picture frame 39
errors 93
ESI-QuickFORM 46
expansion coefficients 177–9, 180
EXPRESS 145, 161

fabric area, normalisation by 311, 312
fabric-based TPCs 257, 258–9, 260–1
  co-moulding 273–4
  hot stamping 267–8
fabric reinforced fluid (FRF) model 30
fabric thermoset prepreg 241–7
fabrics, simplified dynamic equation for 60–1
failure analysis
  composite design and development 289
  simulation of compression moulding 166–9
failure factor 167–8, 170
failure strain 201–2, 207
fatigue
  resistance for FML 199, 200, 202–3
  stress intensity factors for FML 214–15
FiberSIM 46
fibre breakage 128, 253
fibre bridging 202–3
fibre distribution function 153, 154, 156, 160–1
fibre extension (stretch) 30
  thermoset prepreg 246, 254

fibre interaction coefficient 156–7
fibre metal laminates (FML) 197–219
  development of 199–201
  modelling 213–17, 218
    analytical modelling of mechanical properties 213–15
    FE modelling of FML structures 215–16
    forming processes 216–17
    lay-up processes 217
  production processes 205–13, 218
    cutting processes 205–7, 218
    formability aspects 207–8
    forming processes 208–11
    lay-up techniques for FML shells 211–13
  properties 201–5, 218
    mechanical 201–3
    physical 204–5
fibre orientation
  design variables for 286
  distribution 128
  reorientation 22
  simulation and composite design 279, 280
  simulation of compression moulding 153–61
    examples of use 164–5
    fibre pre-orientation 160–1
    fibre transport 153–5
    model for calculation of fibre rotation 155–8
    solution of differential equation 158–9
    thermoplastic discrete layer calculation 159–60
  thermoset prepreg 241
fibre volume fraction 14–17
  gradients 181, 182–3
fibre wrinkling, spring-in due to 190–1
filament wound tubes 180
finite difference method 150–1
finite element analysis (FEA) 25–6, 46–79, 81–2, 193–4, 277–8
  approaches to 49–50
  bi-component model of dry fabric forming 28
  commercial FE code 56–7
  comparison of model of uniaxial and biaxial tension of woven fabric with 90–3
  component design analysis and optimisation 285–91
  composite development process 279
  continuous approaches 49–57

322   Index

discrete approach 50, 57–9
forming simulation 284
  comparison with kinematic simulation 284
  mechanism and specificity of composite forming 48–9
  modelling of FML structures 215–16
  multi-ply forming and re-consolidation simulations 70–5
  multi-scale materials 48
  semi-discrete approach 59–70
fire resistance 204–5
fishnet model 23–4
FIT (Fibre Imprégnée de Thermoplastique) process 261
flame treatment 233–4
flat plates/parts 178
  distortion 181–6, 194
    due to asymmetric lay-ups 181–2
    due to resin bleed and volume fraction gradients 182–3
    due to tool-part interaction 183–6
flattening coefficient 99, 100, 111
flow conductivity 146–7
flow moulding 264–7
flow simulation 145–9
  examples of use 161–4
  isothermal 146–8
  non-isothermal 148–9
Fokker 199
Folgar-Tucker fibre orientation model 155–7
footwear, safety 235
formability 207–8
forming boxes 251–2
forming limit diagram (curve) 129–30, 131, 132, 133, 216
four-node finite element 65–7
frame angle 37–8, 301
frame length 310, 311
friction 2, 10–12, 56–7
  between glass rovings and steel 99, 100
  moment 94, 95
  stress gradients due to frictional effects before gelation 185–6
fully constituted TPCs *see* unidirectional (UD) TPCs

gelation 188, 190
generators (fibre paths) 282–3
geodesic generators 283
geometric model 23–4
GLARE laminates 200–1, 202, 204, 214
glass fabric TPCs 257, 258–9

glass fibre-based GLARE 200
glass fibre-reinforced polymers 39–41, 224–5
glass mat thermoplastic (GMT) 1, 144
  forming trials for self-reinforced polypropylene using a commercial GMT production line 230–1
  TPCs 257, 258–9
    co-moulding 273–4
    flow moulding 264–7
glass rovings, properties of 97–100
glass transition temperature 181–2
glass UD TPCs 257, 258–9
global failure criterion 167
global ply modelling 279, 280
GMText 274
golden section method 123
gradient techniques 285
Green-Naghdi approach 51, 52

Halpin-Tsai model 166
hand lay-up 240–1
heat sources 250–1
heat transfer simulation 149–53
  energy transport after filling stage 149–51
  energy transport after part ejection 152–3
  energy transport during filling stage 149
  four stress layers 151–2
Hele-Shaw flow model 145, 146
helper points 150, 152–3
hemispherical forming
  matched hemispherical mould experiments 226–9
  optimisation of process parameters 136–40
  unbalanced fabric 67–70
high temperature polymer diaphragms 251, 270
homogenisation 51
honeycomb structure 199
Hong Kong University of Science and Technology (HKUST) 294, 297–313, 314
hoop stresses 177, 178
hot compaction 220–38
  commercial exploitation 224–5
  commercial products 232–5
  future developments 235–6
  postforming studies 225–32
    forming trials using a commercial GMT production line 230–1

Index   323

picture frame and bias extension tests 229–30
tensile tests and matched hemispherical mould experiments 226–9
thermoforming strategies 232
process 220–4
hot compaction temperature 222
hot stamping process 266, 267–8, 295–7
hybrid drape algorithm 25
hybrid materials 197
 development of composite/metal hybrids 198–201
 FML *see* fibre metal laminates
hydrostatic stress 32
hypoelastic model for fibrous materials 51–4, 55

impact resistance 203, 233
implicit method 158–9
in-plane deformations
 FML 207
  modelling 216
 semi-discrete approach and in-plane shear 62–5
 see also intra-ply shear
inclined plane method 10–11
indirect (iterative) optimisation methods 134–42
initial zone sizing 284–5
injection 48
input data 111–12
interacting particle model 24–5
interaction between fibres 155–7
interface traction 35, 42–3
interior elementary work 65–7
internal stresses *see* residual stresses
interply shear (interply slip) 71
 woven fabric prepreg 241–3
intertow slip 246
intra-ply shear 2, 3–9, 19, 207
 benchmarking 294, 297–316
 bias extension test *see* bias extension test
 constitutive modelling 36–42
 picture frame test *see* picture frame test
 thermoset prepreg 243–4, 247–8
intra-ply tensile loading 2, 9–10
Invar 190
isothermal bladder moulding 271
isothermal diaphragm forming 270
isothermal flow simulation 146–8
isothermal forming 262, 263
isothermal vacuum forming 268–9

isotropic fibre distribution 153, 154
isotropic polypropylene sheet 224–5
iterative optimisation method 134–42

Jaumann approach 51
Jeffrey model for fibre orientation 155

Katholieke Universiteit Leuven (KUL) (Belgium) 294, 297–313, 314, 315
Kawabata Evaluation System for Fabrics (KES-F) 3, 36
kinematical models 46
 continuum-based laminate modelling 30–4
 discrete model 23–4
 draping simulation 282–4
  compared with finite element forming simulation 284
Kriging method 125–6

Laboratoire de Mécanique des Systèmes et des Procédés (LMSP) (France) 294, 297–313, 314
LAM option 285–6
laminates 178
 compression of 86–8
 zone and ply descriptions 277–8
laser jet cutting 206–7
lay-up processes
 FML 211–13, 218
  modelling 217
 hand lay-up 240–1
Lie derivative 54, 55
linkages, amplification 305–9
liquid composite moulding (LCM) processes 49
locking 60, 102, 107, 180, 245
locking angle 4–5, 8, 107, 129, 230, 282
 benchmarking 310, 314–15
long discontinuous fibre prepreg 254
long fibre thermoplastic (LFT) 273–4
looseness factor 100–1
luggage, protective 234

macromechanical models
 compression moulding simulation 166, 167–8
 used in a commercial FE code 56–7
macroscopic scale 48
magnetic properties 204
mandrels 251–2
manufacture, composite design and 279–80, 281
mapping approaches 23

drape modelling 23–4
mapping function 133–4
masses and springs 58, 59
matched mould processes 18
  matched hemispherical mould experiments 226–9
material data measurement 172–4
material property sets 289
materials characterisation 1–21
  axial loading 2, 9–10
  benchmarking see benchmarking
  compaction 2, 14–17, 19
  consolidation 2, 14, 17–19
  constitutive modelling 35–43
  intra-ply shear 2, 3–9, 19
  ply bending 2, 12–14
  ply/tool and ply/ply friction 2, 10–12
matrix phase 221–3
mechanical conditioning of samples 301, 309–10, 313, 315
mechanical fastening 207
mechanical modelling 23
mechanical properties
  FML 201–3
    analytical modelling 213–15
  TPCs 258–9
  see also under individual properties
mechanical simulation of compression moulding 166–9, 170
mechanical work of shear deformation 93–6
melt impregnation 259
melted and recrystallised phase 221–2, 227–8
melting peaks 221–2, 227–8
mesoscopic (discrete) approach to finite element analysis 50, 57–9
mesoscopic scale 48
metal matrix composites (MMCs) 197–8
metal sheet 129, 130
  optimisation 140–2
metal volume fraction (MVF) approach 214
micromechanical models 43, 166–7
microscopic scale 48
minimum bend radius 208
minimum energy, principle of 80–1, 84, 90
moisture uptake 181, 182
molecular weight 224
mould filling 47
MSC.Nastran 279, 287–8
  PCOMP card 285, 286
Mullins effect 251
multi-component continuum models 27–9

multi-layered composites 24
  continuum-based laminate modelling 34–5
  drape models 29
multi-ply forming 70–5
  shell element with pinching 72–4
  simulation of CFRTP forming processes 71–2
  simulation of forming and re-consolidation stage of a Z profile 74–5
multi-scale materials 48

natural element method (NEM) 46
nesting 86–8
Newton-Raphson (NR) method 134
Newtonian flow 145, 146–7
nodal interior load 65–7
non-crimp fabrics (NCFs) 5, 6, 57
  meso-mechanical model 59
  optimisation of process parameters for NCF with tricot stitch 136–40
non-isothermal diaphragm forming 270–1
non-isothermal flow simulation 148–9
non-isothermal forming 262–4
non-isothermal vacuum forming 269–70
non-Newtonian flow 145
non-orthogonal constitutive models 54–6
normalisation 9
  benchmarking studies 300, 310–13
NSF Composite Sheet Forming Workshop 293
numerical forming investigation 313–15
nylon twill 67–70

objective derivatives 51–4, 55
objective functions 118, 132–4, 135, 142
  in composite forming 127–31
  sheet metal and optimisation of BHF 141–2
optical measurement of shear angle 302–3
optical strain analysis 63, 64
optimisation 117–43
  component design analysis 284–91
  composite forming 126–42
    objective functions 127–31
    optimisation of process parameters using iterative optimisation method 134–42
    shape optimisation using a direct method 131–4, 135
  methods 118–24
  using a surrogate model 124–6
orientation see fibre orientation

Index 325

orientation tensors 153
oriented fibre phase 221–2, 227–8
out-time (age) 240–1
over-injection moulding 273
overlap splice 212

PAM-FORM code 56–7
parallel simulations 121, 124
part ejection, heat flow after 152–3
partially impregnated materials 17–18
particle-based drape models 24–5
performance analysis 117
phenomenological models 16
physical properties of FML 204–5
picture frame test 3–7, 63–4
　benchmarking 294, 297–313
　continuum-based laminate modelling
　　36–41
　　equilibrium 39
　　kinematics 37–9
　hot-compacted polypropylene 229–30
　pre-consolidated laminates 39–41
　types of frame 298–9
pinching 72–4
plain weave 295, 296
ply bending 2, 12–14, 245
ply descriptions 277–8
ply details 279, 281
ply drop-off 279
ply lay-up 279, 281, 287, 288
ply/ply friction 2, 10–12
ply pull-out tests 11, 42–3
ply/tool friction 2, 10–12, 43
poly crystal diamond (PCD) tipped tool
　bits 206
polyetherimide (PEI) 261
polyethersulphone (PES) 261
polyethylene 221–2, 223, 235–6
polyethylene naphthalate (PEN) 236
polyethylene terephthalate (PET) 222, 223,
　236
polyoxymethylene (POM) 236
polypropylene (PP)
　comparison of mechanical properties of
　　PP-based materials 224–5
　self-reinforced composites 221, 222,
　　223–5, 236
　　commercial products 232–5
　　postforming studies 225–32
pores 71–2
post yield modulus 226, 227
powder impregnated fabrics 260–1
　hot stamping 267–8
preconditioning 301, 309–10, 313, 315

preform shape *see* blank (preform) shape
preforming 22, 240
pre-orientation distribution factor 160–1
prepreg 1, 239
　consolidation behaviour 17–19
　picture frame test 5, 7
　thermoplastic 127–30, 131
　thermoset *see* thermoset prepreg
press brake bending 209
press forming 210, 211, 249–50
Press rheometer 172–4
pressing force, calculated vs measured
　163–4
pressure
　autoclave pressure 192
　distribution and hot stamping 267–8
　temperature-pressure-time profiles 262–3
process parameters 127
　optimisation 117, 131–42
　　iterative optimisation method 134–42
　　shape optimisation 131–4, 135
projected generators 283
Propex Fabrics 220, 224
protective cases and luggage 233, 234
pull-out tests 11, 42–3

quality control 169–71
quenching 262, 263

radial stresses 177, 178
random short glass fibre filled
　polypropylene 224–5
rate of deformation tensor 32–3, 38
reaction moulding 274–5
re-consolidation 70–5
　see also consolidation
reinforcement compaction 2, 14–17, 19
relaxed state of a woven fabric, model of
　82–4
release film 268
repeatability 309–10, 313, 315
representative volume elements (RVEs)
　25–6
residual strength 203
residual stresses 177–96
　distortion in flat parts 181–6
　distortion in more complex parts 192–4
　FML 204, 208
　double curved laminates 213
　fundamental mechanisms causing
　　177–81
　spring-in of curved parts 186–92
resin bleed 181, 182–3, 191–2
resin infusion 18, 211

resin-poor regions 183
resin-rich regions 181, 183
responsive surface method (RSM) 124–5
rhombus test *see* picture frame test
riveted joints 215–16
roll bending 210
roll forming 210, 211
  TPCs 272
Rosenbrock functions 119–20, 121, 123, 125
routing 206–7
rubber press forming 249–50

safety footwear 235
sample preparation 300–1
sandwich materials 197, 199
sandwich moulding 274
saturation 16
scissor and slide draping mechanisms 282
self-reinforced composites *see* hot compaction
semi-discrete approach 59–70
  experimental and virtual test for tensile and in-plane shear behaviour 62–5
  four-node finite element made of woven cells 65–7
  hemispherical forming of an unbalanced fabric 67–70
  simplified dynamic equation for fabrics 60–1
semi-preg materials 17–18
shape
  composites development process 278–9, 281
  objective functions and shaping 127–31
  optimisation 129–31
    direct method 131–4, 135
  process capabilities of thermoset prepreg 253, 254
shear 81–2
  composite design and development 288–9
  in-plane shear behaviour 62–5
  intra-ply shear *see* intra-ply shear
  model of shear of woven fabrics 93–6
  parametric description of fabric behaviour under simultaneous shear and tension 96–111
    comparison with experiments 105–11
    parameterisation of the shear diagram 100–5
    properties of glass rovings 97–100
    shape of the shear diagrams 101–3
shear angle 5–7, 31, 38, 137, 138, 301–3

  benchmarking 306, 309, 315
  vs shear force 5–7, 39, 40–1, 309–13
shear couple 61, 63–4
shear edges 265–6
shear force 5–7, 217
  benchmarking 304–5, 306–8, 309
    normalisation 310, 311, 312, 313
  vs shear angle 5–7, 39, 40–1, 309–13
shear lag analysis 189–90
shear modulus 103–5, 106–7
  benchmarking study 313–15
shear resistance 103–5
shear strain interpolation 66
shear stress 314
  ply/tool friction 11–12
shearing 206
sheet forming 117
  optimisation *see* optimisation
sheet moulding compounds (SMCs) 1, 144
  see also thermoset prepreg
shell-based drape models 26–7
shell elements 290
  with pinching 72–4
short fibres 48
shrinkage 229
  chemical *see* chemical shrinkage
silicone rubber diaphragms 251, 270
simplex 119
  see also downhill simplex method
simplified dynamic equation 60–1
single curved laminates 212
single-diaphragm forming 248, 249
skin panels for aircraft structures 211–12
sled runner 171
slip factor 35, 42
slip law 35
slurry deposition (wet laying) 259
smooth particle hydrodynamics (SPH) 46
soccer shin guards 233, 234
solid element generation 290–1
solvent impregnated TPCs 261
splices/splicing 212, 279
sports protection 233, 234
spring back 208
spring-in 179, 180
  of curved parts 186–92, 194
    effect of consolidation 191–2
    fibre wrinkling and tool–part interaction 190–1
    shrinkage after vitrification 187–8
    shrinkage before vitrification 188–90
    thermoelastic spring-in 186–7
squeeze flow rheometers 172–4
stack forming behaviour 247

stamp forming 127–8
  hot stamping 266, 267–8, 295–7
  TPCs 264–8
  see also compression moulding
standardisation 36, 315
stationary layers 159, 160
stiffening index 16–17
stiffness analysis 166–9
stochastic design improvement (SDI) 289–90
strain interpolation coefficients 66–7
strain tensor 53
stress gradients 184–6
stress intensity factors 214–15
stress layers 151–2
stress-strain relations
  FML 201–2
    modelling 213–14
  hot compacted polypropylene sheet 226, 227
  non-orthogonal constitutive models 55–6
  silicone rubber 251
  woven fabric prepreg 244–5
stress tensor 50, 53
stretch forming 210, 211
structural analysis 279
suitcase shell 162, 163–4
surface activity 233–4
surface profile functions 86
surrogate model 124–6

tape prepreg 247–8
tearing 129–30, 131
Technical University of Delft 199–200
temperature
  hot compaction process
    hot compaction temperature 222
    melting peaks of two phases 221–2, 227–8
    postforming studies 226–32
  thermoset prepreg 246
temperature-pressure-time profiles 262–3
tensile behaviour 2, 9–10
  fibre metal laminates 201–2
  glass rovings 98, 99–100
  model of uniaxial and biaxial tension of woven fabric 89–93
  parametric description of fabric behaviour under simultaneous shear and tension 96–111
  self-reinforced polypropylene 226–9
  semi-discrete approach 62–5
tensile modulus 90–1, 201, 202, 257, 258–9

tensile strength 257, 258–9
tensile surfaces 60–1, 62, 63
tension tensor 61
Tensylon 236
thermal contraction, differential 177–9, 180, 181–2
thermal history 246
thermoelastic spring–in 186–7
thermoforming 127–8, 240
  thermoforming studies of self-reinforced polymers 225–32
thermoplastic composites (TPCs) 144, 256–76
  basic principles of TPC forming technologies 262–4
  discrete layer calculation of the fibre orientation 159–61
  drape models 28
  forming methods 264–72
  recent developments 273–5
  TPCs for forming 256–62
thermoplastic prepreg 127–30, 131
thermoset composites
  hot compaction process see hot compaction process
  see also thermoset prepreg
thermoset prepreg 49, 144, 239–55
  deformation mechanisms in woven fabric prepreg 241–7
  diaphragm forming tooling 251–2
  forming processes 248–50
  future trends 253–4
  potential problems 252–3
  practicalities of forming thermoset prepreg stacks 240–1
  process capabilities 253, 254
  tape prepreg 247–8
  tooling and equipment 250–1
thermostamping process 266, 267–8, 295–7
thickness
  design variables for 286
  diversity 128
  glass rovings 97
three-dimensional (3D) fabric, compression of 88–9
through-thickness contraction 188–90
through-thickness expansion coefficients 179, 180
tool-part interaction 180
  distortion in complex parts 192–3
  distortion in flat parts 183–6
  spring-in due to 190–1
tool/ply friction 2, 10–12, 43

torsion, mechanical work of 94, 95, 96
total Cauchy stress 31–4
tow bending 245
tow buckling 245–6
transversal forces 89–90, 93, 94
transverse stretching 248
trellising *see* intra-ply shear
tricot stitch, NCF with 136–40
truss elements 25–6, 58
twist 178, 179
two-dimensional (2D) laminates, compression of 86–8

ultimate tensile stress 201, 202
unbalanced fabric 67–70
  twill weave 295, 296
unbending of yarns, mechanical work of 94, 96
uniaxial tension model 89–93
  comparison with finite element simulation and experiment 90–3
  outline of algorithm 89–90
unidirectional (UD) laminates 207
unidirectional (UD) TPCs 257, 258–9, 261–2
  hot stamping 267–8
unit cells 25–6
University of Massachusetts Lowell (UML) (USA) 294, 297–313, 314, 315
University of Twente (UT) (Netherlands) 294, 297–313, 314

vacuum consolidation 18
vacuum forming 268–70
vacuum infusion 16
vacuum pump/compressor 250
vacuum table 269
Vantage Polymers Ltd 220
variable layers 159–60
velocity profile 154, 159–60
vertical displacement, mechanical work of 94, 96
virtual testing 19, 80–116
  creating input data for forming simulations 111–12
  mechanical model of the internal geometry of the relaxed state of a woven fabric 82–4
  model of compression of woven fabric 84–9

model of shear of woven fabric 93–6
model of uniaxial and biaxial tension of woven fabric 89–93
parametric description of fabric behaviour under simultaneous shear and tension 96–111
  comparison with experiments 105–11
  parameterisation of the shear diagram 100–5
  properties of glass rovings 97–100
viscosity 57, 172–4, 246
viscous material continuum models 27
viscous stress 31, 32–3
viscous traction 35, 42–3
vitrification
  chemical shrinkage after 187–8
  chemical shrinkage before 188–90
volume fraction *see* fibre volume fraction
von Mises strain distribution 302–3
von Mises stresses 177–8

wall adhesion 173, 174
wall slippage effects 172–4
warp yarns 101
weave patterns 82–3, 100
weaving density 100–1
weft yarns 101
wet laying (slurry deposition) 259
width, roving 97
WiseTex 82
woven cells 65–7
wrinkles/wrinkling 4, 22, 181, 310
  finite element analysis 68–70
  optimisation of composite forming 128–30, 131, 138–40
  sheet metal optimisation 129, 130
  spring-in due to fibre wrinkling 190–1
  thermoset prepreg 252–3
  woven prepreg 245–6

yield stress 201, 202
Young's modulus 90–1, 201, 202, 257, 258–9

Z reinforcement 71
  simulation of forming and re-consolidation stage 74–5
zone descriptions 277–8, 285–6, 287–8
zone lay-up 279, 281
zone orientation 287